W9-BEA-370

SAUNDERS GOLDEN SERIES IN ENVIRONMENTAL STUDIES

ECOLOGY, POLLUTION, ENVIRONMENT by Amos Turk, Jonathan Turk, and Janet T. Wittes

ENVIRONMENTAL SCIENCE by Amos Turk, Jonathan Turk, Janet T. Wittes and Robert Wittes

ECOSYSTEMS, ENERGY, POPULATION by Jonathan Turk, Janet T. Wittes, Robert Wittes and Amos Turk

SCIENCE, TECHNOLOGY and the ENVIRONMENT by John T. Hardy

IMPACT: SCIENCE ON SOCIETY edited by Robert L. Wolke

SCIENCE, MAN AND SOCIETY (*Second Edition*) by Robert B. Fischer

OUR GEOLOGICAL ENVIRONMENT by Joel S. Watkins, Michael L. Bottino and Marie Morisawa

Friends of the earth, we too care. This book has been printed on 100% recycled paper. It is estimated that in the total printing of 30,000 copies, 100 trees have been saved.

JONATHAN TURK, Ph.D.
Naturalist

JANET T. WITTES, Ph.D.
Hunter College

ROBERT WITTES, M.D.
Department of Medicine,
Memorial Sloan-Kettering Medical Center, New York

AMOS TURK, Ph.D.
Professor, Department of Chemistry,
The City College of the City University of New York

ECOSYSTEMS, ENERGY, POPULATION

W. B. SAUNDERS COMPANY / Philadelphia / London / Toronto

W. B. Saunders Company: West Washington Square
Philadelphia, PA 19105

1 St. Anne's Road
Eastbourne, East Sussex BN21 3UN, England

1 Goldthorne Avenue
Toronto, Ontario M8Z 5T9, Canada

Library of Congress Cataloging in Publication Data

Main entry under title:

Ecosystems, energy, population.

(Saunders golden series)

Includes index.

1. Ecology. 2. Human ecology. I. Turk, Jonathan.

QH541.E32 301.31 74–12919

ISBN 0–7216–8930–2

Cover photography courtesy of GEORGE SILK

Ecosystems, Energy, Population ISBN 0-7216-8930-2

Last digit is the print number: 9 8 7 6 5 4 3 2

PREFACE

Our previous textbooks dealing with environmental studies* have been directed at an overview of biological, chemical, and physical aspects of the subject. The wide acceptance of these books has prompted us to apply their style and pedagogical approach to *Ecosystems, Energy, Population.*

This text is for the student who is interested primarily in ecology and the biological aspects of environmental science and is intended for use in a one-semester course devoted to such subject matter. It is also suitable for use as a supplement to the general biology course. The first half of the book discusses the principles of natural ecology, population ecology, and speciation in natural systems in sufficient detail to provide a comprehensive introduction for the beginning student. The second half probes three vital problems of man's interaction with his environment: human population growth, energy, and agriculture.

The discussion of natural ecology is nearly as complete as that found in many small textbooks that are devoted exclusively to the subject. The somewhat detailed chapters on human population, energy, and agriculture supplement the subject matter and constitute an introduction to the wider study of the biological aspects of the environment.

Scientific, social, and economic issues are integrated into the discussions on human interactions with the environment. A wide variety of study problems is provided for each chapter, and answers are given for all the computational exercises. Suggested assignments are listed for each chapter section.

---

*Turk, Turk, and Wittes: *Ecology, Pollution, Environment,* 1972; Turk, Turk, Wittes, and Wittes: *Environmental Science,* 1974. Both published by W. B. Saunders Company, Philadelphia.

Acknowledgments

Professor Peter W. Frank of the Department of Biology, University of Oregon, who reviewed the entire manuscript of our previous textbook, *Environmental Science,* offered the major suggestions for direction, emphasis, and selection of new material, on which the structure of the present text is largely based. In addition, other colleagues examined particular chapters concerning their fields of special experience and expertise. We gratefully acknowledge these reviews in the following list:

CHAPTER	REVIEWER
1. Natural Ecosystems	Richard S. Miller, Professor of Wildlife Ecology, Yale University Eugene P. Odum, Alumni Professor of Zoology, University of Georgia Professor Armando de la Cruz, Mississippi State University Professor Stanley Wecker, Department of Biology, The City College of The City University of New York
2. Population Ecology	Professor Richard S. Miller
3. Species Diversity	Professor Richard S. Miller
4. Extinction of Species	Professor Richard S. Miller Dr. Beryl Simpson, Associate Curator, Department of Botany, Smithsonian Institution
5. Human Population	Dr. Judith Goldberg, Research Statistician of the Health Insurance Plan of Greater New York Dr. Elizabeth Whelan, Executive Director, Demographic Materials, Inc.
6. Energy	Professor John Fowler, Department of Physics, University of Maryland
7. Agriculture	Mr. Robert Josephy, Proprietor of Blue Jay Orchards, Bethel, Connecticut; Vice-Chairman of the Connecticut State Board of Agriculture

We are also indebted to Mrs. Pearl Turk, who gathered and classified source material, to Mrs. Evelyn Manacek, who typed the manuscript, to Mr. Grant Lashbrook and his staff at W. B. Saunders Company, whose art illustrates the text, and to the many others at Saunders who helped to complete the project.

JONATHAN TURK
JANET T. WITTES
ROBERT E. WITTES
AMOS TURK

CONTENTS

1

INTRODUCTION TO NATURAL ECOSYSTEMS

1.1 WHAT IS ECOLOGY?

No machine can perform as many diverse functions as a living organism. Animals and plants, unlike machines, can feed and repair themselves, adjust to new external influences, and reproduce. These abilities depend on intricate interrelationships among the separate parts of the organism. Each mammal, for instance, is far more than an independent collection of brain, heart, liver, stomach and other organs. What affects one part of the body affects its entirety.

Even with all its built-in mechanisms of life, however, an individual plant or animal cannot exist as an isolated entity; it is dependent upon its environment, including other organisms to which it relates. It must convert some source of energy (food or light), ingest water and minerals, dispose of wastes, and maintain a favorable temperature.

Plants and animals occurring together, plus that part of their physical environment with which they interact, constitute an **ecosystem.** An ecosystem is defined to be nearly self-contained, so that the matter which flows into and out of it is small compared to the quantities which are internally recycled in a continuous exchange of the essentials of life. The dynamics of the flow of energy and materials in a given geological environment, as well as the adaptation made by the individual and the species to find a place within the environment, constitute the subject matter of the ecology of natural systems.

Ecosystems differ widely with respect to size, location, weather patterns, and the types of animals and plants included. A watershed in New Hampshire, a Syrian desert, the Arctic ice cap, and Lake Michigan are all distinct ecosystems. Common to them all is a set of processes. In each there are plants which use energy from the sun to convert simple chemicals from the environment into complex, energy-rich tissues. Each houses various forms of plant-eaters, predators who eat the plant-eaters, predators who eat the predators, and organisms that cause decay.

Throughout this book we shall sometimes refer to ecosystems as geographically distinct areas with unique characteristics. But ecosystems are of course interconnected, so we shall also discuss the nature of their linkages.

The definitions of ecology and ecosystem are discussed in more detail in the marginal box on page 18.

1.2 ENERGY

(Problems 2, 4)

After reviewing some general concepts of energy, we shall apply them to the study of ecosystems.

Energy is the capacity to do work or to transfer heat. In ordinary speech we refer to "physical work" or "mental work" to describe a variety of activities that we think of as energetic. To the physicist, however, "work" has a very specific meaning: work is done on a body when a body is *forced to move*. Merely holding a heavy weight requires force but is not work, because the weight is not being moved. Lifting a weight, however, is work. Climbing a mountain or a flight of stairs is work. So is stretching a spring, or compressing a gas in a cylinder or in a balloon, because all of these activities move something.

When heat is transferred to a body, its temperature rises or its composition or state is changed, or both. For example, when heat is transferred to air, the air gets warmer; when ice is heated enough, it melts; when sugar is heated enough, it decomposes. All of these changes require energy.

Now imagine that you see an object lying on the ground. Does it have energy? If it can do work or transfer heat, the answer is yes. If it is a rock lying on a hillside, then you could nudge it with your toe and it would tumble down, doing work by hitting other objects during its descent and forcing them into motion. Therefore, it must have had energy by virtue of the potential to do work that was inherent in its hillside position. If it is a lump of coal, it could burn and be used to heat water, or cook food, and therefore it too has energy. If it is an apple, you could eat it and it would enable you to do work and to keep your body warm, and so an apple has energy (see Table 1.1).

Of course, if you roll an apple down a hill, it has energy as a falling body. It is proper to assign a quantity of energy to an apple as a food (one small apple contains 64 Calories) rather than as a falling body because people eat apples; they don't build power plants in orchards to generate electricity from the falling fruit. This means that the energy you assign to a body depends on the process you have in mind. A person who looks at a warm lake and says, "There is energy in all that water," may be thinking of its ability to spill over a dam and force turbines to generate electricity, or its ability to melt ice, or its potential for use in a fusion reacter (see Chapter 8), where some of its hydrogen would be converted to helium with accompanying release of energy. Therefore, we refer to the energy of a body only in relation to a specific process. The potential energy of the water to be used in the hydroelectric plant is the *difference* between its energy at the foot of the dam and its energy at the top of the dam. It doesn't matter what absolute value we assign to any one of the two states—it is the energy difference between them that counts.

The energy released when an apple falls 15 feet would be only about 1/1000 Calorie, Sir Isaac Newton's head notwithstanding.

Let us return to the apple. The apple has 64 Calories because that is the amount of energy released when it is converted by oxidation to carbon dioxide and water. It does not matter whether this oxidation is performed by metabolism in a human being, or in a worm, or by combustion in a fire. It is only the difference between the final and initial states that matters. But if the apple is left in a closed vessel where there is insufficient oxygen to convert it all to carbon dioxide, it may ferment to yield alcohol. It will thus have reached a different state and the amount of energy transferred will be different—only about two Calories. Of course, the alcohol thus produced may also be considered to have energy, because it could be burned completely to carbon dioxide and water, releasing another 62 Calories. So, either path yields 64 Calories (Fig. 1.1).

Energy is necessary to all living organisms because many biochemical reactions are energy-requiring. In complex ecosystems such as those that exist today, all organisms are interrelated with other organisms by the food- (energy-) gathering process. Therefore, the study of the energy relationships within an ecosystem are of primary interest to the ecologist.

TABLE 1.1 ENERGY: WHAT IT CAN DO

	HEAT A BODY ABOVE THE SURROUNDING TEMPERATURE	*WORK*
Apple	Eating it helps you to maintain your body temperature at 98.6°F, even when you are surrounded by air at 70°F.	Eating it helps you to be able to pull a loaded wagon up a hill.
Coal	Burning it keeps the inside of the house warm in winter.	Burning it produces heat that boils water that makes steam that drives a piston that turns a wheel that pulls a freight train up a hill.

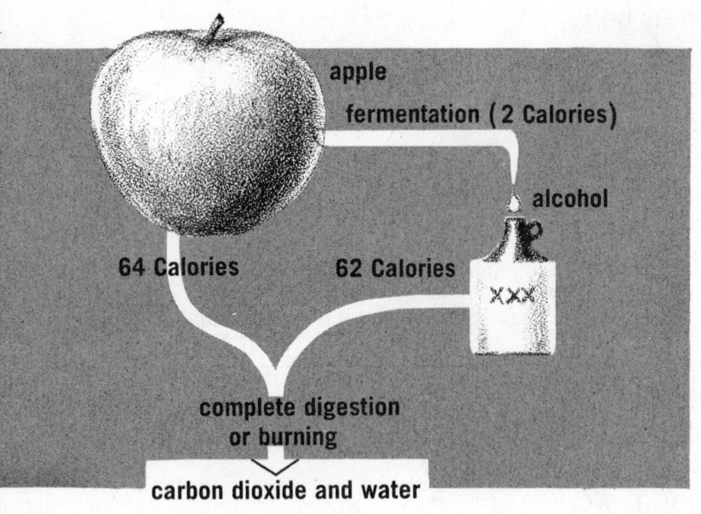

FIGURE 1.1 Energy from an apple.

1.3 ENERGY RELATIONSHIPS WITHIN AN ECOSYSTEM

(*Problems 3, 5, 6, 7, 8, 9, 10, 23*)

An auto- (self) troph (nourish) obtains energy by *itself* from the sun. A hetero- (other) troph obtains energy by eating *other* organisms.

A carnivore **is an organism that eats the flesh of animals (living or dead). Thus, carnivores occupy the third or higher trophic levels. The definition of** predator **is ambiguous. Some authors consider predator to be synonymous with carnivore, although this definition would include carrion-eaters, and thus would eliminate the sense of "attack" or "pillaging" which is the essence of the older meaning of predation. This text, as well as many others, uses a much broader definition: A predator is an organism that eats other living organisms. By this definition, a cow is a predator of grass.**

Plants are able to trap the sun's energy and transform it into chemical energy to build molecular structures such as those of sugars, starches, proteins, fats, and vitamins. For this reason plants are called **autotrophs,** meaning self-nourishers. All other organisms obtain their nourishment (energy) from other sources and are called **heterotrophs** (other-nourishers). This broad classification includes such widely diverse species as cows, grasshoppers, mountain lions, sharks, maggots, and amoebas. There are many ways to classify the heterotrophs, depending on the type of study one wishes to conduct. A comparative anatomist would make separate categories based on evolutionary and morphological similarities. An ecologist wishes to focus attention on function, specifically on the position of the organism in the energy flow. He is therefore interested in levels of nourishment, or **trophic levels.** Autotrophs occupy the *first trophic level.* All heterotrophs that obtain their energy directly from autotrophs are known as **primary consumers** and are said to occupy the *second trophic level.* They may be as different from each other as a grasshopper is from a cow, but both have similar ecological functions: they are grazers. Praying mantises eat grasshoppers, and owls eat field mice; therefore, both of these predators are secondary consumers. That is, they obtain energy from the plants only indirectly, in two steps, and are said to occupy the *third trophic level.* Let's take another step: shrews eat praying mantises, and martens eat owls; therefore, both are tertiary consumers. Owls who eat shrews are quaternary consumers, and martens who eat the owls who have eaten shrews are still another step removed from the original plant. Now just a minute—how can an owl be both a secondary and a quaternary consumer? The answer is that these

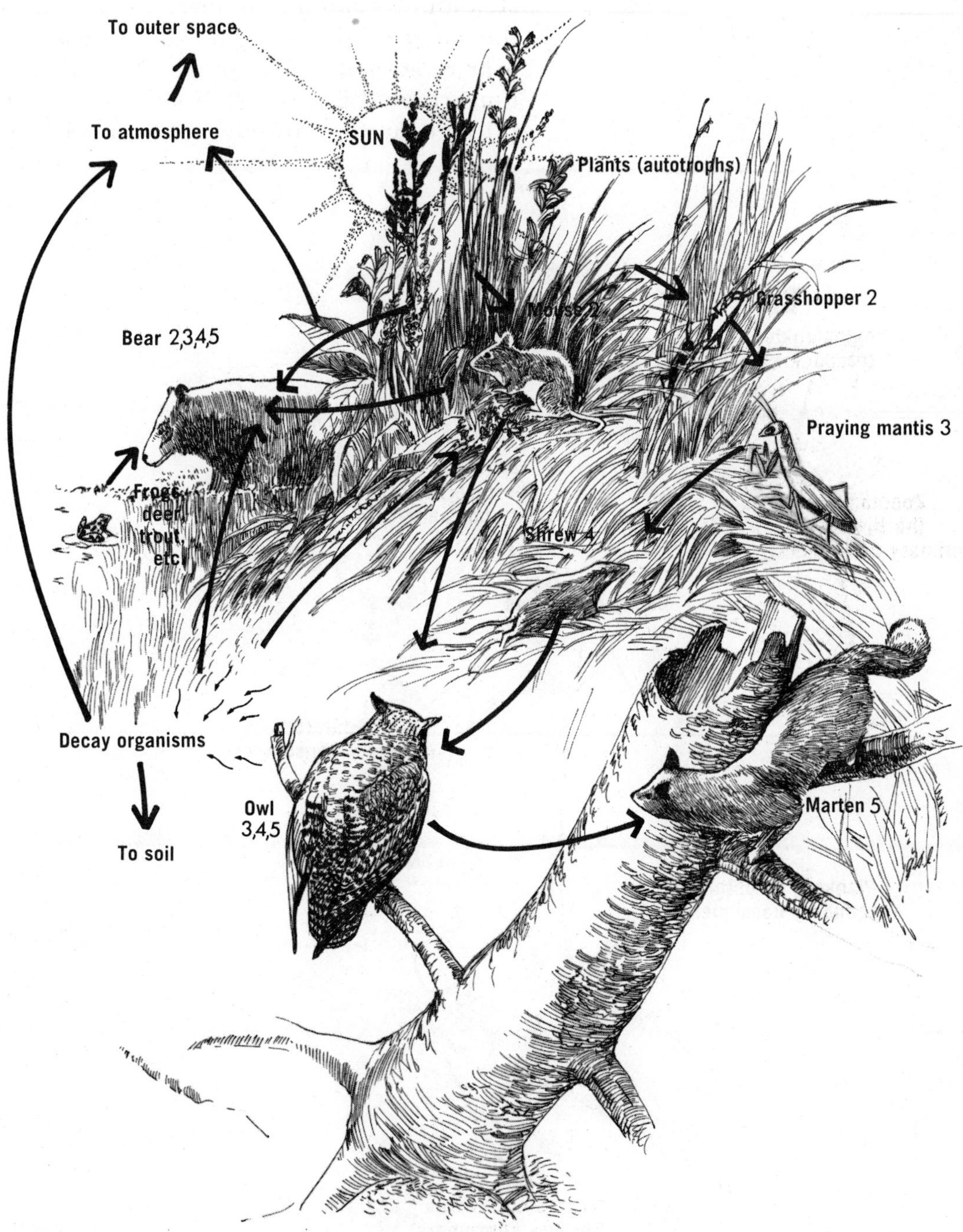

FIGURE 1.2 Land-based food web. Arrows are in the direction of progressive loss of energy available for life processes. Numerals refer to trophic levels.

categories are not mutually exclusive; a hungry owl doesn't care about human definitions, but dives at whatever is little and running about, regardless of how we classify him.

In very simple ecosystems the flow of food energy is said to progress through a food chain in which one step follows another. In most natural systems, such as the one partially outlined above, the term **food web** is a more accurate description of the observed interactions (Figs. 1.2 and 1.3). Food webs are further complicated by the presence of **omnivores,** species that

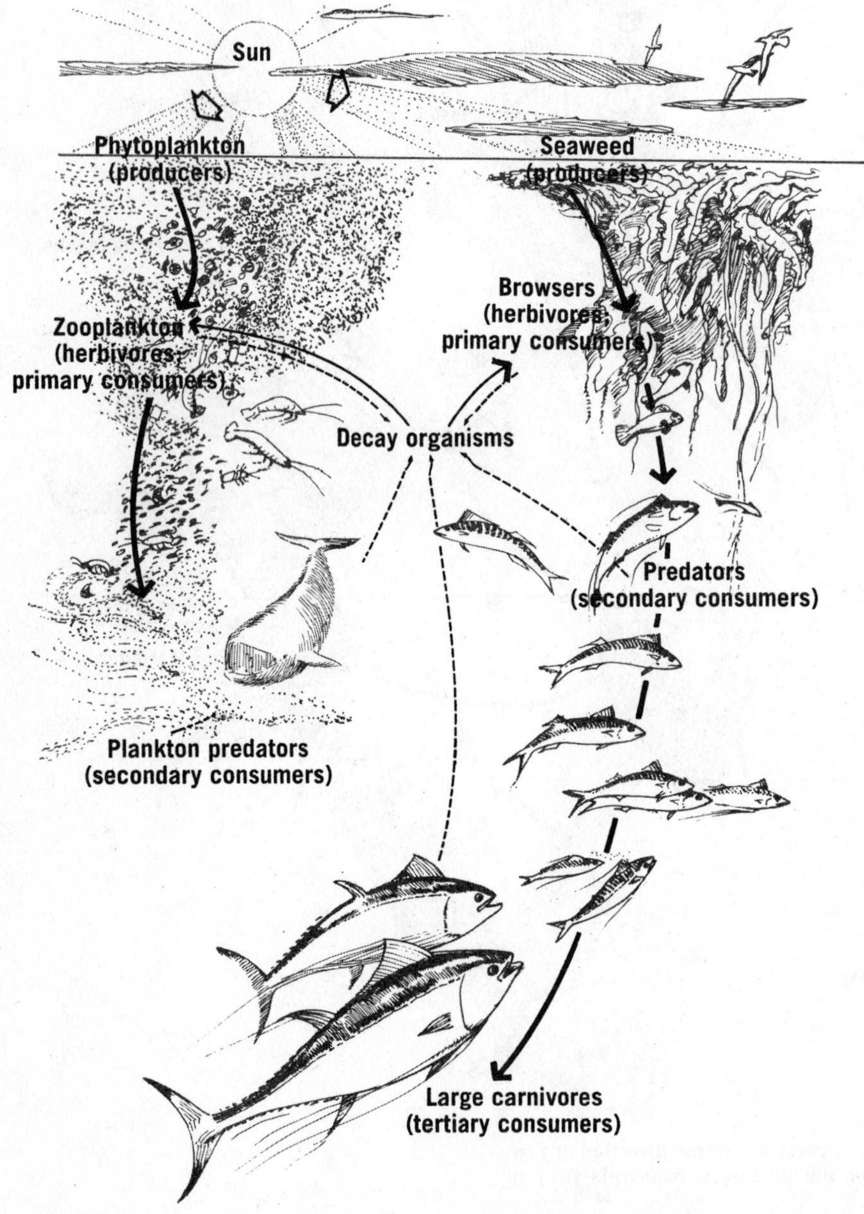

FIGURE 1.3 Aquatic food web.

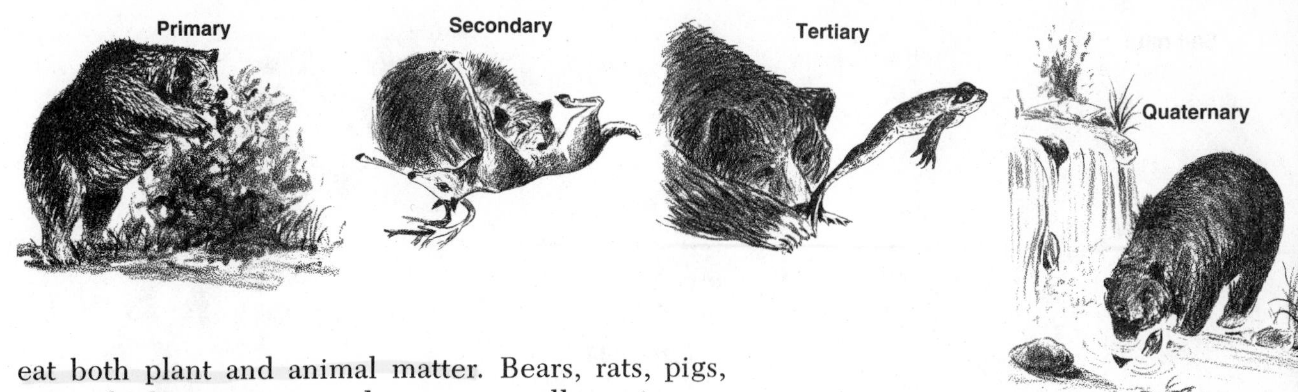

eat both plant and animal matter. Bears, rats, pigs, man, chickens, crows, and grouse are all omnivores. These species are often very difficult to place within any simple scheme. When a grizzly bear eats roots and berries, he is a primary consumer; when he eats a deer, he is a secondary consumer; eating a frog makes him a tertiary consumer; and eating a trout qualifies him as a quaternary consumer.

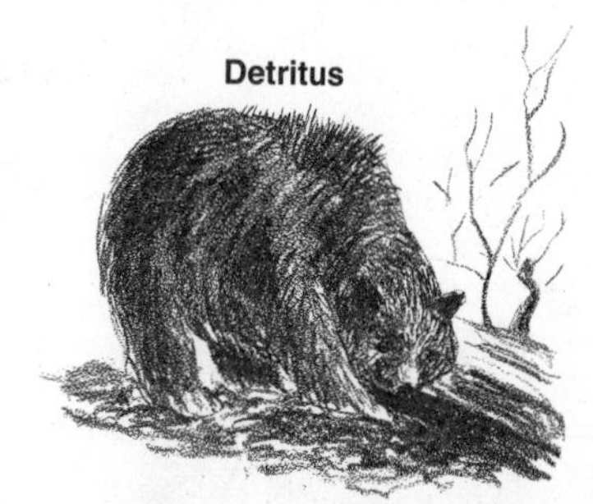

Of course, individuals often die without being immediately consumed, but instead remain to decay. The decay process is performed by many widely varying types of species, such as fungi, bacteria, soil mites, nematodes, ostracodes, and snails. These organisms, both plants and animals, are also classified by function and are called **saprophytes** (Fig. 1.4). Since saprophytes are also subject to predation, and some are omnivorous, the food web within a forest floor or on a lake bottom is quite complex. All the non-living organic matter in an ecosystem is known collectively as **detritus.** This debris is consumed by a group of organisms known as **detritus feeders.** Thousands of different species of plants and animals, and billions of individuals, feed on detritus. Although the complex interactions among the organisms are poorly understood, we do know that all stable ecosystems depend on the detritus food web to maintain stability.

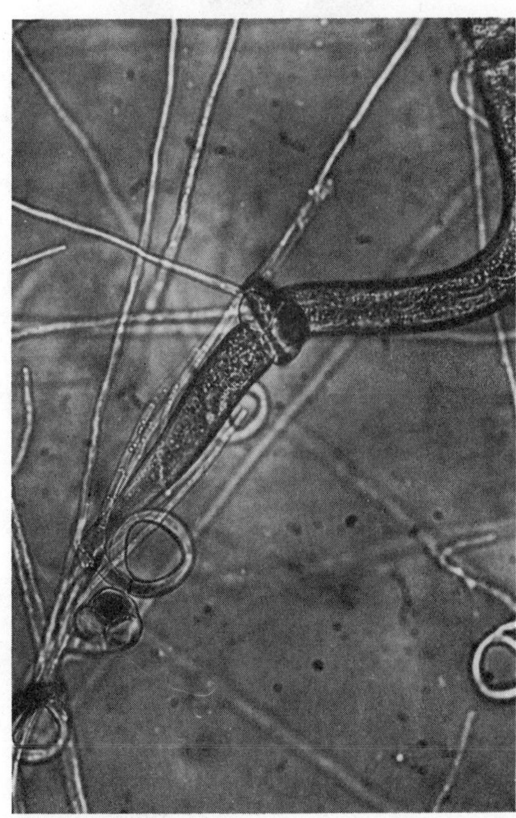

Nematode captured by the predaceous fungus *Dactylella dreschleri* **(× approx. 560). (Courtesy of Dr. David Pramer.)**

One characteristic of ecosystems must be reemphasized: Subclassifications are helpful in focusing attention on a particular aspect of study, but the lines are not sharply drawn in nature. Consider the bear (clearly not a decay organism) who rips open a rotting log in search of termites. In consuming them, he derives energy. Some energy has now been removed from the detritus food web and has entered the food web of the larger animals.

Energy is continuously received from the sun at a constant rate. Ultimately this heat is radiated off into space. Because we observe that the average annual

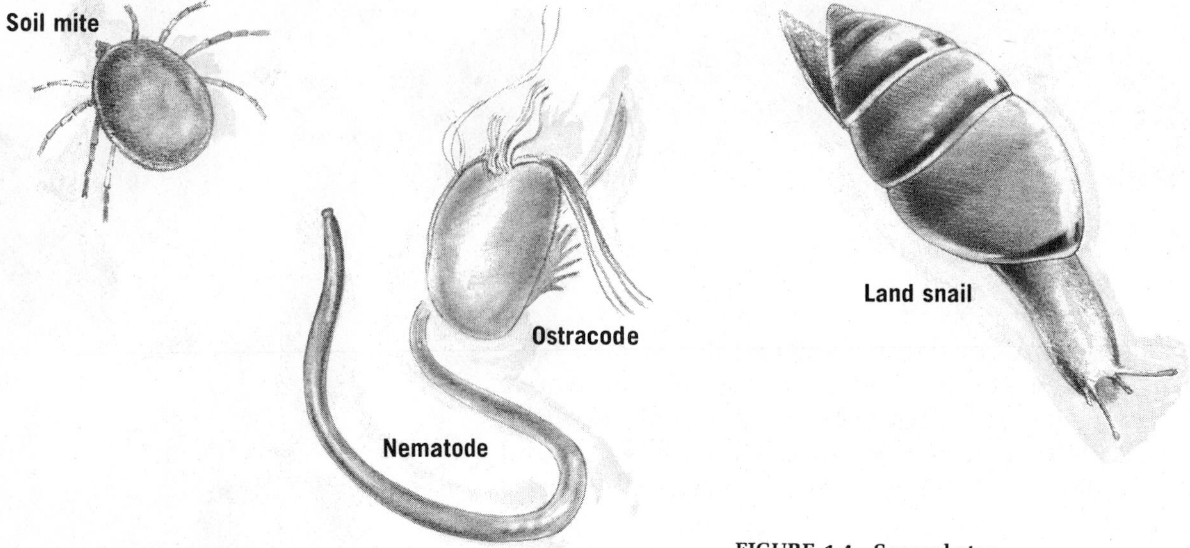

FIGURE 1.4 Saprophytes.

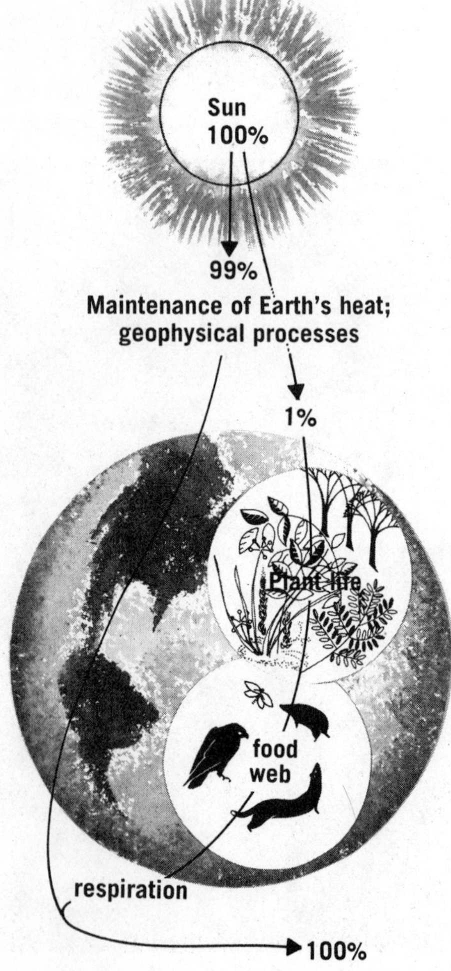

FIGURE 1.5 The flow of energy on Earth.

temperature of the earth remains relatively constant from year to year, we know that the energy gain and loss of the Earth must be in balance. The condition in which the inflow equals the outflow is called a **steady state.** The flow and ultimate fate of energy from the sun is represented schematically in Figure 1.5.

The chemical energy available for life processes constantly decreases through the food chain.* To understand the workings of an ecosystem, one must unravel the energy flow through the food web. Therefore, the ecologist is interested in the total production of organic tissue and the total consumption, which can be measured as respiration. The ratio of total (gross) production to total respiration is an indicator of the energy balance of the system; it tells us whether the system is growing, aging, or maintaining a steady state. Another important variable that is related to both the present and past production/respiration ratios is the **biomass** – the total mass of organic matter present at one time in an ecosystem. Finally, the *biomass at each trophic level* is of fundamental importance. For terrestrial ecosystems, a large primary production is required to support a small weight of predators, so that a bar graph of biomass of a self-contained ecosystem usually resembles a pyramid (Fig. 1.6).

We shall now follow the energy flow through a

*See Chapter 6 for a discussion of the First and Second Laws of Thermodynamics.

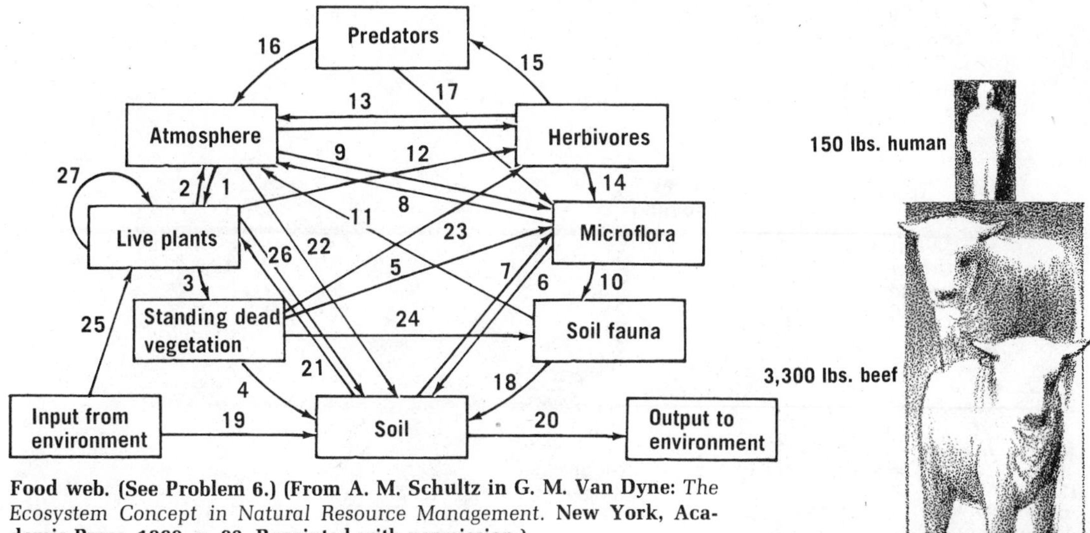

Food web. (See Problem 6.) (From A. M. Schultz in G. M. Van Dyne: *The Ecosystem Concept in Natural Resource Management.* **New York, Academic Press, 1969, p. 83. Reprinted with permission.)**

FIGURE 1.6 Food pyramid. The mass shown in each box is the amount required to produce the mass of tissue in the box above it. The areas in the boxes are proportional to the masses.

simple food chain. Consider a plant that receives 1000 Calories of light energy from the sun in a given day (Fig. 1.7). The efficiency of conversion of the sun's energy to chemical energy depends on the species of plant and the conditions of growth, but, in any case, most of the energy is not absorbed; instead, it is reflected or transmitted through the tissue. Of the energy that is absorbed, most is stored as heat and used for evaporation of water from leaf surfaces and other types of physical processes. Of the energy that is used in the life processes of the plant, most is expended in respiration. The remainder, about five calories, is stored in the plant tissue as energy-rich material, suitable as food for animals.

Now suppose a herbivore, say a deer, eats a plant containing five calories of food energy. She would dissipate about 90 per cent of the energy received to maintain her own metabolism as well as to retain some energy for muscle action to move around. As a result, the deer would convert only about ½ calorie to weight gain. A carnivore eating the deer is likewise inefficient in converting food to body weight, so the energy available to each succeeding trophic level is progressively diminished. The energy advantage of the herbivores is one important reason why there are so many more herbivores than carnivores. It is obvious that man, who can occupy primary, secondary, or tertiary positions in the food chain, uses the sun's energy most efficiently when he is a primary consumer, that is, when he eats plants.

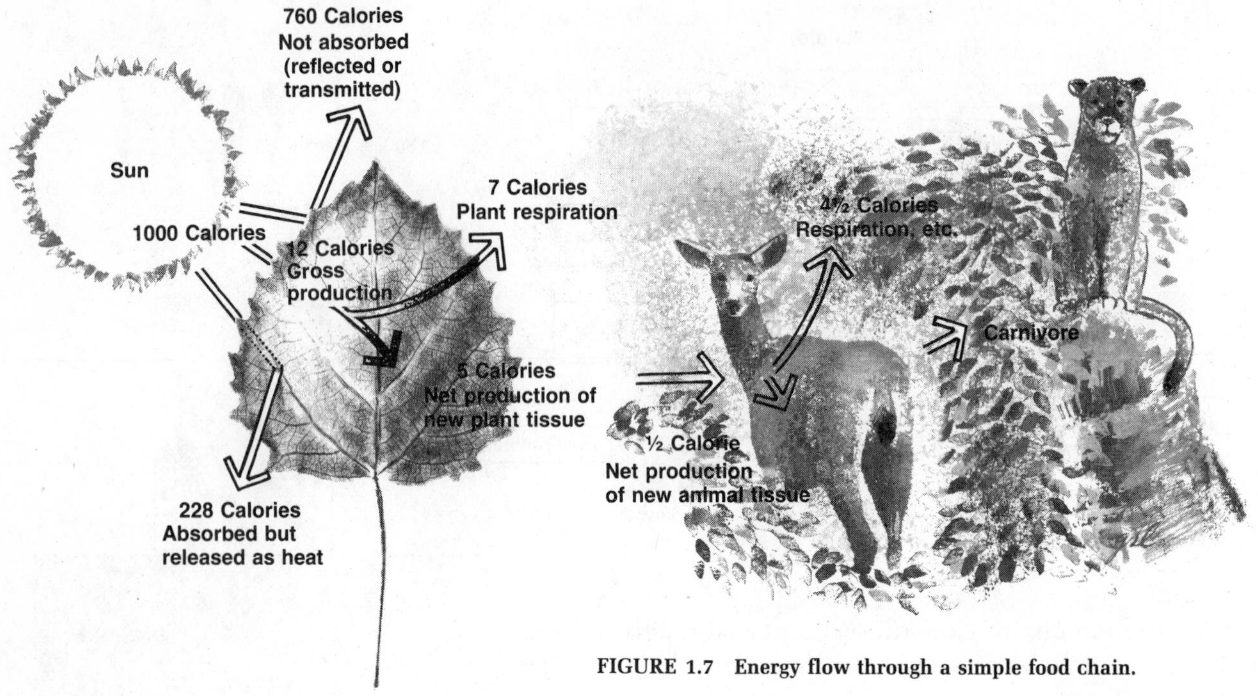

FIGURE 1.7 Energy flow through a simple food chain.

1.4 NUTRIENT CYCLES

(*Problems 11, 12, 13, 14, 15, 16, 17, 18*)

Energy alone is insufficient to support life. Imagine, for example, that some aquatic plants such as algae were sealed into a sterilized jar of pure water and exposed to adequate sunlight. The process of photosynthesis which utilizes the atmospheric carbon dioxide and releases oxygen would not be balanced by the plants' own respiration, and soon the plants would starve for lack of carbon dioxide. If the experimenter maintained the proper atmosphere through some gas-supply and exhaust lines, the plants still would not survive. They would starve for lack of the chemicals necessary for life. Suppose, then, that the experimenter fertilized the jar with the proper inorganic chemicals, which had been sterilized to insure that no additional living organisms entered the system. The plants would grow, and the cells which died in the normal processes of life would accumulate. Soon there would be no more room to introduce new fertilizer, and, unless the jar were enlarged, the fertilization would have to be stopped and the algae would die. The jar might be stabilized indefinitely, however, if some plant-consumer, a snail for example, were introduced, and the pure water were replaced by pond water to supply inorganic nutrients and saprophytes. Assuming an initially balanced ratio of snails to algae, the jar would now perhaps become a balanced, stable ecosystem in which the nutrients would be continuously recycled. The algae use water, carbon dioxide, sunlight, and the dissolved nutrients to support life and build tissue. Oxygen is one major waste product of algae and at the same time is an essential requirement of snails and other heterotrophs.

In general, plants produce more food than they need. This overproduction allows animals to eat parts of the plants, thus obtaining energy-rich sugars and other compounds. These compounds, in turn, are broken down during respiration, a process that uses oxygen and releases carbon dioxide. Other animal waste products, such as urine, though relatively energy-poor, are consumed by microorganisms in the pond water. The waste products of the microorganisms are even simpler molecules which can be reused by the algae as a source of raw materials. The consumption of dead tissue by aquatic microorganisms is essential both as a means of recycling raw materials and for waste disposal. The interaction among the various living species permits the community of the jar to survive indefinitely. In fact, biology laboratories have sealed aquariums in which life has survived for a decade or more. We would like to stress that the picture presented of this cycle has been greatly simplified. Cycles which occur in nature are so complicated that they are not fully understood today.

The biosphere itself can be considered to be a sealed jar, for although it receives a continuous supply of energy from outside, it exchanges very little matter with the rest of the universe. Thus, for all practical purposes, life started on this planet with a fixed supply of raw materials. There are finite quantities of each of the known elements. The chemical form and physical location of each element can be changed, but the quantity cannot.

This statement does not apply to radioactive elements.

A generalized nutrient cycle can be visualized as follows (Fig. 1.8): Suppose substance A is a vital raw material for a certain species of organism 1, and substance B is one of the organism's products. Substance B must then be a vital raw material for a second organism, which then produces a new product. This alternation of products and the organisms that consume them must continue until the cycle is complete; that is, until some organism produces our original substance A. When averaged over a long period of time, each prod-

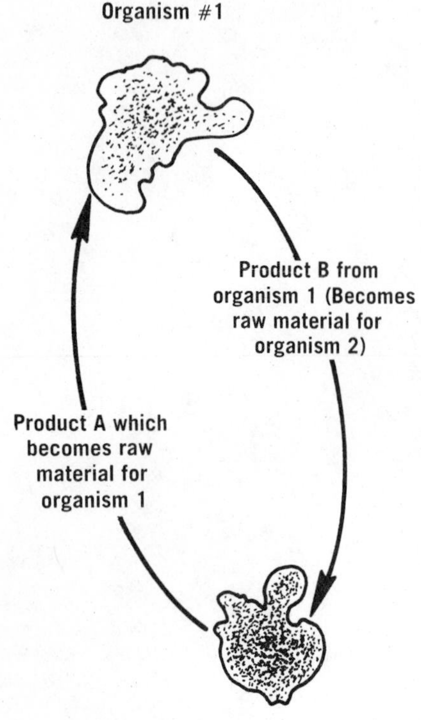

FIGURE 1.8 Generalized food cycle.

uct must be produced and consumed at the same rate; consumption must balance production. If any substance in the cycle, for example, substance B, were produced faster than it was consumed, it would begin to accumulate. The population of organism 2 would then increase in response to its greater food supply, and it would begin to consume more of substance B. Eventually the consumption of B would be increased enough to match its production, and the cycle would thus have regulated itself. Conversely, if B were produced at a rate *less* than the demand for it, the population of organism 2 would decline because of this insufficiency. The resulting decreased demand would enable production to catch up with consumption, and balance would be reestablished. Any permanent interruption of the cycle, for example by extinction of one of the species of organisms and the failure of any other organism to fulfill its function, would necessarily lead to the death of the entire system.

The substances produced by living organisms are widely varied. First, there are wastes associated with the act of living. Plants produce more oxygen than their life processes require; animals exhale carbon dioxide. However, organisms themselves can be viewed as products. A tree produces leaves, twigs, a trunk; a lion produces bones, a mane, a tail. All these parts of an organism, indeed the entire organism, when it is either living or dead, is a product which can be consumed by other organisms.

When studying nutrient cycles, it is helpful to separate those elements that can be found in large quantities in the air from those which are dissolved in water or located in the soil or the earth's crust. The continuous state of turbulence of the atmosphere insures a constant composition (omitting water vapor and air pollutants) throughout the earth. Moreover, gases in the air are relatively quickly assimilated and released by living organisms. The net result is the existence of a large, constant, readily available pool of nutrients. This does not mean an infinite, unchanging pool, for atmospheric composition has, in fact, changed considerably in the course of time.

The elements that occur naturally on Earth range from hydrogen (atomic number 1) to uranium (atomic number 92), but four atomic numbers are missing: technetium (43), promethium (61), astatine (85), and francium (87).

Of the 88 natural elements, about 40 are needed by living systems to maintain life. Many of these 40 are needed in only trace amounts, while others such as carbon, oxygen, hydrogen, and nitrogen constitute a large proportion of the mass of a system.

Oxygen Cycle

Oxygen atoms are present in the Earth in the following forms: (a) As molecular oxygen, O_2, in the atmosphere; (b) in water, H_2O; (c) in gaseous carbon dioxide, CO_2; (d) in many organic compounds, such as sugars, starches, and proteins; (e) in ions such as nitrate, NO_3^-, and carbonate, $CO_3^=$, that are dissolved in water; and (f) in many kinds of rocks, minerals, and other geological formations in the Earth's crust, such as limestone, $CaCO_3$, quartz, SiO_2, and metallic ores like bauxite (aluminum ore), Al_2O_3, and hematite (iron ore), Fe_2O_3 (Fig. 1.9). This last category accounts for most of the oxygen present in the biosphere. Most of the rapid *exchange of oxygen,* however, occurs by the action of living organisms (Fig. 1.10).

A plant synthesizes carbohydrates by combining carbon dioxide with water in the presence of sunlight and discharging oxygen as a byproduct:

$$6CO_2 + 6H_2O \xrightarrow{\text{sunlight}} \underset{\text{sugar, a carbohydrate}}{C_6H_{12}O_6} + 6O_2 \quad (1)$$

This process is called **photosynthesis.** Most molecules in plant tissue contain the element oxygen. Heterotrophs, which are unable to build sugars from carbon dioxide and water, consume plants and assimilate their complex chemicals, including the oxygen contained therein. As these chemicals move through the food web to consumers of other orders, they are all ultimately converted to carbon dioxide and water by

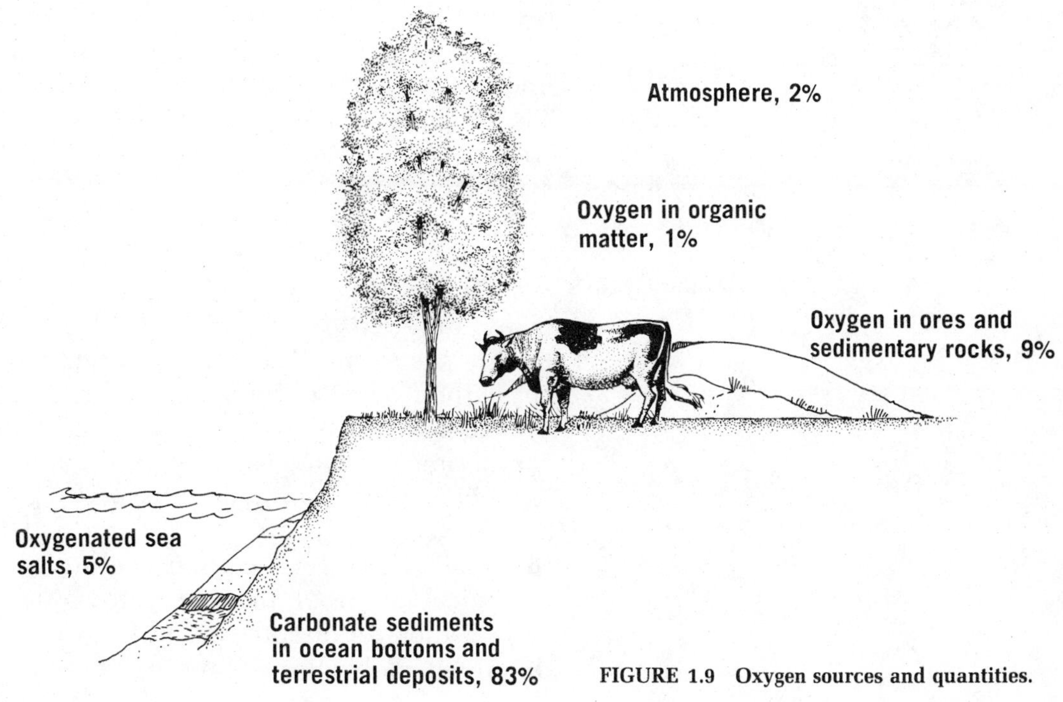

FIGURE 1.9 Oxygen sources and quantities.

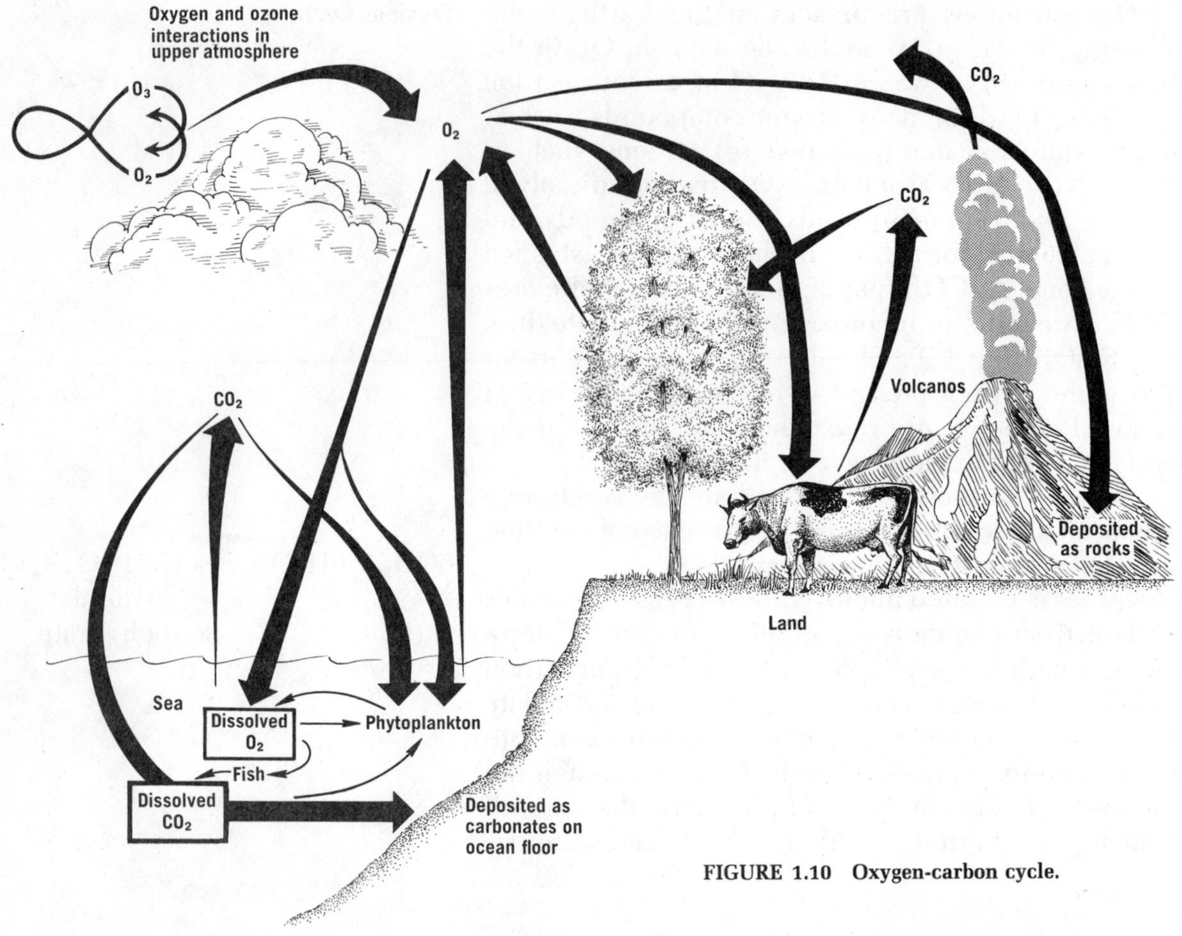

FIGURE 1.10 Oxygen-carbon cycle.

respiration or other oxidative processes. The organic oxygen participates in these conversions and so it, too, finds its way into CO_2 and H_2O molecules:

$$6O_2 + C_6H_{12}O_6 \xrightarrow{\text{respiration}} 6\ CO_2 + 6H_2O \qquad (2)$$

This completes the cycle, for the carbon dioxide and water may then be reused as raw materials for new synthesis.

Carbon Cycle

Carbon, like oxygen, is present in the Earth in many different forms. (See Table 1.2.) It can be readily seen from Equations 1 and 2 that the *biological* carbon cycle is intimately related to the oxygen cycle, for whenever oxygen is cycled by life processes, carbon must accompany it. (See Figure 1.10.) Carbon is incorporated into organic tissue by photosynthesis and released by respiration and decay. However, less than half of the total carbon cycling occurs through biological or other organic pathways. Atmospheric carbon dioxide dissolves readily in water, and some dissolved molecules escape from the sea to the air.

This exchange occurs as a dynamic equilibrium, that is, atmospheric carbon dioxide molecules are constantly being dissolved into the ocean, while those that were previously dissolved are constantly escaping from the sea to the air. There is little net change in the ratio of dissolved carbon compounds to atmospheric carbon compounds. This geochemical cycling is independent of any living process; it is an inherent property of the chemistry of carbon dioxide and water.

Some of the dissolved carbon dioxide reacts with sea water to form carbonates, which settle to the ocean floor as calcium carbonate either in the form of inorganic precipitates (limestone) or as skeletons of various forms of sea organisms. This loss is partially balanced by the action of inland water which slowly dissolves limestone deposits on land and carries the carbonates to sea.

Currently there is about 15 times as much carbon locked into fossil fuel deposits as there is present in the atmosphere. These deposits were caused by past imbalances in the carbon cycle. The burning of coal and oil since the Industrial Revolution has released much carbon, permitting it to reenter biotic cycles. In playing an active role in the long-term recycling processes of the biosphere, industry is measurably changing man's environment.

The carbon from fossil fuels is released into the atmosphere primarily as carbon dioxide. The atmospheric concentration of carbon dioxide before the Industrial Revolution is difficult to establish. Our best estimate, however, is that the worldwide CO_2 concentration has increased by 10 to 15 per cent during the past century, that is, from about 280 ppm to its present level of 315 ppm, and that this rate of increase is rising.

Not all the additional CO_2 released into the atmosphere by the burning of fossil fuels remains there—about two-thirds of it is reabsorbed into the biosphere. Some of this reabsorbed carbon is incorporated into plant tissue, some dissolves into the ocean, and most of the remainder is deposited into the Earth's crust as limestone and other sedimentary rocks. As man's use of fossil fuels continues to increase, the carbon dioxide concentration of the atmosphere also increases; we do not know, however, how the biosphere will respond to this change.

TABLE 1.2 SOURCES AND QUANTITIES OF CARBON IN THE BIOSPHERE

SOURCE	*QUANTITY (BILLIONS OF TONS)*	*PER CENT OF TOTAL*
Limestone, oil shale, and other sedimentary deposits	20,000,000	99.75
Oceans	35,000	.17
Coal and oil	10,000	.05
Dead organic matter	3,700	.018
Atmosphere	700	.003
Plants on land	450	.002
Plankton	10	.00005

Burning of Fossil Fuels

$$\underset{\text{coal}}{C} + O_2 \rightarrow CO_2$$

$$\underset{\text{octane}}{2C_8H_{18}} + 25O_2 \rightarrow 16CO_2 + 18H_2O$$

Perhaps most of the additional carbon will dissolve into the oceans, precipitate out as limestone, and lie relatively inert on the ocean floor. On the other hand, it is possible that more and more of the additional carbon dioxide will remain in the atmosphere. The consequences of this potential situation are hard to predict. We do know that plant growth is stimulated by small increases in the carbon dioxide concentration. Such enrichment is like fertilizing the air. We also know that carbon dioxide is a more efficient absorber of infrared (heat) rays than are most other atmospheric gases such as N_2 or O_2. Therefore, if the carbon dioxide concentration in the air increases, the Earth's atmosphere will absorb more heat, and its temperature is likely to rise. A significant rise in the temperature of the Earth could conceivably melt the polar ice caps and flood the coastal areas of the continents.

Scientists estimate that, if the rate of removal remains constant, a doubling of the present CO_2 concentration, which is an increase of about 300 ppm, would require 400 years and would warm the surface of the earth by an average of about two Celsius degrees (3.6 Fahrenheit degrees). Fluctuations of this magnitude have occurred in recorded history, and the glaciers didn't melt, so don't sell your beach property on this account.

Nitrogen Cycle

Nitrogen is an important constituent of proteins, and therefore necessary to both plants and animals. Although nitrogen is roughly four times as plentiful in the atmosphere as oxygen, it is chemically less accessible to most organisms. Almost every plant and animal can utilize atmospheric oxygen, but relatively few organisms can utilize atmospheric nitrogen (N_2) directly. The nitrogen cycle (Fig. 1.11) must therefore provide various bridges between the atmospheric reservoir and the biological community. Lightning, photochemical reactions, modern fertilizer factories, and specialized bacteria and algae (called nitrogen-fixers) transform molecular nitrogen into forms usable by living organisms. Usually, fixed nitrogen is assimilated by plants and then travels into the heterotrophic chain. It may now cycle for a considerable time within the food web. Organic decay might return it to the soil, or it might be digested and returned to the soil through feces or urine. Once in

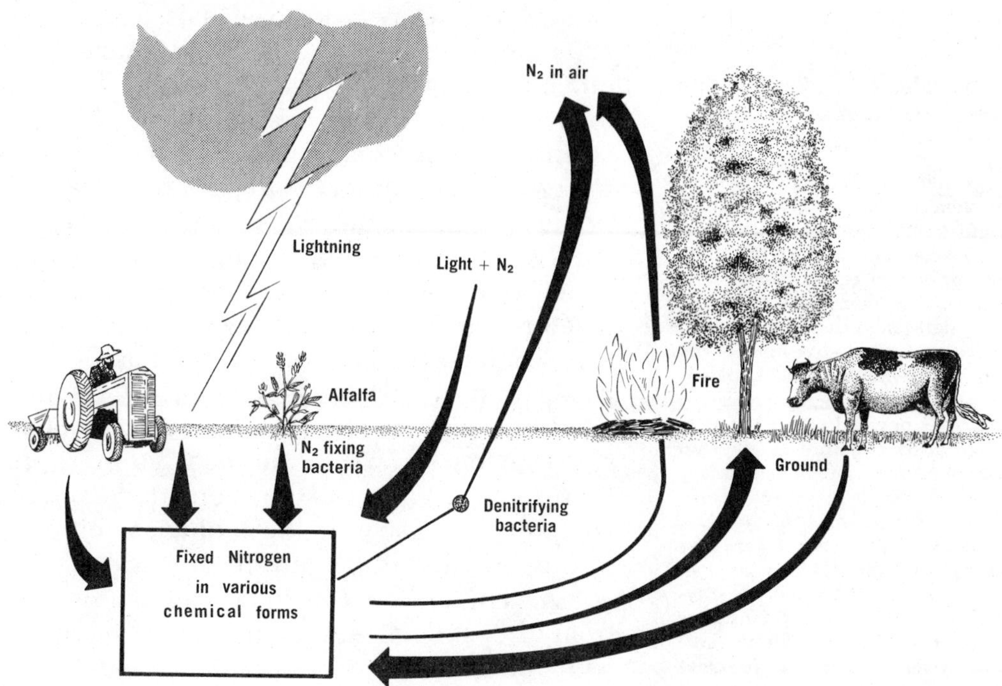

FIGURE 1.11 Pathways whereby nitrogen is removed from and returned to the atmosphere (nitrogen cycle).

the soil, it may reenter plant systems. Denitrifying bacteria and fire provide mechanisms for the return of nitrogen to the atmosphere. The rates of assimilation and denitrification are generally slow compared to the rates of nitrogen cycling. Therefore, unless some process removes nitrogen from an ecosystem, a given nitrogen atom may cycle many times before it is returned to the atmosphere. As a result, a large natural biomass can be supported by a small rate of nitrogen assimilation.

In agricultural systems, on the other hand, nitrogen-rich crops are harvested and generally trucked to distant cities. If the soil is to remain fertile, either the biological rate of nitrogen fixation must be high or man must artificially return nitrogen to the soil. Only a few organisms—for instance, the bacteria on the roots of legumes such as alfalfa, peas, and beans—are able to fix atmospheric nitrogen, that is, to convert N_2 to ammonium ion, NH_4^+. Most grain crops lack this ability, and therefore must be supplied with fixed nitrogen from some outside source if the fields are to be harvested annually. Nitrogen fertilization can be effected by any of three different techniques. The sim-

To "fix" means to make firm or stable. In this sense, a gas (which is not firm) can be fixed by binding it in some form of solid or liquid.

plest, and the method used most commonly by the ancients, is to recycle plant and animal matter. Another technique is by crop rotation, that is, planting legumes and grain in alternate years and thus maintaining soil nitrogen. Finally, man has recently learned to convert atmospheric nitrogen to plant fertilizer by producing ammonia ($N_2 + 3H_2 \rightarrow 2NH_3$), which is then readily converted to ammonium ion.

Today, enormous quantities of ammonia are manufactured in this manner; by some estimates, industrial fixation accounts for *one-third* of the total annual production of nitrogen compounds in the biosphere.

Many different types of nitrogen compounds exist simultaneously within an ecosystem. Figure 1.12 represents a simplified schematic diagram of the complex chemistry of the biological nitrogen cycle. Let us examine this figure in some detail:

A. The roots of the corn plant assimilate nitrogen from the soil as either ammonia, NH_3, as nitrite, NO_2^-, or as nitrate, NO_3^-.

B. The plant uses these simple inorganic compounds as raw materials for the synthesis of proteins. The plant must expend energy to produce proteins. This energy is stored in the molecule of protein and may be released at some later time if the tissue is burned or eaten.

C. In the diagram, some plant tissue is eaten and digested by a cow. Some of the plant proteins are oxidized with the release of energy-poor nitrogen compounds. The released energy is utilized by the cow to meet general metabolic requirements, such as respiration, chemical synthesis, and movement.

D. Some of the plant proteins are only partially decomposed, and the fragments are used for the synthesis of complex, energy-rich animal proteins. If the cow is eaten by a carnivore (not shown in Figure 1.12), these animal proteins will be decomposed with the release of energy and the production of energy-poor organic nitrogen compounds.

E. Energy-poor nitrogen compounds are excreted in the cow's feces.

F. Animals do not utilize all the energy available in plant proteins, and the compounds excreted by the cow can still be oxidized further with the release of additional energy. Certain decay organisms, mostly bacteria, eat the feces, digest the organic nitrogen compounds, utilize the energy which is released, and excrete ammonia as a by-product.

The Problems in Establishing Operational Definitions of Ecology and Ecosystem

Ecology, like economics, derives from the Greek stem eco-, or house. Economics, in its original sense, meant the management of the household, and, in the same spirit, ecology may be called "the science of the economy of animals and plants." When a physiologist studies an organism, he investigates the relationships of its component parts, while an ecologist delves into the relationships of that organism to others and to its physical environment.

Such an abstract definition gives no clue as to how to draw the boundary around an ecosystem. In some sense, every organism on Earth has some relationship, however small, with every other, and Earth itself depends on the sun for light and heat, the moon for tides, and the gravitational fields of the heavenly bodies for its course in space. Yet to call the universe the only ecosystem is frivolous. No ecologist studies every biological and physical phenomenon which impinges upon an organism. Operationally meaningful ecosystems must be smaller units with boundaries defined by the study being performed. If the ecology of an island is investigated, that entire island is regarded as an ecosystem, and the flow of material and energy into and away from the island is considered in the aggregate. If the ecology of a lake on the island is studied, the lake is called the ecosystem, and the relations between the lake and the rest of the island are regarded in the aggregate.

Of course, if the system to be studied is too small, the concept of an ecosystem is irrelevant. An individual organ or organism is not studied by an ecologist except as it relates to other species and to the physical world. A study of the mechanism by which a barnacle attaches to a rock would also not be considered an ecological investigation, even though it relates an organism to its physical environment. Even a study of the interrelationships among organisms is not necessarily ecological. Thus, a study of a virus in man, although it involves two organisms, is not called ecology.

What then, transforms an ordinary biological study into an ecological one? We can say that when a biologist studies an organism as it relates to at least one other or to its physical environment, and when that biologist says he is an ecologist, the study is ecological. Then the ecosystem is the smallest unit of environment that includes the study.

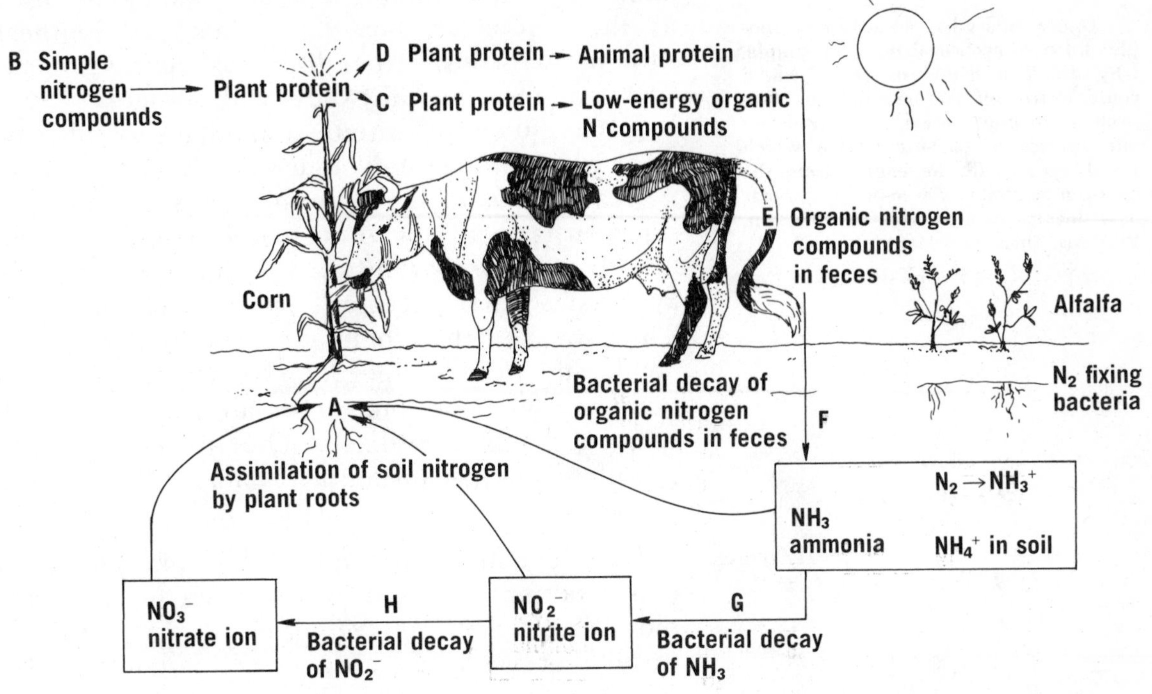

FIGURE 1.12 Chemistry of the nitrogen cycle.

G. The ammonia is usually released into the soil, and from here it may be reassimilated by plant roots. Moreover, ammonia also contains stored energy; it too can be burned in a fire or oxidized in the digestive systems of some organisms. Certain bacteria absorb ammonia, utilize the energy released during oxidation, and release nitrite, NO_2^-, as a waste product.

H. Nitrite ion, NO_2^-, is oxidized to nitrate, NO_3^-, with the release of additional energy. Nitrate ion has the lowest energy in the nitrogen cycle; it cannot be oxidized further.

Back to A. Plants can assimilate either ammonia, nitrite, or nitrate as a source of nitrogen for protein synthesis. Of these three sources, nitrate has the least chemical energy, and ammonia has the most. To understand the concept of chemical energy, consider this analogy. Suppose that you wish to build a house. At the lumber store you are given a choice: you can buy individual boards, or you can buy prefabricated walls and roof trusses. Construction of the house from boards requires more sawing and hammering (work energy) than construction from the prefabricated sections. Thus, the prefabricated pieces contain stored energy. Where did the stored energy originate from? Someone in a factory sawed boards and nailed them together. We can think of nitrate as the boards,

The Indians of the Great Plains used buffalo chips (dried buffalo manure) for fuel. Today, women in India collect cow dung, dry it, and use it in their cooking fires.

If you have ever been in a cow barn, you might recall the strong odor of ammonia, which is released during bacterial decomposition of urine and manure.

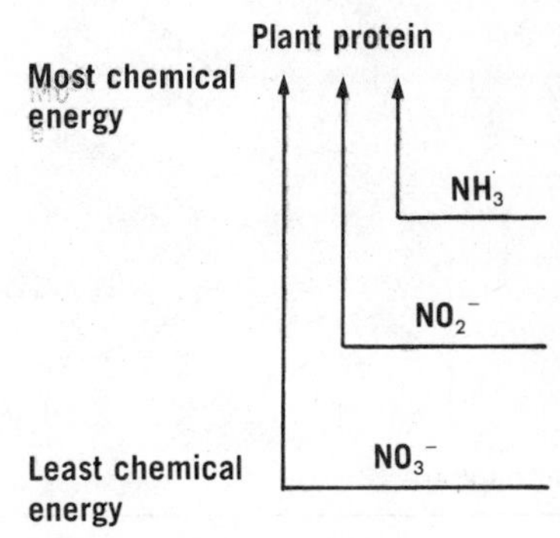

One wonders how or why this complex nitrogen cycle evolved. For example, why didn't a herbivore evolve which could utilize *all* the energy from plant proteins in one process? Or why didn't one species of bacteria evolve which could extract all the energy from the nitrogen compounds in manure? Chapter 3 is devoted to answering the question, Why Are There This Many Species?

ammonia as the prefabricated units, and plant protein as the completed house. A plant can synthesize proteins with less expenditure of energy if it starts with ammonia than if it starts with nitrates.

Why, then, is nitrate assimilated at all? Why don't plants utilize only ammonia? Energy from light is readily available to most plants, whereas soil nitrogen is more apt to be scarce. Plants, therefore, absorb energy-poor nitrogen compounds and then synthesize the necessary energy-rich molecules. Thus, we see that sometimes energy-rich nitrogen compounds release energy in living cells, and sometimes energy-poor compounds are ingested, and the element itself is utilized. The overall result is a complete and self-sustaining cycle.

Mineral Cycles

The cycling of minerals such as phosphorus, calcium, sodium, potassium, magnesium, or iron is much more fragile than the cycling of oxygen, nitrogen, or carbon because of the absence of a large mineral reservoir in a readily available form. (See Figure 1.13.) Let us consider the mineral cycles in a particular watershed in the mountains of New Hampshire. This area includes a ring of mountains and the enclosed valley. Rain is the area's only significant source of

For the details of this example, refer to Chapter 4 of G. M. Van Dyne, *The Ecosystem Concept in Natural Resource Management.* (See references at the end of this chapter.)

FIGURE 1.13 Mineral cycle.

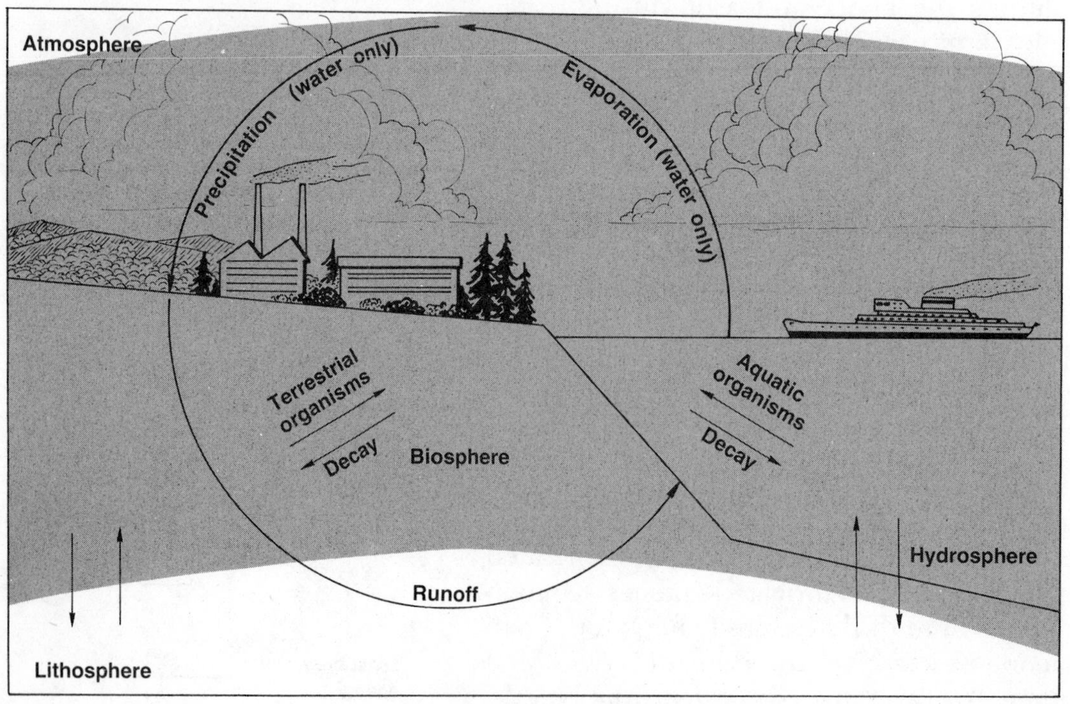

FIGURE 1.14 Calcium balance in a New Hampshire watershed.

water; the only important exit for flowing water in this watershed is through a single stream, because the geology of the area is such that seepage is negligible. In such a system the total input of minerals is limited to two processes: the rain deposits some mineral matter, and weathering of rocks frees some minerals from the earth's crust. This gain is partially balanced by losses.

In the particular watershed that was studied, it appeared that the stream outflow removed fewer minerals than weathering and rainwater were bringing in. (See Figure 1.14.) However, the net gain for the ecosystem was very small compared to the requirement of all the life forms in the valley. This means that the valley life was operating on mineral capital, not mineral income. In fact, the entire biosphere depends for its supply of mineral matter upon the ability of each individual incoming inorganic ion to undergo countless transformations from soil to plant tissue, from plant tissue into the organic detritus or on to animal tissue, and then to the detritus, back to the soil, and around again and again before being washed out of the system. Therefore, input and output rates bear little relationship to the quantity of inorganic matter in the private pool of any given ecosystem.

How important is the presence of healthy vegetation in maintaining this delicate mineral balance? In order to study this problem, all the trees, saplings, and brush in one watershed were cut or chemically killed. No organic matter was removed, so that outright soil erosion could be minimized. In the year following this clearcutting experiment, the stream flow was 40 per cent greater than it would have been had the system remained undisturbed. The water carried away many mineral salts, including calcium, and the net nutrient drain was then greater than the net input.

Transpire (*Latin*, breathe through) means to pass a vapor (usually moisture) through a porous barrier, such as skin or leaves. The word is also used in the sense of leaking out, as of information, or becoming known, or very loosely, coming to pass, or occurring.

In a healthy ecosystem, water and nutrients in the soil are absorbed by plants and carried up to leaf tissues. The nutrients are generally assimilated, but much of the water is lost through **transpiration,** evaporation from leaf surfaces. In a dead system no water is lost through transpiration, and therefore most of the rainwater must ultimately flow into the stream bed. Since all ground water carries nutrients out of the system, a large increase in stream flow disrupts the nutrient balance.

1.5 ECOSYSTEMS AND NATURAL BALANCE

(Problems 19, 20)

We have seen that food webs perpetuate the flow of energy and the cycle of nutrients. Naturally, all processes are being carried on simultaneously. A diagram illustrating all the energy and nutrient transfers of even a simple system would be a blur of arrows showing sugars, mineral ions, nitrogens, oxygens, and carbons going this way and that. How is the ensemble regulated? How can a natural meadow exist in an orderly and relatively unchanged state for many years?

Many opposing forces operate within a natural ecosystem. (See Figure 1.15.) Organisms are born and die. Moisture and nutrients travel out of the soil and they are transferred back into the soil. Furthermore, many of these oppositions are exquisitely protected against disruption. During a dry season, when the mice in a grassland have less food, their birth rate decreases. But their behavioral response to lack of food is to retreat to their burrows, and thus their death rate also decreases because they are less exposed to predation. Their behavior protects their own population balance as well as that of the grasses, which are not consumed by hibernating mice. Such a tendency is called **ecosystem homeostasis.**

The "balance of nature" thus refers to the tendency of natural ecosystems to maintain their existence by appropriate opposition of processes and by regulatory mechanisms which protect these processes against disruptions.

How does this dynamic balance operate; how is it controlled? One clue comes from the observation that in natural, stable ecosystems the biomass is quite large compared with the biomass of less stable systems. Some areas are incapable of supporting a large biomass; in deserts or in tundra the rate of growth and of accumulation of litter is slow, and massive plants such as trees cannot survive. However, the systems of greatest stability (the climax structures, see p. 68) in these harsh areas have a large biomass compared with emergent systems in the same climatic region.

Stable ecosystems are not necessarily in balance all the time, but if they are imbalanced in one direction, they must become imbalanced in the opposite direction at some time in the future if they are to survive with no essential change in character. In fact, all ecosystems naturally fluctuate. For example, climate

Constancy of a system does not, of course, imply constancy for each individual, for balance is achieved not by stagnancy but by an equality of oppositions. The carbon cycle is balanced when the absorption of CO_2 matches the release of CO_2; a scale is balanced when the weights or forces on each pan are equal; population size is considered to be balanced when the opposing processes, the birth rate and the death rate, are equal. What happens when a balance is disturbed? That depends on the nature of the system. Some systems go out of balance easily; others resist change. As an example of the latter type, your blood is balanced between acidity and alkalinity; that is, opposing chemical processes maintain a constant level (one that is very slightly alkaline). Now suppose a certain quantity of acid is added to your blood. The chemical composition of blood is such that it will tend to oppose this disturbance of its acid-alkali balance and to return toward its normal condition. Such resistance to disturbance of acid-alkali balance is called buffering. **Buffering is thus a protective action. An organism tends to maintain a balance of various life processes by feeding itself, keeping itself in repair (healing itself), and adjusting to external changes. This tendency to maintain a stable internal environment is called** homeostasis.

FIGURE 1.15 Moist coniferous forest, a stable ecosystem in the Olympic National Forest, Washington. (U.S. Forest Service photo.)

varies from year to year. Other disruptions, such as migration, drought, flood, fire, or unseasonable frost can cause imbalance in an ecosystem. The ability of an ecosystem to survive depends on its ability to adjust to an imbalance.

Let us consider as an example the relative populations of tigers, grazing animals, and grass in a valley in Nepal. Assume that the mountains surrounding the valley are so high that no animals can enter or leave the valley. One year rainfall is limited and mild drought conditions exist. Because water is naturally stored in pools, the drinking water supply is adequate. However, the grasslands suffer from the lack of rainfall. Therefore, food is scarce for the grazing animals, and many are hungry and weak. Such a situation is actually beneficial to the tigers because hunting becomes easier, and the tiger population thrives.

The following spring, when rains finally arrive, the grazing population is low. This permits the grass to grow back strongly because not much of it will be eaten. However, because the food supply for the tigers is low (and the grazers that are left are the strongest of the original herd), the tigers suffer a difficult time the summer after the drought. The third year there are fewer tigers, so the herd resumes its full strength and balance is achieved again.

Stable ecosystems, if examined superficially, do not seem to go out of balance at all. Actually, their homeostatic mechanisms function so well that slight imbalances are corrected before they become severe. Such sensitive responses are, in fact, the essence of stability, because if severe imbalances are prevented, it is very unlikely that the system will be destroyed and, as a result, it lasts for a long time. Usually, when we refer to a stable ecosystem, we mean one which regenerates itself again and again with little overall change. Sometimes, however, an ecosystem is stable not because it regenerates often without change, but because the dominant plants are long-lived. A wonderful example is a redwood forest. To the hiker entering a redwood forest on a hot summer day, two characteristics of the environment are immediately apparent. It is cool, and there is a thick floor of spongy matter. The coolness, caused by the extensive shade of the tall trees, reduces water loss due to evaporation. The spongy floor is formed by a large bed of decomposed organic matter known as

humus, as well as a liberal supply of fallen needles and other as yet undecomposed litter that catches and holds the rainwater, and thus serves as a water and nutrient reservoir, and as a home for the members of the detritus food chain (Fig. 1.16). Thus, the effect of a drought does not have to be balanced by the death of trees. In addition, the large amount of organic material available ensures a steady functioning of the recycling system. Thick mats of partially decomposed organic matter also retain the heat produced by saprophytes and help maintain a warm environment known as a **microclimate**, which is often more favorable for metabolism than the climate on the surface. Anyone who has ever operated a compost heap has observed this warming effect.

The coolness of the forest offers still another advantage. Clouds passing overhead are more likely to discharge their moisture where it is cool, for cold enhances the tendency of water to agglomerate into raindrops. Thus, the redwood forest tends to cause rain, to restrict evaporation, and to retain a substantial supply of water. Furthermore, decay organisms require moist environments to be effective, so the redwood forest insures the proper recycling of materials by environmental control.

This unusually stable ecosystem also exhibits other highly effective homeostatic abilities. One of these can be observed readily by walking through the forest on the west side of Route 1 as it runs through the town of Big Sur, California. Some time ago a tremendous fire swept through the forest. However, because the bark and the wood of the redwood tree are fire-resistant, the big trees were scarred but not killed. Because the forest environment was maintained, the forest community was quickly regenerated. If a fire of equal intensity had burned through a stand of pine, the trees would have died and many more years would have been needed for the forest to be rebuilt.

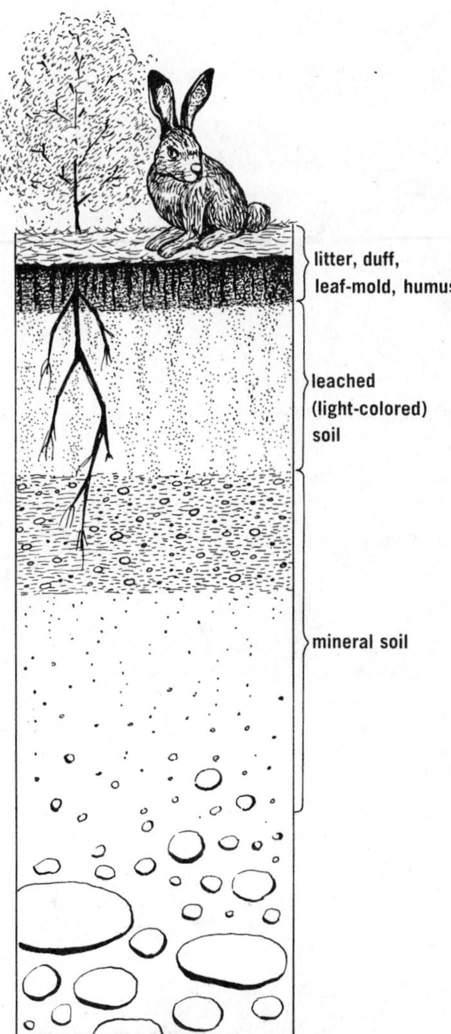

FIGURE 1.16 Cross section of forest soil.

1.6 MAJOR ECOSYSTEMS OF THE EARTH

(Problems 1, 21, 22)

The **biosphere** is the region including all the life-supporting portions of our planet and its atmosphere. This section briefly describes some of the major ecosystems of the biosphere.

Ocean Systems

The ocean holds a varied and intricately interwoven set of ecosystems.

The ocean. (Courtesy of H. Armstrong Roberts.)

Despite the fact that the ocean is known to be more than 19,000 feet (6000 meters) deep in places, light does not penetrate more than about 650 feet (200 meters) in sufficient quantities for photosynthesis to occur. The depth of this illuminated section, or **euphotic zone,** varies considerably with the turbidity of the water, reaching its maximum limits in the central oceans and narrowing to 100 feet (30 meters) in coastal regions. Although the depth of the photosynthetic zone is greatest in the central ocean, the rate of photosynthesis is not greatest there, for the concentration of plant nutrients is low. Indeed, the central ocean has often been likened to a great desert.

FIGURE 1.17 (Courtesy of Professor John Lee, Dept. of Biology, The City College of the City University of New York.)

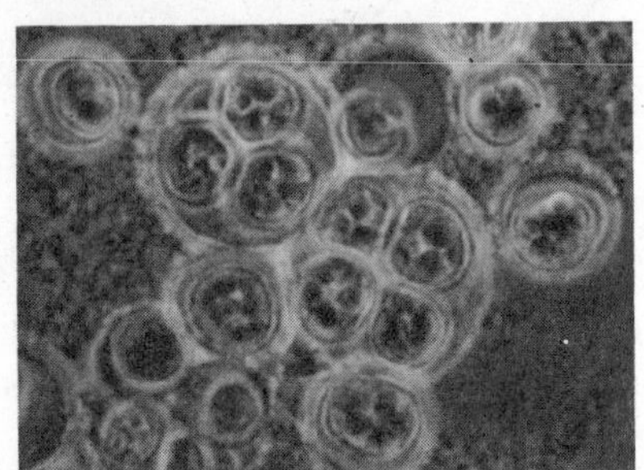

A

A nonmotile marine phytoplankter.

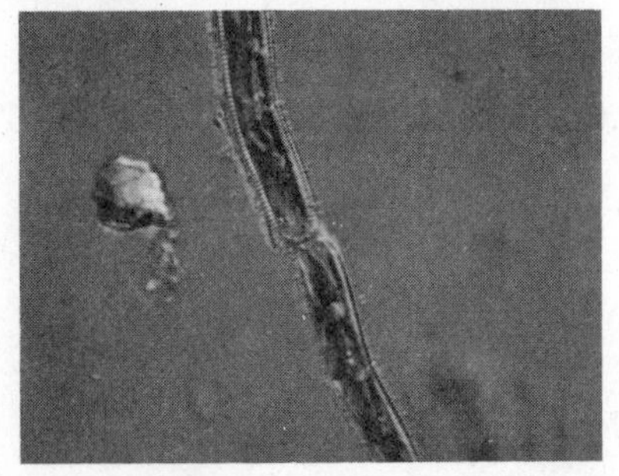

B

A microscopic marine plant.

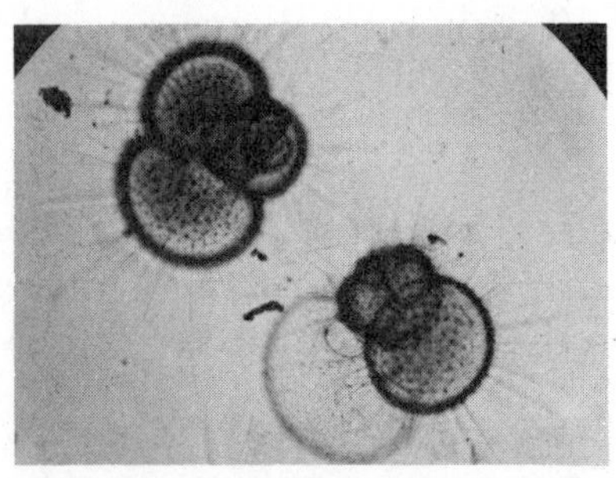

C

A planktonic marine animal *Globigerina bulloides.* Large areas at the bottom of the sea are covered by their remains.

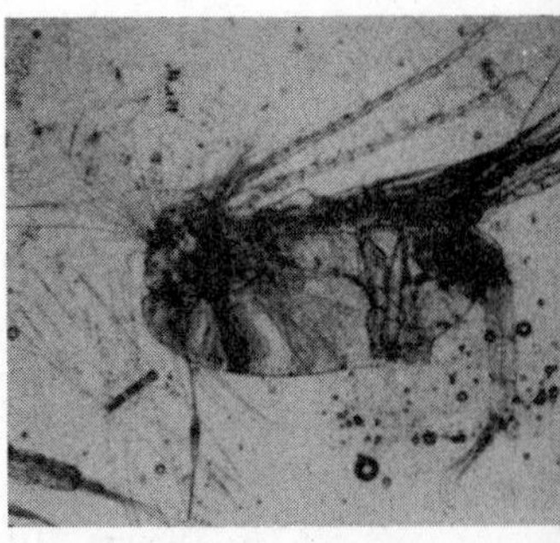

D

Another planktonic animal: ma copepod.

The primary autotrophs in all marine ecosystems are phytoplankton, the one-celled chlorophyll-bearing organisms suspended in the water (Fig. 1.17). In areas near shore, various species of algae, commonly known as seaweed, are abundant and account for a significant portion of the net production of living matter.

A large portion of the primary consumption in the sea is carried out by myriads of species of small

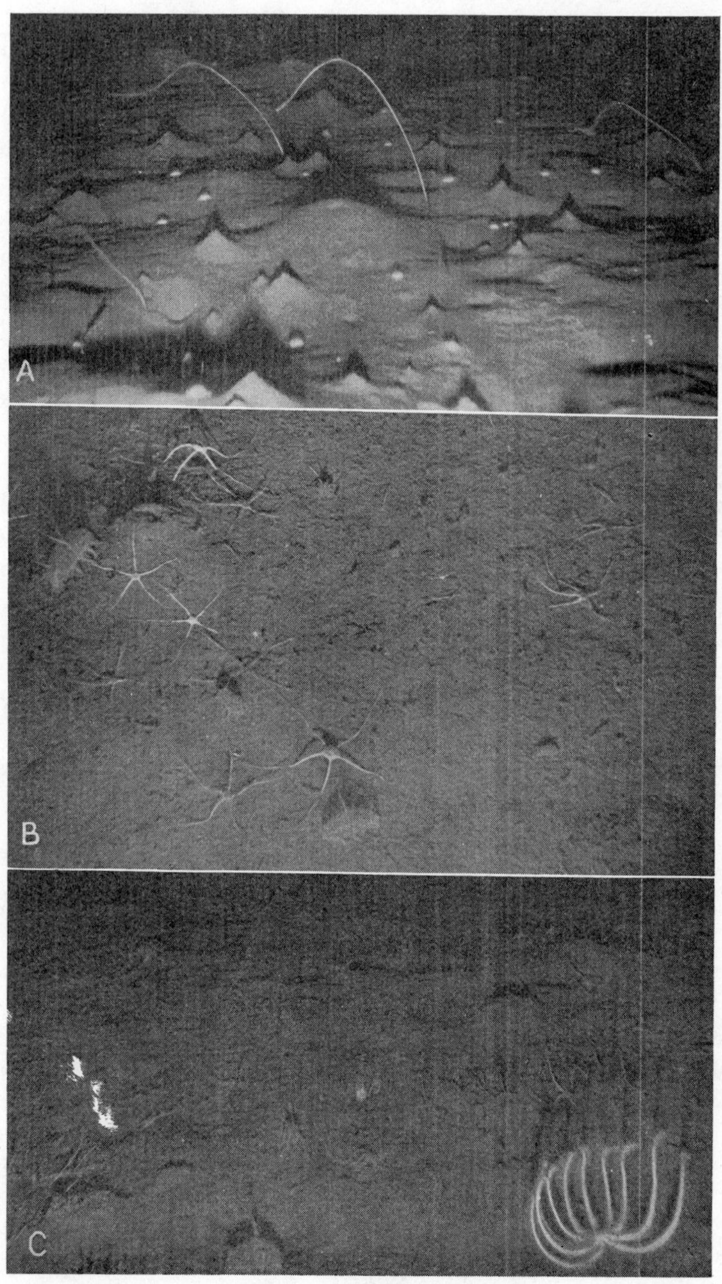

FIGURE 1.18 Benthic species. (From Odum: *Fundamentals of Ecology*. **3rd Ed. Philadelphia, W. B. Saunders Co., 1971.)**

grazers called zooplankton (Fig. 1.17). These animals range in size from roughly 0.2 mm (.008 in.) to 20 mm (0.8 in.) in diameter and consist both of permanent zooplankton and larvae of larger forms of life. The composition by species of a sample of plankton varies with the climate, and plankton occupies several trophic levels. There are few large grazers in the sea analogous to land-based bison, deer, or cattle.

Many omnivores and predators live in the sea. These species represent various biological classifications, including fish, mammals, invertebrates, and reptiles. Organic debris from the life processes on the surface and in the body of the sea are continuously raining down toward the ocean floor. In the deep ocean, some of this food is intercepted, some decomposes, and some falls to the bottom, where **benthic species** (organisms that live in and on the bottom sediments) receive their nourishment (Fig. 1.18). Some of these are semimobile or immobile animals, while others crawl or swim actively in search of food.

While gravity is constantly pulling nutrients to the bottom of the sea, the mechanisms for upward recycling of nutrients are comparatively feeble. Currents and water turbulence serve to move some nutrients back to the surface, but a net loss does occur to the deep sea floor. These chemicals may eventually be recycled by geological upheavals. One net result of poor recycling in the ocean is that the concentration of nutrients dissolved in the bulk of our ocean waters is low and imposes limits on the growth rate of organisms. By contrast, in the food-rich coastal areas, wave actions, nutrient inflow from rivers, and a relative shallowness all combine to support a large biomass.

Estuary Systems

Coastal bays, river mouths, and tidal marshes are all physically contiguous to the open ocean, yet their proximity to land and fresh water affects the salinity and nutrient composition to such a large extent that these areas, known as **estuaries**, are characteristic neither of fresh nor of salt water (Figs. 1.19 and 1.20). The combination of such factors as (a) easy access to the deep sea, (b) a high concentration and retention of nutrients originating from land and sea, (c) protective shelter, and (d) rooted or attached plants supported in shallow water makes estuaries very productive areas indeed. Estuaries provide nurseries for many deep-water fish which could not produce viable young in the harsher environment of the open sea.

Freshwater Systems

Freshwater systems involve ecological relationships similar to those which operate in the oceans. Again, the food web starts with plankton and culminates in large predators. A major difference between the two systems is that there are more trophic levels in salt water than in fresh. Freshwater systems, unlike

FIGURE 1.19 Shoreline features: coastal bay, estuary, tidal marsh, coral lagoon behind corl reef.

FIGURE 1.20 Salt marsh. (U.S. Forest Service photo.)

other aquatic systems, are continuously fertilized by nutrients leached from the nearby soil. Because bodies of fresh water are shallower than the oceans, rooted plants, marsh grasses, and lilies, as well as algae, are much more important in the food webs. Finally, as one would expect, ponds, lakes, mudpuddles, springs, creeks, and rivers all have unique and characteristic species.

Terrestrial Systems

The characteristics of large, stable, terrestrial ecosystems, known as **biomes**, result from the interactions of many environmental, biological, and evolutionary

FIGURE 1.21 Tundra in July on the coastal plain near the Arctic Research Laboratory, Point Barrow, Alaska. (Photos by the late Royal E. Shanks, E. E. Clebsch, and John Koranda. From Odum: ***Fundamentals of Ecology*****. 3rd Ed. Philadelphia, W. B. Saunders Co., 1971.)**

FIGURE 1.22 Natural temperate grassland (prairie) in central North America. *A*. Lightly grazed grassland in the Red Rock Lakes National Wildlife Refuge, Montana, with a small herd of pronghorns. *B*, Short-grass grassland, Wainwright National Park, Alberta, Canada, with herd of bison. (From Odum: *Fundamentals of Ecology*, 3rd Ed. Philadelphia, W. B. Saunders Co., 1971.)

factors. Rainfall, average and seasonal temperatures, altitude, and soil conditions all have profound influences. Specialized physical factors such as seasonal changes or ocean mists can help to create unique biomes.

A general characterization of the major terrestrial biomes of North America is outlined in the following paragraphs.

In the coldest regions, where summer temperatures average under 10°C (50°F) and trees cannot survive, stable plant cover is known as **tundra** (Fig. 1.21). Some of the grazer-plant and predator-prey relationships in this area will be discussed later.

The region between the tundra and the northern forest is characterized by tundra grasses growing between dwarfed, gaunt, windblown conifers. (Such border ecosystems, known as **ecotones**, will be discussed further on page 72.) Animals from both the tundra and the forest systems live here.

Still farther south, where the weather is warmer, is the **taiga**, or northern evergreen forest, which supports deer, elk, moose, caribou, wood buffalo, and rodents as the primary grazers and wolves, lynx, mountain lions, and man as the largest predators.

South of the taiga are forests consisting largely of deciduous trees. Such forests need 30 to 60 inches (75 to 150 cm) of rain per year and an average temperature during the growing season of about 60° to 65° F (15° to 18° C). Therefore, they do not exist farther north than central Maine or southwestern Quebec or farther south than central Georgia or southern Louisiana.

In the temperate areas in which the rainfall is too low to support forests, the stable ecosystems are **prairies** or **grasslands** (Fig. 1.22). In wet areas, the virgin grasses of North America were as tall as a man; elsewhere, they were short enough that the early scouts could "see the horizon under the bellies of millions of buffalo."

In very dry areas, where rainfall is less than 10 inches (25 cm) per year, the climax is a **desert**, either barren or able to support only scrub brush and cactus (Fig. 1.20). Deserts can be hot, as in Nevada, or relatively cold, as in eastern Washington.

In addition to the above ecosystems are some not found in North America. **Tropical rain forests** are stable systems of the Amazon basin, parts of Africa, Central America, and parts of Southeast Asia, Malaysia, and New Guinea. Here the seasonal temperature changes are less pronounced than daily ones. Vegetation is thick, and most of the available nutrients are found in the biomass. Decay and recycling are rapid.

Tropical **savannas** are fire-dependent systems occurring in areas with annual wet and dry seasons where the yearly rainfall is 40 to 60 inches (100 to 150 cm). Fires are common during the dry season. The flora consists of a few fire-resistant trees and annual vegetation **blooms**, or periods of rapid growth. Many of the animals are migratory.

When studying biomes of the world, it is of interest to compare similar climatic regions which have been separated for millennia. The classic comparison between North American plains and those of Australia shows similar trophic structure but dissimilarities be-

FIGURE 1.23 **Two types of deserts in western North America.** *A.* **A low altitude "hot" desert in southern Arizona.** *B.* **An Arizona desert at a somewhat higher altitude with several kinds of cacti and a greater variety of desert shrubs and small trees. (From Odum:** *Fundamentals of Ecology.* **3rd Ed. Philadelphia, W. B. Saunders Co., 1971.)**

tween species. Thus, kangaroos and bison, both important grazers, are morphologically quite different.

PROBLEMS

1. **Vocabulary.** Define ecology, ecosystem and biosphere.
2. **Energy.** Classify each of the following substances as energy-rich or energy-poor with specific reference to its ability to serve as a food or fuel: sand, butter, paper, fur, ice, marble, and paraffin wax.
3. **Vocabulary.** Define autotroph, heterotroph, and trophic level.

4. **Energy.** Organisms have evolved to obtain energy from foods or from the process of photosynthesis. Would it have been possible for life forms to evolve which obtain their energy by some other process, such as rolling down hills? Defend your answer.

5. **Food web.** What is a food web? Sketch a diagram of a food web that primarily involves life in the air, such as birds and insects. Will you need links to terrestrial or aquatic systems?

6. **Food web.** Consider the food web in the diagram on page 9. Arrow number 15 can be explained by the relationship: mountain lion eats deer. Write similar statements for each of the other 27 arrows.

7. **Food pyramid.** The discussion presented in Section 1.3 and illustrated in Figures 1.6 and 1.7 shows that more energy from the sun is used to nourish a human being who eats meat than a human being who eats vegetables. Explain why this fact, by itself, is *not* an argument for or against vegetarianism, taking into account the following questions: Does the choice of human diet affect the total energy flow through the food web? Does it affect the total biomass of plant matter on Earth? Does the choice between the alternatives of a large human population living largely on vegetable food or a smaller human population living largely on meat and fish significantly affect the total biomass of animal matter on Earth? In your answer do not take into account the possible destruction of species; this topic is treated separately in Chapter 3.

8. **Food pyramid.** An experiment has shown that the total weight of the consumers in the English Channel is five times the total weight of the plants. Does this information agree with the food pyramid shown in Figure 1.6? Can you offer some reasonable explanations for the findings?

9. **Food pyramid.** It is desired to establish a large but isolated area with an adequate supply of plant food, equal numbers of lion and antelope, and no other large animals. The antelope eat only plant matter, the lions, only antelope. Is it possible for the population of the two species to remain approximately equal if we start with equal numbers of each and then leave the system alone? Would you expect the final population ratio to be any different if we started with twice as many antelope? Twice as many lions? Explain your answers. (Assume that lions and antelope have the same body weight.)

10. **Consumer levels.** Explain why a beer drinker is to some extent a secondary consumer.

11. **Nutrient cycles.** We speak about nutrient cycles and

energy flow. Explain why the concepts of nutrient *flow* and energy *cycle* are not useful.

12. **Nutrient cycles.** Give three examples supporting the observation that nutrient cycling hasn't been 100 per cent effective over geological time.

13. **Oxygen cycle.** Trace an oxygen atom through a cycle that takes (a) days, (b) weeks, (c) years.

14. **Nutrient cycles.** Why don't farmers need to buy carbon at the fertilizer store? Why do they need to buy nitrogen?

15. **Carbon cycle.** The carbon dioxide concentration in the air just above trees varies considerably between night and day. From your knowledge of the biochemistry of carbon, predict whether the atmospheric carbon dioxide concentration will be higher during the night or during the day.

16. **Nutrient cycles.** Certain essentials of life are abundant in some ecosystems but rare in others. Give examples of situations in which each of the following is abundant and in which each is rare: (a) water, (b) oxygen, (c) light, (d) space, and (e) nitrogen.

17. **Nitrogen cycles.** The conversions shown in the margin represent transformations that occur in living systems. For each conversion, state whether the organism gains or expends energy.

a. $NH_3 \rightarrow NO_3^-$
b. $NH_3 \rightarrow$ protein
c. amino acid $\rightarrow NH_3$
d. $NO_3^- \rightarrow NH_3$
e. protein $\rightarrow NO_2^-$

18. **Mineral cycles.** Loggers can harvest timber either by clearcutting (removing all the trees from an area) or by selective cutting (removing only the most desirable trees). Explain why clearcutting is an ecologically unsound practice in most woodlands.

19. **Ecosystem.** Could a large city be considered a balanced ecosystem? Defend your answer.

20. **Homeostasis.** Consider two outdoor swimming pools of the same size, each filled with water to the same level. The first pool has no drain and no supply of running water. The second pool is fed by a continuous supply of running water and has a drain from which water is flowing out at the same rate at which it is being supplied. Which pool is better protected against such disruptions of its water level as might be caused by rainfall or evaporation? What regulatory mechanisms supply such protection?

21. **Vocabulary.** Define euphotic zone, phytoplankton, and benthic species.

22. **Estuary systems.** Copy Figure 1.19 and indicate which bodies of water are salty, fresh, and brackish.

The following question requires arithmetical computation:

23. **Energy.** Assume that a plant converts 1 per cent of

the light energy it receives from the sun into plant material, and that an animal stores 10 per cent of the food energy that it eats in its own body. Starting with 10,000 calories of light energy, how much energy is available to a man if he eats corn? If he eats beef? If he eats frogs that eat insects that eat leaves? Of the original 10,000 calories, how much is eventually lost to space?

(*Answer:* 100 cal; 10 cal; 1 cal; 10,000 cal)

BIBLIOGRAPHY

Three basic textbooks on ecology are:

Edward J. Kormondy: *Concepts of Ecology.* Englewood Cliffs, N.J., Prentice-Hall, Inc., 1969. 209 pp.

Charles J. Krebs: *Ecology.* New York, Harper and Row, 1972. 694 pp.

Eugene P. Odum: *Fundamentals of Ecology.* 3rd Ed. Philadelphia, W. B. Saunders Co., 1971. 574 pp.

A periodical issue devoted in its entirety to "The Biosphere" is:

Scientific American. September, 1970. 267 pp.

Three books dealing with specific areas of natural ecology are:

R. Platt: *The Great American Forest.* Englewood Cliffs, N.J., Prentice-Hall, 1965. 271 pp.

B. Stonehouse: *Animals of the Arctic: the Ecology of The Far North.* New York, Holt, Rinehart and Winston, 1971. 172 pp.

G. M. Van Dyne, ed.: *The Ecosystem Concept in Natural Resource Management.* New York, Academic Press, 1969. 383 pp.

A classic study of ecology and conservation as seen through the eyes of a naturalist is:

A. Leopold: *A Sand County Almanac.* New York, Sierra Club/ Ballantine Books, 1966. 296 pp.

2

NATURAL GROWTH AND REGULATION OF POPULATIONS

2.1 POPULATION GROWTH AND CARRYING CAPACITY

(Problems 1, 2, 3, 4, 5)

We have previously noted that old ecosystems are typically well balanced because the interactions within such natural environments regulate the population levels of all native species. Of course, oscillations in population size do occur even in stable systems, but these oscillations are generally small, and the total population of each species as well as the *ratio* between populations fluctuate only slightly. In contrast to this observed stability, calculations show that the **biotic potential** of any species is extremely high. For example, a bacterium can split into two bacteria in approximately 20 minutes. If enough food is available and there is no predation, these two bacteria can grow into four after another 20 minutes; by the end of an hour the original bacterium will have become eight. By the end of a day and a half, a growing colony would have increased through 108 generations (36 hours at 3 generations per hour). Since each generation leads to a doubling of the number of individuals, the colony would consist of 2^{108}, or roughly 10^{32} individuals. This number of bacteria could cover the entire surface of the Earth to a uniform depth of 1 foot. Such a growth pattern is known as a geometric rate of increase and can be described mathematically as follows: If there are x organisms

The biotic potential of a species is defined as the maximum rate of population growth which would result if all females bred as often as possible, and all individuals survived past their reproductive age. To grow at the biotic potential, a population must have ample food and living space and be free from disease and predation.

at time 0 and ax organisms (where a is greater than 1) after one generation, there would be $a(ax)$, or a^2x after a second generation, a^3x after a third, and a^nx after n generations. Most other organisms have longer generation times than 20 minutes, and consequently, their biotic potential is smaller than that of bacteria. Nevertheless, the breeding capacity of *all* species is large, and no ecosystem could support the geometric growth rate of any species for very long.

Some combination of environmental pressures must, therefore, act to inhibit the potential growth of every species. Examples of many of these environmental pressures are easy to observe: you swat a mosquito, a bird dies during a cold spell immediately following a winter rain, a puppy dies of diphtheria, you mow your grass before it goes to seed, a cat catches a rat, a bluejay chases a sparrow away from a crust of bread and the smaller bird loses a meal, or you eat an apple and throw the core (including the seeds) into a garbage disposal unit. Other pressures are less easy to observe, and yet are occurring all around us: a paramecium eats a bacterium, a small crustacean eats a paramecium, a wild oat seed competes with a wild barley seed for an almost microscopic hole in the earth favorable for growth, a parasite infects a beetle, a seed fails to germinate during a drought, an acorn rots during a particularly wet spring, or a hailstone hits a caterpillar on the head. The sum of all the environmental interactions which collectively inhibit the growth of a population is known as the **environmental resistance.** Individual components of the environmental resistance will be discussed in Sections 2.3 through 2.7, but for the moment we will be concerned with the effect that the environmental resistance has on growth. The question we wish to discuss can be stated simply: We know that natural populations cannot grow at geometric rates, therefore how do they actually change in size as time goes on?

Consider some population which is initially very small. The very fact that it is small places it in danger of extinction, because it may not be able to recover from such setbacks as epidemics, famine, or poor breeding. Even if such factors do not totally destroy the population, they will limit its growth rate, and therefore the population will increase only slowly at first. However, once the population is established, its

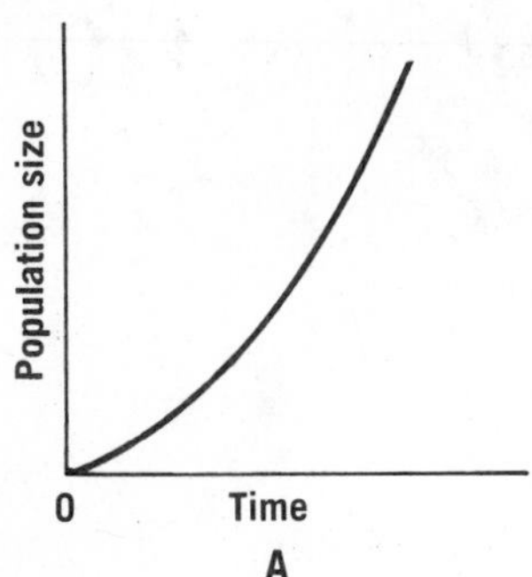

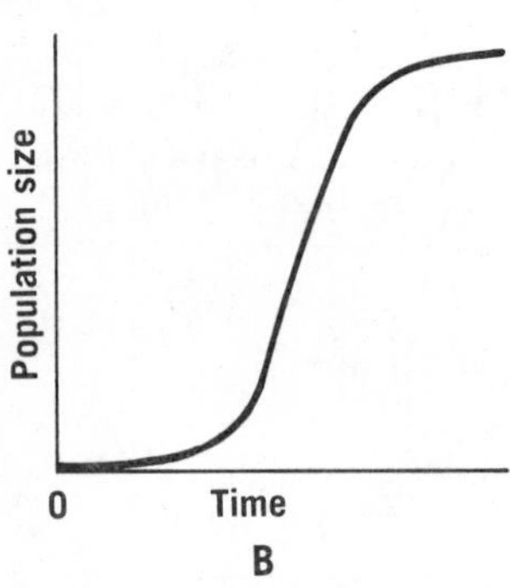

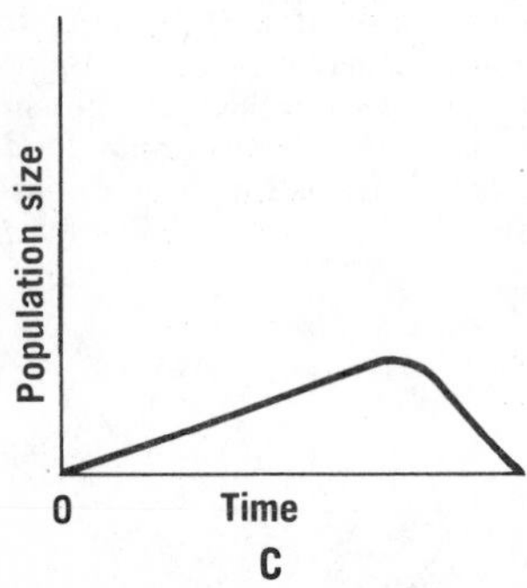

FIGURE 2.1 Schematic growth curves. *A.* **Geometric pattern of growth.** *B.* **Sigmoid curve of growth.** *C.* **Growth curve of a population that becomes extinct.**

size will rise more rapidly as long as there are adequate food sources, relatively few predators and favorable living conditions. When the population becomes very large with respect to its food supply, availability of shelter, and vulnerability to predators, and becomes so dense that disease spreads rapidly, the environmental resistance increases and the growth rate decreases. The entire curve of growth looks like an S and is said to be **sigmoid** or S-like, as shown in Figure 2.1. The rate of growth shown by the upper right portion of the sigmoid curve is very nearly zero. A zero growth rate does not mean that there are no births and no deaths. It simply means that the total number of births plus immigrations equals the total number of deaths plus emigrations. When this equilibrium is reached, the biotic potential is balanced by the environmental resistance. The magnitude of this upper population level is characteristic for a given species in a given ecosystem; thus, we say that each ecosystem has a given **carrying capacity** for each species. When the carrying capacity has been reached, the system cannot continue to support any more individuals of that species. The carrying capacity is not constant from region to region; for example, a wheat field has an inherent ability to support more locusts than a short-grass prairie. Indeed, the carrying capacity may vary from time to time: the carrying capacity of a region can be altered by some calamity such as fire or by annual variations in temperature, rainfall, etc.

zero growth rate

The **carrying capacity** is the maximum number of individuals of a given species which can be supported by a particular environment.

Do real populations in natural ecosystems grow in a smooth sigmoid fashion and eventually approach

Sigmoid Curve

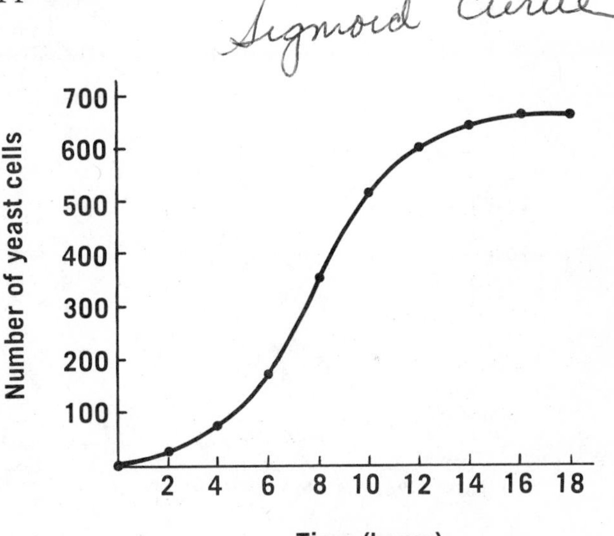

FIGURE 2.2 Growth function of yeast cells in the laboratory. (From *The Biology of Population Growth*, by Raymond Pearl. Copyright 1925 by Alfred A. Knopf, Inc., and renewed 1953 by Maude de Witt Pearl. Reprinted by permission of Alfred A. Knopf, Inc.)

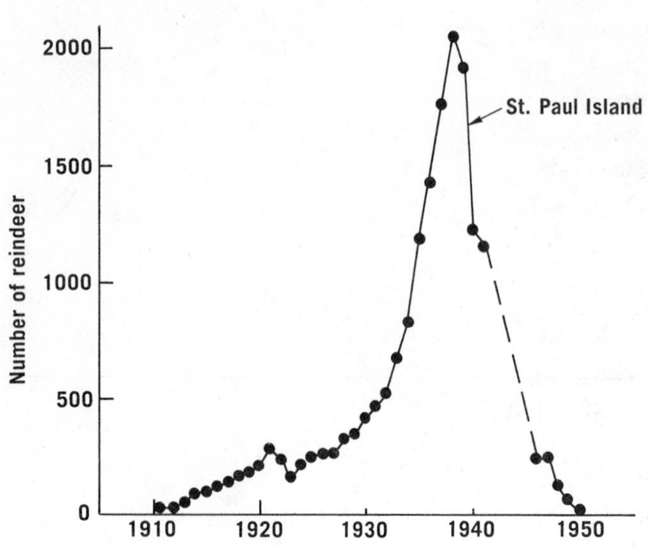

FIGURE 2.3 Growth and death of reindeer on St. Paul Island. (Redrawn from C. J. Krebs: *Ecology*. New York, Harper & Row, Publishers, 1972, p. 197.)

a constant size? Yes, sometimes. For example, yeast cells experimentally introduced into a culture do exhibit a sigmoid growth pattern (Fig. 2.2). On the other hand, when 4 male and 21 female European reindeer were experimentally introduced onto St. Paul island near the coat of Alaska in 1912, a different result was observed. St. Paul island was free of predators and environmentally favorable to the deer, and the population growth rate initially increased according to a sigmoid pattern, but the expected orderly approach to equilibrium was not observed. (See Figure 2.3.) Instead, a nearly geometric growth rate continued long after the carrying capacity of the island had been exceeded, but then the population declined abruptly. Why did this occur? Initially, the food supply on St. Paul was abundant, and since there was no predation, the animals were healthy and their fertility was high. As the population continued to

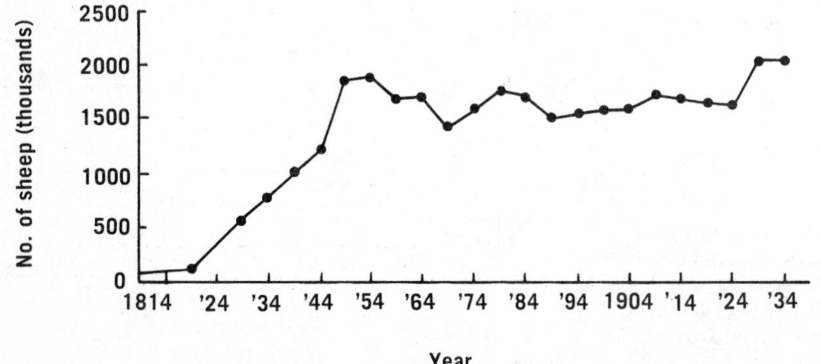

FIGURE 2.4 Growth of a sheep population introduced into a new environment on the island of Tasmania. (Redrawn from Odum: *Fundamentals of Ecology*. 3rd Ed. Philadelphia, W. B. Saunders Co., 1971.)

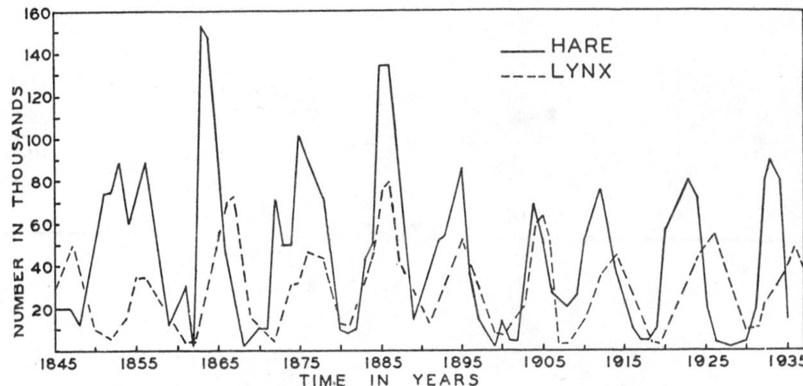

FIGURE 2.5 Population levels for the lynx and the snowshoe hare, as indicated by the number of pelts received by the Hudson Bay Company. (From Odum: *Fundamentals of Ecology.* **3rd Ed. Philadelphia, W. B. Saunders Co., 1971.)**

increase, the reindeer were eating plant matter faster than the island could replace it by photosynthesis. The reindeer, however, were unaware of the instability of their situation; their food was still abundant and they continued to multiply. Then quite suddenly almost all the food was gone. The island was barren, and mass starvation and death occurred, until in 1950, only eight animals remained alive on the island.

Although the pattern exhibited by the reindeer on St. Paul island is somewhat extreme, the tendency for any population to oscillate is quite common. Thus, when sheep were introduced into Tasmania the population grew to about two million individuals, then declined slightly, rose again, and continued to vary (Fig. 2.4). In this case the observed fluctuations

The snowshoe or varying hare, famous in ecological annals for its spectacular cyclic abundance (see Figure 2.5). The individual shown is in its white winter pelage. The change from the brown summer pelage to the white one of winter has been shown to be controlled by photoperiodicity. (U.S. Soil Conservation Service Photo.)

were small, and the sheep population was nearly constant over a long period of time.

We can see that the oscillations of the population of one species in an ecosystem may create oscillations in the populations of other species. Specifically, a large predator population causes a decline in the number of prey (reindeer can be considered to be "predators" of the moss they eat), and an increase in a prey population will generally be followed by an increase in the population of its predators. A classic predator-prey oscillation is displayed in Figure 2.5, which is a graph of population levels for the snowshoe hare and its predator, the lynx. We see that the predator cycle is closely correlated with the prey cycle, but the points of the maxima and minima do not generally coincide. These discrepancies can be explained by reasoning that if rabbits are abundant in a given year, say 1863, the lynx will be healthy, and many successful matings will occur. However, a pregnant female counts as only one individual, and therefore does not contribute to an increase in the lynx population until her kits are born, one season later. Thus, the maximum in the lynx population is not observed until 1864.

Extinction occurs when the population of a species reaches zero. When the population of a species fluctuates dramatically, it is possible that one drastic decline in population will not be reversed in time and that all individuals will die. This phenomenon will be discussed further in Chapters 3 and 4.

2.2 SPORADIC AND CYCLIC GROWTH CURVES

(Problem 6)

The preceding section described four patterns of population growth:

Pattern	*Example*
Sigmoid	Yeast in experimental culture
Sharp peak followed by sharp decline	Reindeer on St. Paul island
Oscillating-low amplitude	Sheep in Tasmania
Oscillating-high amplitude	Hares and lynx in Canadian woods

An explanation of these patterns and a theory that could predict the population growth of a species would be valuable. Unfortunately, no completely satisfactory theory is yet at hand.

The conceptual problem, however, is not completely hopeless. Some oscillating growth patterns are explicable and predictable. For example, many plants and animals are unable to survive such periods of hardship as winter cold, summer dryness, or spring floods. These species survive by producing hardy seeds, spores, or pupae which can grow anew when favorable conditions return. A population curve for an annual plant or animal, therefore, shows a rapid increase every spring followed by the death of all adults by winter. Other species can survive throughout the year, but conditions for growth and reproduction are much more favorable in some seasons than in others. This situation is exemplified by thrips, which are small plant-eating insects. One species of these creatures feeds primarily on roses. As shown in Figure 2.6, the thrip population increases dramatically each spring when the roses bloom and food is plentiful, and then decreases in winter; but the population never reaches zero.

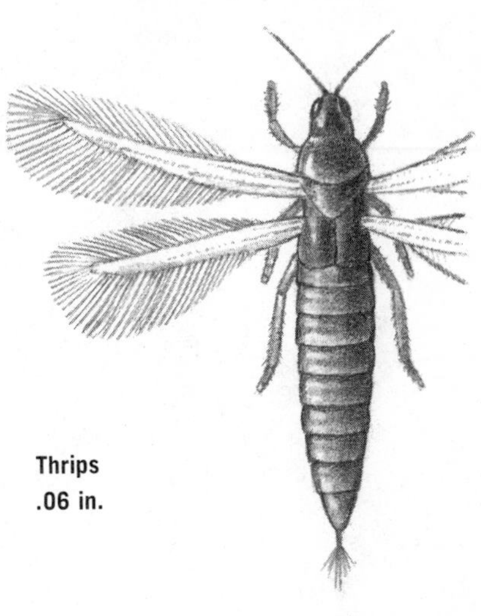

The research was done in Australia, so that October is seasonally equivalent to April in the Northern Hemisphere, November is equivalent to May, etc.

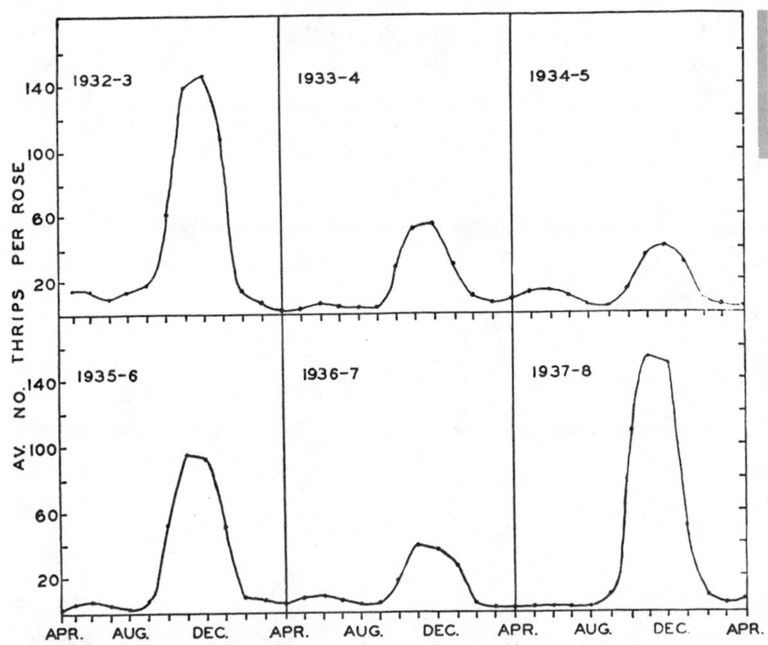

FIGURE 2.6 Seasonal changes in a population of adult thrips living on roses. (Data from Davidson and Andrewartha, 1948. From Odum: *Fundamentals of Ecology.* **3rd Ed. Philadelphia, W. B. Saunders Co., 1971.)**

Conversely, some cyclic population curves are not at all easy to explain. For example, no one can understand why the hare and lynx populations oscillate in a regular 10 year cycle. Neither rabbit and coyote populations in the southwestern United States nor wolf and deer populations in the Canadian forests show such dramatic population cycles.

There are many examples of inexplicable growth functions. We have accounted for the observed behavior of reindeer on St. Paul island, but if we use our hypothesis to predict that other newly introduced species will follow a similar growth pattern, we may be embarrassed by the facts. For example, at the same time that reindeer were introduced onto St. Paul island, a similar herd was introduced onto nearby St. George island, which resembled St. Paul in size and ecological properties. The size of the herd on St. George, however, grew in a more controlled fashion than the St. Paul herd. (See Figure 2.7.) Other introductions of reindeer into Alaska have exhibited population growth patterns intermediate between these two examples. No general theory can be offered to explain the observed discrepancies.

North of the Canadian forest lies the Arctic tundra, and one common rodent of the tundra is the lemming. Lemmings exhibit predictable three or four year population cycles. One summer the population will be extremely high; the next year the population will rapidly decline, or crash. For another year or two, the population recovers slowly; then it

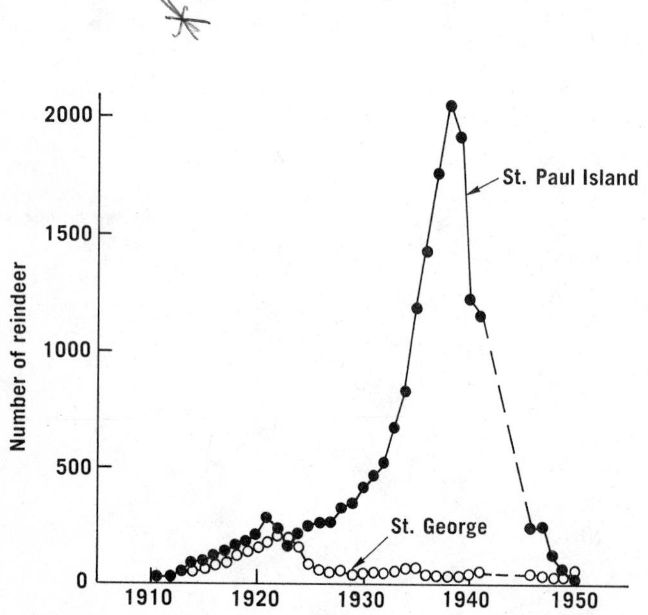

FIGURE 2.7 Growth function for reindeer on two similar islands. (From C. J. Krebs: *Ecology*. New York, Harper & Row, Publishers, 1972, p. 197.)

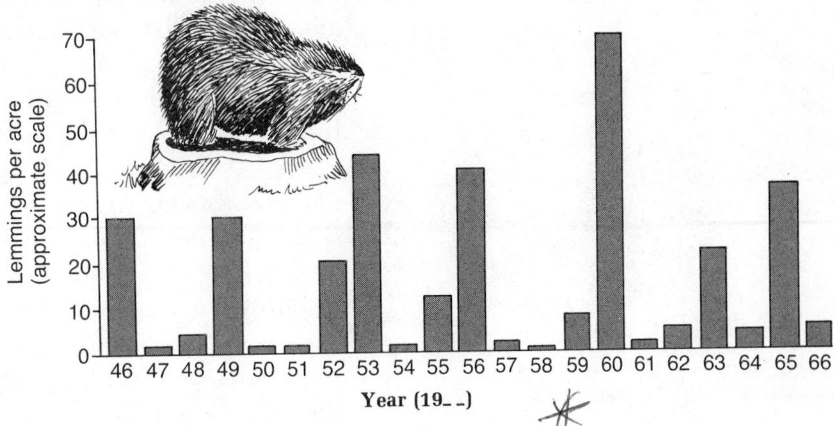

FIGURE 2.8 Lemming population cycles at Point Barrow, Alaska, from 1946 to 1966.

skyrockets for a season and the cyclic pattern repeats itself.

Lemming abundance is associated with forage cycles (Fig. 2.8). During an abundant lemming year, the tundra plants are plentiful and healthy. Most of the available nutrients exist in plant tissue, and there is little stored in the soil reservoir. In the spring following the overabundance of lemmings, the heavily overgrazed plants become scarce, and most of the nutrients in the ecosystem are locked in the dead and dying lemmings. The process of decay and return to the soil requires another year or two, during which time the population and health of both plants and rodents increase.

Unfortunately, recognition of the relationship between lemming and forage cycles does not explain why the cycles exist at all, or why the amplitude of the variation does not increase or decrease appreciably in time. We know that populations of other grazers in the Arctic do not fluctuate as much as the population of the lemmings.

Erratic **population blooms** occur even in old, seemingly stable systems. Since recorded history, a species of red phytoplankton has often been observed to enter a period of rapid growth in coastal areas and turn the ocean red. There seems to be no predictable cyclic pattern associated with this phenomenon, but instead, the plankton grow very rapidly at unexpected times—seemingly in response to a new influx of nutrients. Traditionally, these "red tides" were often observed to occur after flood waters washed large quantities of soil nutrients into the sea. This influx created a plentiful food source. The red-colored plants were best able to take advantage of it. Unfortunately,

A bloom is a period of very rapid growth or sudden development. We use the word most frequently to describe the growth of a flowering plant in the spring, but other usages are also correct. For example, "The skinny awkward teenager bloomed into a robust adult." When a population of any species, plant or animal, expands *quickly*, it is said to bloom.

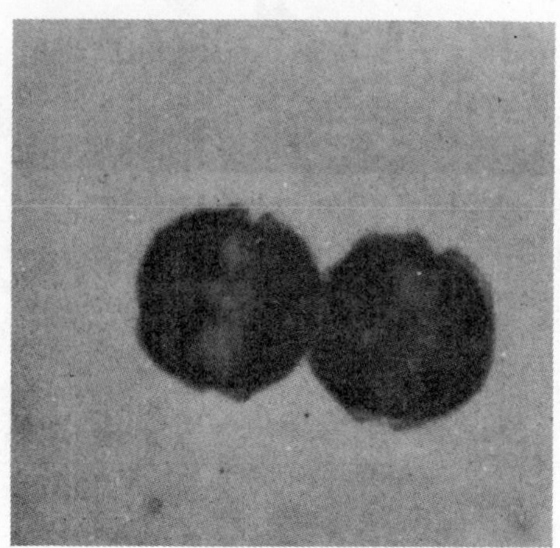

Red tide organism.

these organisms discharge toxic substances into the sea, killing fish and aquatic mammals, and are therefore considered a great hazard. Recently, a series of red tides apparently unrelated to natural phenomena have been tentatively attributed to the nutrient content of pollutants discharged into the sea.

Let's review what we have been discussing. In Chapter 1 we emphasized that ecosystems are stable and balanced. In this section we have shown that the populations of many species regularly overshoot the carrying capacity of their environment as part of a recurrent oscillation, while other species bloom and then decline erratically. How can we reconcile this apparent contradiction?

First, erratic fluctuations of populations, such as red tides and the death of lemmings, are dramatic and attract attention out of proportion to the frequency of their occurrence. More significant is the fact that even ecosystems that are superficially regarded as "well behaved" do reveal perturbations when they are examined more intimately. For example, if you walk through a section of the Canadian forest, the scenery will look nearly the same whether the hares or the lynx are scarce or abundant. Tall evergreen trees will rise above you, and trout will swim in the brooks and lakes. The pine needles on the forest floor will be moist and spongy in spring, harder and noisier when you walk through in autumn, and covered with snow in winter. Tracks of deer, moose, caribou, wolves, coyotes, hares, and lynx will be apparent during every year. Moreover, most biologists believe that this landscape has been maintained for many years and will continue to exist far into the future. In this long-term sense the system is stable; however, we have already learned how hare and lynx populations fluctuate from year to year. Thus, for example, a logger might consider the system stable, while a trapper, whose interests are differently focused, might consider it unstable.

2.3 ENVIRONMENTAL RESISTANCE

(*Problems 7, 8, 9, 10, 11, 12, 13*)

If we study the population of the phytoplankton near the surface of central oceans it becomes immediately obvious that there is a lot of sunlight available but relatively few plankton. Of course, there is sufficient water to support life; why, then, aren't there more plankton? The central ocean is an old ecosystem and one would except that population equilibrium has been reached; that is, one would not expect average phytoplankton populations to rise appreciably in the next few years. Imagine that you tried to answer this question by studying the size of the grazing population in the deep sea, thinking that perhaps the low level of plant population is due to a high degree of predation. If, however, you determined the total biomass of the system (the weight of plants plus all the heterotrophic organisms) and compared it to the biomass of coastal aquatic ecosystems, or to the theoretical biomass that all the available sunlight could support, you would again find the phytoplankton population of the central oceans to be abnormally low. Why? The answer is that many nutrients such as nitrates and phosphates are scarce in the deep sea. The shortages of nutrients exist because of the nature of the physical environment. On land, nutrients are generally retained within an ecosystem for a long time, but in the sea small pieces of debris settle below the euphotic zone and become unavailable. Moreover, nutrients moving into the sea from sources on land are diluted by the vastness of the oceans. Thus, the phytoplankton populations are limited by a shortage of a few nutrients, even though water, light, and many mineral salts are available in quantity.

Recall that phytoplankton are the microscopic free-floating plants that are responsible for most of the primary food productions in aquatic systems (page 27).

In general, the population level of any species in any ecosystem is regulated by those essentials of life which are available in the *minimum quantity.* This **Law of Limiting Factors** applies not only to nutrients but also to other physical factors. Consequently, an organism cannot survive without sufficient light, heat, moisture, and space. Additionally, many organisms are limited by the turbulence of their environments. For example, benthic organisms in streams cannot tolerate an excessive water flow; ocean dwellers are limited by the force of waves; and many plants cannot survive on exposed windswept areas.

Populations may be limited not only by too little of certain physical factors, but also by too much of these factors. Desert plants cannot survive where there is too much water, polar bears cannot survive in the tropics, and even saltwater fish would die in the Great Salt Lake where salt concentrations are unusually high.

Suppose we were to use laboratory observations

to study the growth limits of the prairie grasses. We could plant several batches of seeds under conditions which differ only with respect to acidity of the soil. By measuring the rate of growth of the grass seedlings in each pot, we could draw a graph of growth as a function of soil acidity. The experiment could then be repeated under different conditions. Nutrients, light intensity, duration of darkness, moisture, and temperature could all be varied. For each factor, one might observe three important experimental points. There is a minimum value for each variable below which no seeds will sprout. There is also a maximum limit above which no seeds will sprout. Finally, there is an optimum growth condition. If we collect the results of our individual experiments, we can construct a reasonable approximation of the growth conditions of the grasses. We know, for example, not to expect prairie grasses to grow in marshes, in Northern Canada, or in old mineshafts. Unfortunately, an exact description of the limits and of the optimal environment is elusive. A major complication is that many factors do not operate independently of each other. For example, when soil nitrogen is in short supply, plants can grow well only if all other nutrients are found in abundance, but their resistance to drought is lowered. Grass cannot grow if nitrogen *and* water are available in limited supplies simultaneously.

Some species of plants and animals thrive in a wide range of physical habitats. One species of jelly-

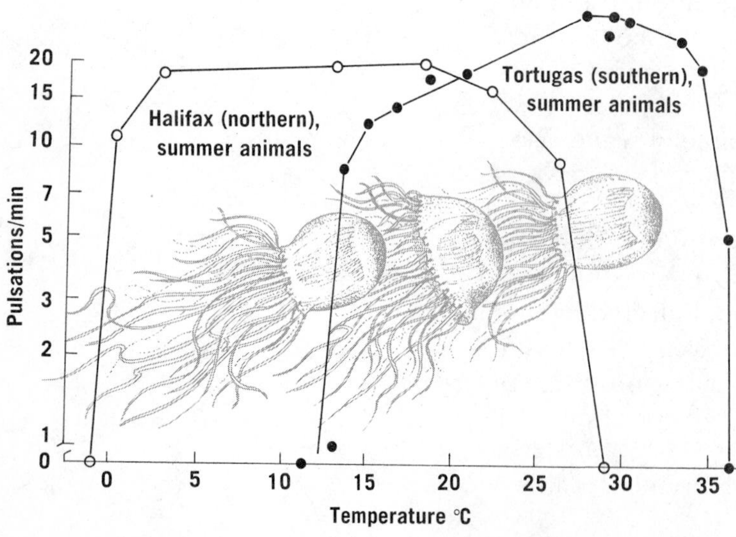

FIGURE 2.9 Swimming speeds for two ecotypes of jellyfish at different temperatures. (After Bullock, J. H.: Compensation for Temperature in the Metabolism and Activity of Poikotherms. Biol. Rev. 30: 311–342, 1955.)

fish, *Aurelia aurita*, grows both in northern seas near Halifax, Nova Scotia, and in southern waters near Tortugas, Haiti. Those individuals living in the north must survive and operate efficiently both in cold winter water and in the warmer temperatures of the summer, while all the southern jellyfish live in a more constant environment. Is one population of this species living at or near its optimum growth condition most of the time, while the other struggles along near the limits of its tolerance? Experiments tell us that this is not so; on the contrary, we find that isolated populations of the same species have evolved different limits and different optimal conditions for growth. These different subgroups within a species are called **ecotypes.** Figure 2.9 shows the swimming motion of jellyfish as a function of temperature for two ecotypes of jellyfish. The southern jellyfish die when water temperatures are below 11° C, whereas their northern brethren can survive water temperatures below 0° C; on the other hand, northern animals die in waters above 29° C, but the southern ecotype operates most efficiently at about 29°C and can survive temperatures up to 36° C. Interestingly, the northern ecotype has a much broader temperature optimum then the southern jellyfish.

FIGURE 2.10 The arctic winter covers all plants and is a density-independent factor.

Of course, the physical environment is not an unchanging background. Storms arise; rivers shift position; floods, earthquakes, volcanoes and other natural calamities occur. Since natural disasters bring death to many plants and animals, these phenomena must also be included as components of the environmental resistance. Natural disasters often act abruptly and lethally on an entire population; they are said to be **density-independent** factors. As an extreme example, an erupting volcano on a small Pacific island can destroy an entire population of nesting birds irrespective of whether 6 or 600 individuals live there. Physical calamities are not limited to rare occurrences such as volcanic eruptions. Two rather common examples are: (a) Every year, the winter destroys the adult populations of some plants, insects, and other small animals. The species survive because seeds, spores, or pupae endure through the cold months, but the adult population has been reduced to zero. (b) Hurricanes occur regularly in some parts of the world, and the wave action produced by these high winds eliminates whole colonies of intertidal organisms.

The growth of organisms is limited not only by the physical environment, but also by the growth of other organisms in the ecosystem. Thus, a population is partially controlled by the abundance of its prey and by the pressures of its predators. In addition, if an essential nutrient is scarce, then different species will compete for it; if winds, water, currents, waves, or changes in temperature make survival difficult, then the plants and animals will compete for shelter. The biological component of the environmental resistance consists largely of pressures from predation and competition. These two topics will be discussed in greater detail in Section 2.5.

If the population of one species of an ecosystem changes, the environmental resistance from predation, competition, and availability of food will also change. To understand this relationship, imagine that a small population of beetles is introduced into a new environment. It would be immediately subject to pressures from general hunters—animals that pursue whichever individuals they see or sense first. Many insect predators, however, specialize (e.g., certain wasps parasitize the larvae of specific beetles). If the initial beetle population were small, these specific hunters would probably not be attracted to the area, but as the beetle population increased, specific predators might migrate into the area, adding to the pressure produced by the indigenous generalists. Competition also becomes increasingly severe as the population increases. For example, suppose that there are n ideal sites suitable for beetle burrows. As long as there are fewer than n beetles there should be no serious competition for real estate. When the beetle population exceeds n, however, problems arise, and one beetle will have to move to some inferior building site. Inferior homes provide less protection from the weather and from predators than prime homes do; consequently, as the beetle population increases, the average life expectancy of beetles decreases. As the population increases, the environmental resistance increases and brakes the rate of growth. The result is that the population tends to approach the carrying capacity gradually, and a sigmoid growth curve is observed. The components of the environmental resistance which vary with the population size are said to be **density-dependent** factors.

In some environments, such as a desert or an

exposed section of seacoast, density-independent calamities are likely to occur regularly, while in others such as a tropical jungle or the deep sea, natural disasters are rare. Ecologists recognize that different plants and animals have evolved different mechanisms to deal with these two types of situations. Most organisms that live in variable environments are able to grow and reproduce very rapidly during short periods of favorable conditions. They compete with their neighbors by quickly grasping the essentials of life and then reproducing before the supplies run out. Such organisms often have dormant phases such as spores, seeds, or pupae. The population curves for these types of organisms are almost never sigmoid, rather they are characterized by periods of rapid growth and precipitous decline. On the other hand, most populations indigenous to more constant environments exist at or near the carrying capacity of the system all the time. They have evolved mechanisms to grow and reproduce within a constant, crowded, and complex world; they compete by an ability to crowd out their neighbors, to extract nutrients most efficiently all year round, or to find a unique mode of life.

The speed with which some organisms take advantage of a favorable situation is truly remarkable. For example, alpine flowers must sprout, bloom, and go to seed in an extremely short growing period. In the early summer when the snow is melting rapidly from mountain meadows, several feet of new earth often appear each day, but some flowers do not wait for the snow to melt before sprouting. These species sense the time when the snow cover is only a few inches thick and sprout under the snow a day or two before the snow melts off. In this way, they can be the first to take advantage of the spring growth period without being forced to live under the snow for too long.

Populations are regulated not only by external factors such as weather, predation, or competition, but sometimes also by internal factors. Internal regulations of population occurs when an individual, a family, or a group spontaneously restricts its birth rate. A human female may choose to bear two children even though she has enough money to feed, clothe, and shelter many more, and is biologically capable of bearing a larger family. Patterns of voluntary birth control are difficult to analyze for human populations (see Chapter 5), but when we study animal populations the patterns and motivations are even more obscure. For example, a wolf pack typically consists of a dominant male (the leader), his mate (the dominant female), and a number of subordinate males and females. The dominant pair mate every year, but other members of the pack generally do not copulate even though they are sexually mature. In one study of a wolf pack kept in a large fenced enclosure, the dominant male became aroused and tried to mate with a subordinate female, but she would not allow it, and with tail between her legs, she cowered and avoided his advances. It is easy to understand that this voluntary birth control is beneficial to a wild wolf population, for otherwise the wolves, who are not themselves subject to predation, would overpopulate their range and face mass starvation. It isn't easy, however, to understand why the wolves act this way or what drives individuals to abstain. The female in the experimental enclosure was well fed all the time and was not acting in direct response to a food shortage. Self-regulation has been observed in many systems. The phenomenon is not limited to mammals, for some species of insects, birds, fish, and reptiles are also known to limit their own populations even if food is plentiful and predation is minimized.

2.4 THE ECOLOGICAL NICHE

(*Problems 14, 15, 16, 17*)

There are roughly one and a half million different species of animals and one-half million species of plants on Earth. Each species performs unique functions and occupies specific habitats. The combination of function and habitat is called an **ecological niche.** To describe a niche fully, one would first have to describe all physical characteristics of a species' home. One might start with specifying the gross location (for example, the Rocky Mountains or the central floor of the Atlantic Ocean) and the type of living quarters (for example, a burrow under the roots of trees). For plants and the less mobile animals, one would have to describe the preferred micro-environment, such as the water salinity for species living at the interfaces of rivers and oceans, the soil acidity for plants, or the necessary turbulence for stream dwellers. An animal's trophic level, its exact diet within the trophic structure, and its major predators are also important in the description of its niche. Mobile animals generally have a more or less clearly defined food-gathering territory or **home range** which is another factor in establishing the physical niche.

A niche is not an inherent property of a species, for it is governed by factors other than genetic ones. Social and environmental factors contribute to the choice of niche. Recall from Section 2.3 that each organism has an optimum growth condition for any given physical factor. In particular, recall that jellyfish swim fastest (and presumably forage more efficiently) when the water is maintained at a certain ideal temperature. Of course, it would be unreasonable to expect all southern jellyfish to be living in waters of exactly 29° C all the time, or to expect all northern jellyfish to be equally favored by their particular optimum water temperature. The weather changes, cold and warm spells occur, yet organisms survive. Imagine a situation in which a warm, sunny, sheltered bay provided an optimal physical environment for jellyfish but exposed them to abnormally high predatory pressures. Since the conditions in the bay would not be optimal, many individuals might migrate to a less favorable physical environment to find more favorable biotic surroundings. Thus, the observed niches of a bay-dwelling jellyfish and of a migrating jellyfish are different from each other and from the theoretical optimal niche. The niche of a given species in a given ecosystem is not a set of conditions that would be best suited to the genetic makeup of the organism, but rather it is the best accommodation that the organism can make to the realities of its environment.

2.5 INTERACTIONS AMONG SPECIES

(*Problems 18, 19, 20, 21, 22, 23, 24, 25*)

Food webs are complex. Populations overlap in place and function. Many homeostatic equilibria exist simultaneously in the same system. These complexities, the interplay of the checks and balances, the interactions between species, maintain population levels and thereby stabilize the system. Interactions among species are made up of separate events; an individual of one species interacts with one individual of another species at a given time. Of course, every isolated encounter has importance to the two interacting species. But when analyzing the system, we wish to focus attention upon the effects that the encounter has on the total community. In a larger sense, we are ultimately concerned with the effects of the sum of encounters on the total community. For example, knowing the relationship between deer and aspen saplings is necessary, but is only one prerequisite to the more comprehensive knowledge of the relationship of the community of grazers toward the community of plants.

When two individuals come into contact, the interactions are beneficial, harmful, or neutral to either one or both of them. Two-species interactions can be catalogued into eight major types: neutralism, competition, amensalism, predation, parasitism, commensalism, protocooperation, and mutualism.

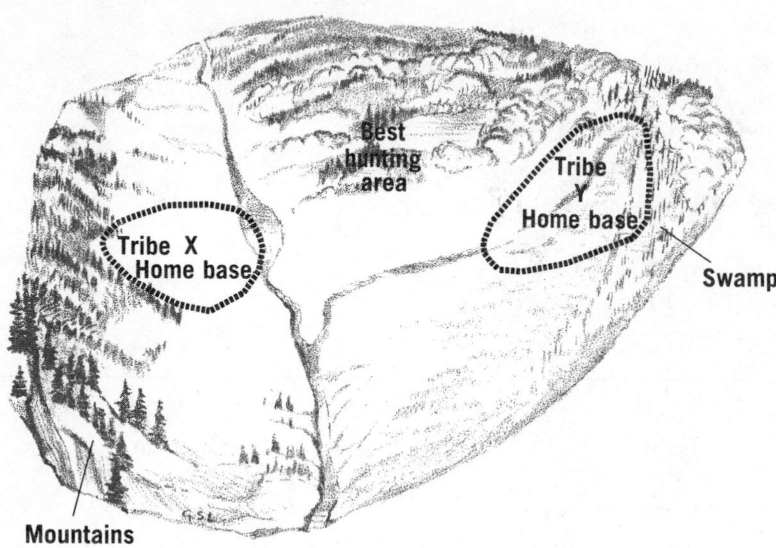

FIGURE 2.11 A model of competition.

Neutralism is the inconsequential case of very little interaction. Wild rose bushes and lynx have almost no direct relations with each other.

Competition is an interaction in which two or more organisms try to gain control of a limited resource. When competition for a given commodity is severe, individuals often alter their demand somewhat. Thus, competition is a driving force toward diversity in natural ecosystems. To understand the interplay between competition and diversity, imagine two tribes, X and Y, of nomadic hunters living in the area of the fictitious map in Figure 2.11. Suppose that the best hunting grounds are found along the flood plain of the river. Naturally, both tribes would compete for this prime hunting area, perhaps by sporadic warfare. One probable result of this competition is that tribe X and tribe Y would maintain different secure territories on the edge of the flood plain. Between these two home bases, some of the hunting area would be visited by the hunters of both tribes. If tribe X had sole possession of the best hunting areas, the people could be fed easily from the population of lowland animals, but in the face of competition from tribe Y, the hunters from tribe X must obtain part of their game by hunting the mountainous area near their home base. Similarly, the hunters in tribe Y must search the swamp for a portion of their food. If some catastrophe, such as fire, destroys the lowland hunting areas, members of both tribes, experienced in hunting alternate sources of food, could probably feed their people.

Competition between animal species can be considered analogous to the competition between the two tribes. If the words "niche" and "species" are used in place of the words "home territory" and "tribe," the example just given illustrates a simplified case of competition between two animal groups.

These relationships can be further illustrated by several examples taken from natural systems. Consider two species of barnacles that live on the Scottish coast. The larger species, *Balanus*, dominates between the high and low tide marks where the individuals can be assured of daily contact with the sea. The smaller species of barnacles, *Chthamalus*, lives predominately higher up along the beach where they must survive frequent and more prolonged periods of desiccation. Experiments have shown that in the absence of competition, the smaller barnacles can survive over a larger range (see Figure 2.12), while the larger species survives only in the region of plentiful moisture. The smaller species, however, is unable to compete with the stronger one in a wetter environment, and any larvae of the smaller species that attempt to grow outside of their observed range are competitively displaced or dislodged and eaten by predators. One species survives by virtue of its ability to adapt to extreme physical pressure, the other by virtue of its competitiveness and its ability to survive predation, and thus diversity is built into the system.

We can imagine that if the niches for two separate

***Desiccate*, which means to deprive thoroughly of moisture, is from the Latin word meaning "to dry completely." An obvious cognate is the French word for dry, "sec." Sec is used in English to describe certain wines.**

FIGURE 2.12 **Competition among barnacles.**

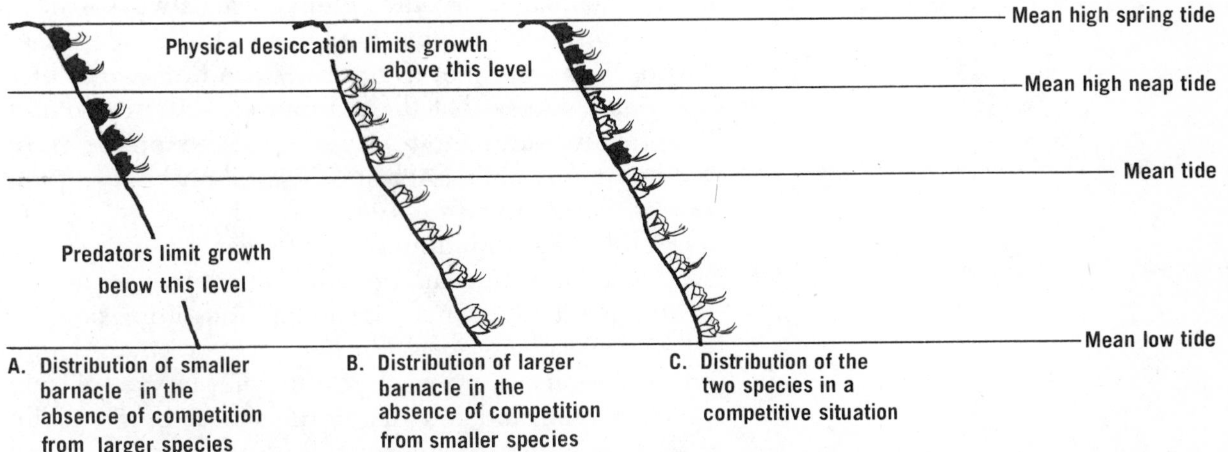

species were identical, competition for food and shelter would be particularly intense. In fact, both experiment and field observation have led to the generalization that two species in the same ecosystem cannot both survive in the same niche. If there are two species with the same preferred niche, competition will either lead to the elimination of one of them or to the adaptation of one of them to fit a new niche. A corollary to this rule is the generalization that in an ecosystem which houses many species, not only is the diversity of niches very large, but species are coadapted to overlapping niches so that elimination through competition is reduced.

It is reasonable to ask whether or not two species of plants living side by side, with their roots in the same soil and their tops touching, are not, in fact, occupying the same niche. Naturally, not every set of plants living near each other has been studied, but those studies that have been conducted have shown that if there isn't a difference in nutrient or light requirements, then there is usually some difference in root depth or in timing of life cycles. Of two species of clover that coexist, one was observed to grow faster and spread its leaves sooner than the other. The second clover species ultimately grows taller than the first. Thus, each has its period of peak sunlight, and they are both able to survive.

One laboratory experiment showed that if two species of flour beetles were placed in a jar of flour, and an ample supply of food maintained, only one species would survive. Yet, in the absence of competition, either one could survive in the jar indefinitely. Further studies showed that if the jar were kept warm and moist, one of the species always won, while if the jar were cool and dry, the other species always won. It is easy to imagine what happens to these beetles in nature. For example, if grain is stored in a large elevator, it is likely that the conditions at the bottom of the pile are warm and moist while at the top the grain is drier and cooler. Each species of flour beetles then can fit into its own niche.

The differentiation of niches can be quite delicate; competition can be minimized by seemingly insignificant differences in adaptation. For example, when two species of grain beetles were placed in a homogeneous supply of grain, the larvae of one species lived and fed inside the wheat berry, while the other larvae lived and fed on the outside of the

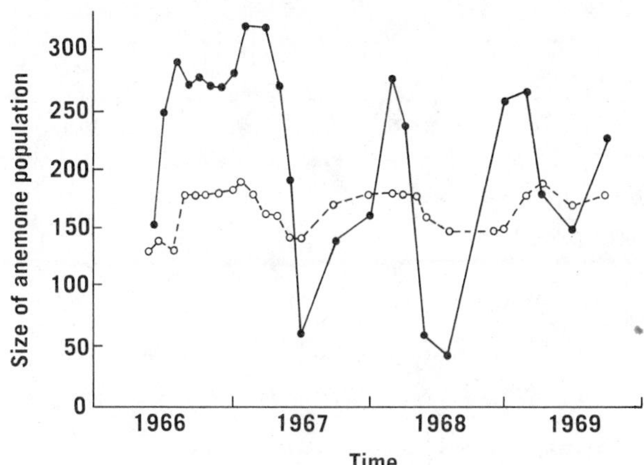

FIGURE 2.13 Effect of removal of barnacles from intertidal region. Solid circles: Anemone population in an area where barnacles were removed. Open circles: Control; anemone population in an undisturbed area. (From P. K. Dayton: Ecolog. Mongr., *41*(4), 1972, p. 373. Reprinted with permission of author.)

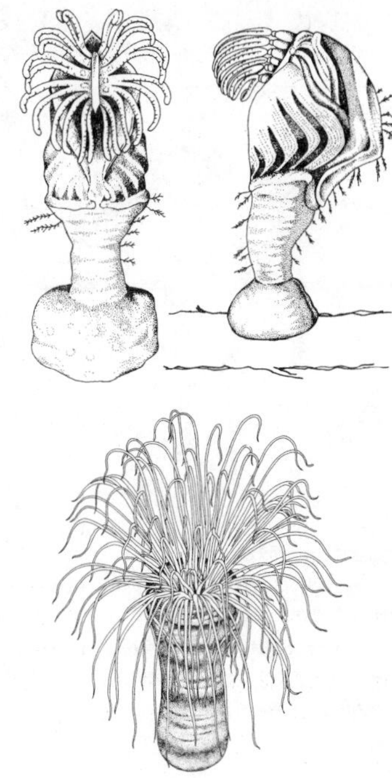

berry. Thus, direct competition was minimized and the species coexisted.

In almost all cases, competition between two species retards the growth of both competitors, and therefore it is tempting to say that one species would be "better off" somehow if its competitors were eliminated. Natural ecosystems, however, can be fascinatingly complex. In one section of the shoreline in the state of Washington, barnacles and sea anemones competed for suitable living space in the intertidal regions. When all the barnacles were removed during an experiment, the anemone population grew rapidly, as one might have expected, but this bloom was followed by a rapid decline, as shown in Figure 2.13. The decline did not result from a food shortage, rather it was found that the anemones died of desiccation. In an unperturbed system anemones grow close to their competitors, the barnacles, thereby finding shelter from the hot summer wind.

Niche competition and the resultant diversity are not merely a two-species interaction. Of the 10 principal mammalian predators living in the southern tundra and northern forests, no two species have identical hunting habits, yet significant overlap is apparent, as illustrated in Table 2.1. Let us see how this system operates. All the predators but the shrew hunt the lemming, a rodent whose population is susceptible to periodic fluctuations (page 45). During abundant lemming years, all the predators can eat lemmings. This prevents lemmings from becoming

TABLE 2.1 NICHE COMPETITION IN THE TUNDRA*

MAMMALIAN PREDATORS	*FOODS*							
	Rodents	*Insects*	*Large Grazers*	*Eggs*	*Birds*	*Fish*	*Carrion*	*Vegetables and Berries*
Bear	X	X	X	X		X	X	X
Wolf	X		X	X	O		O	
Coyote	X		O	X	X		O	
Fox	X			X	X			
Wolverine[a]	X		X	X	X	X	X	
Otter	X			X	X	X		
Marten[b]	X			X	X			
Shrew		X						
Lynx[c]	X			X	X		O	
Primitive man	X		X	X	X	X		X

*Legend: X = regular or staple part of diet; O = occasional part of diet.
[a]Can gnaw bones unchewable by other species.
[b]Can climb trees and hunt tree-dwelling animals more effectively than other species.
[c]Migrates when food is scarce.

so populous that they destroy their food supply. Then, during years with few lemmings, the wolves can eat more caribou, the foxes can subsist on birds, the lynx can move south, and so on. Thus the predator populations survive. If the diets of predators were limited

Migrating herd in Africa. (From C. A. Spinage: *Animals in E. Africa*. Boston, Houghton Mifflin Co., 1963.)

to lemmings, some species might become extinct, and, in that case, there might not be sufficient control mechanism during peak lemming years.

We have shown some relationships among niches that occur when several species gather food in a single geographic area, or when several species of stationary organisms, such as barnacles, live along a gradient of climate. Sometimes niche interactions are defined by patterns of migrations. Of the numerous grazers in the Serengeti grasslands of Africa, the relationship between the zebra and the Thompson gazelle is illustrative. The former is a large, horse-like animal which does not chew its cud. Its physiology is therefore quite different from that of the cud-chewing gazelle. Since the zebra is able to eat large quantities of food and excrete the unnecessary carbohydrates, it can live on plants with a low protein content. The gazelle's digestive system is unable to handle such large quantities of food, and it must subsist on protein-rich grasses. During the abundant wet season both animals graze together on the hillsides, but when the yearly drought arrives, the zebra cannot subsist on the sparse growth and moves down into the floodplain. There it eats the tall grasses which are plentiful but nutritionally poor. The gazelle, meanwhile, remains to wander about the hillsides picking out the small, rich surface plants which the zebra did not consume. Because the gazelle is small and needs less total food than the zebra, he can afford the time to search for food. The zebra would have starved had he stayed. After some time, even the gazelle must migrate. The tall grasses of the floodplain have been cut down by this time by the zebras, leaving the lower, smaller, richer surface vegetation for the gazelle. The zebra continues to migrate in a path that eventually brings him back to the hillsides at the start of the next rainy season.

Amensalism is an interaction in which the growth of one species is inhibited, while the growth of another species is unaffected. As an example, certain shrubs native to southern California secrete toxic chemicals which kill nearby grasses. (See Figure 2.14.)

Predation is an interaction in which certain individuals eat others. Since all heterotrophs must eat to survive, predation is an integral part of the function of any ecosystem. In stable ecosystems,

Top, **Zebras in Africa;** *Middle and bottom*, **gazelles in Africa. (Courtesy of Dr. C. A. Spinage, College of African Wildlife Management, Mweka, Box 3031, Moshi, Tanzania.)**

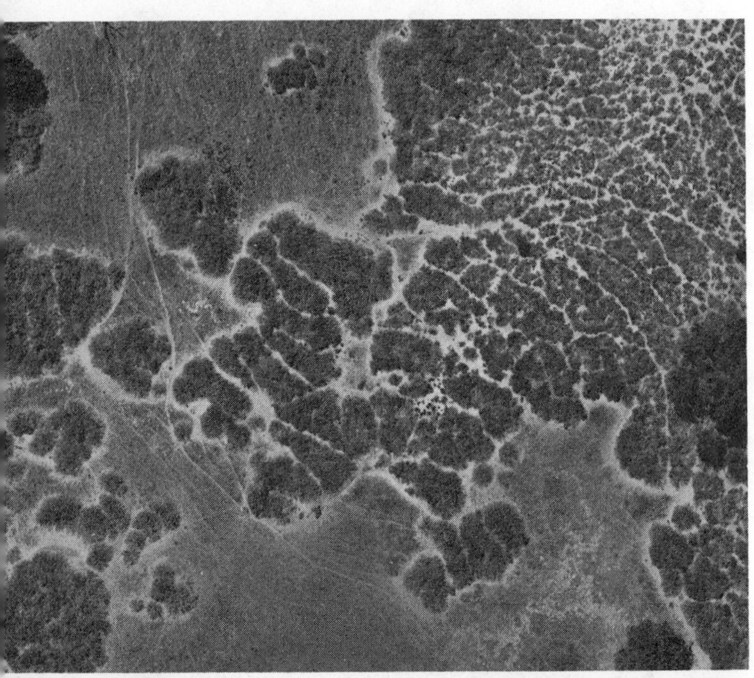

FIGURE 2.14 *Top,* **Aerial view of aromatic shrubs** *Salvia leucophylla* **and** *Artemisia californica* **invading an annual grassland in the Santa Ynez Valley of California and exhibiting biochemical inhibition.** *Bottom,* **Closeup showing the zonation effect of volatile toxins produced by** *Salvia* **shrubs seen to the center-left of A. Between A and B is a zone two meters wide bare of all herbs except for a few minute, inhibited seedlings (the root systems of the shrubs which extend under part of this zone are thus free from competition with other species). Between B and C is a zone of inhibited grassland consisting of smaller plants and fewer species than in the uninhibited grassland seen to the right of C. (Photos courtesy of Dr. C. H. Muller, University of California, Santa Barbara.)**

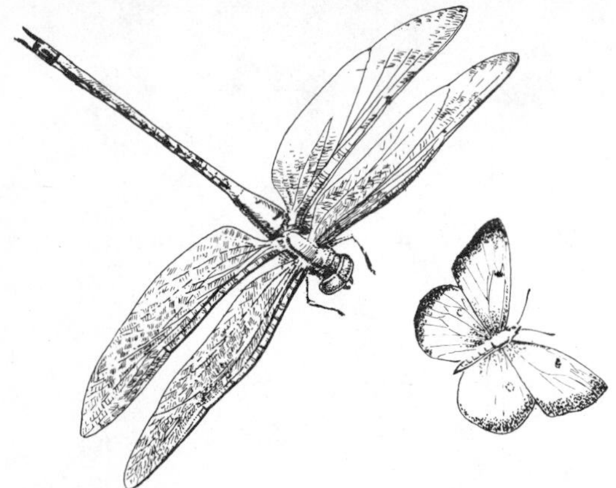

growth and predation are balanced in such a way that all species maintain viable populations. In the example on page 24, we saw that the grazer population in our hypothetical valley in Nepal was regulated by both the availability of its supply of food (the grass) and the size and vitality of its predator population (the tigers). In turn, the tiger population was regulated mostly by the size and health of the herds of grazers.

Of course, not all ecosystems are so finely regulated, and predators have destroyed certain prey populations in some regions. The homeostasis of an ecosystem is not dependent on individual restraint by a particular predator, but rather depends on the statistical balance between offense and defense. Such natural controls do not always function smoothly. We saw in Section 2.1 that dramatic population fluctuations occur, and we will see in Chapter 4 that sometimes the predator population will cause a species of prey to become extinct. In relatively stable ecosystems, however, consumption and growth are nearly balanced.

Predation is a delicate and fascinating process. One of us (Jon) was recently fortunate enough to watch a lone timber wolf stalk a moose by a river in the northern Canadian forest. The wolf followed the moose along a river bank, always maintaining a separation of several hundred yards. While the moose frequently looked toward her predator, she never broke into a run, but continued to feed and move slowly downstream. After about five minutes the wolf turned abruptly, trotted over the hill, and disappeared from view. Both animals knew that a wolf was no match for a young, healthy cow moose. Empirical analyses of caribou killed by wolves have shown that the old, the crippled, the sick, and the very young are killed in disproportionately high numbers. A healthy adult caribou or moose is attacked only very rarely by a wolf or wolf pack. By selecting the less fit animals as victims, predation is a force in the natural selection of the hunted species.

This story is only a small part of the total predation picture, however. We have already mentioned that two species cannot occupy exactly the same niche because one always dominates or displaces the other. It has been shown that in the absence of predation, two species with different but similar niches will often be unable to coexist. Yellowstone Park has a food web that was once balanced and rich with diverse species. Man's initiation of the destruction of the predator population has led to a rapid rise in the number of elk. The abundance of elk has put so much competitive pressure on the deer that they are threatened with elimination from the area.

In other systems it has been shown that predatory pressures *increase* the number of prey species in a system. In one study of intertidal communities on the Pacific coast, a typical section of shoreline harbored 15 resident benthic species. There were

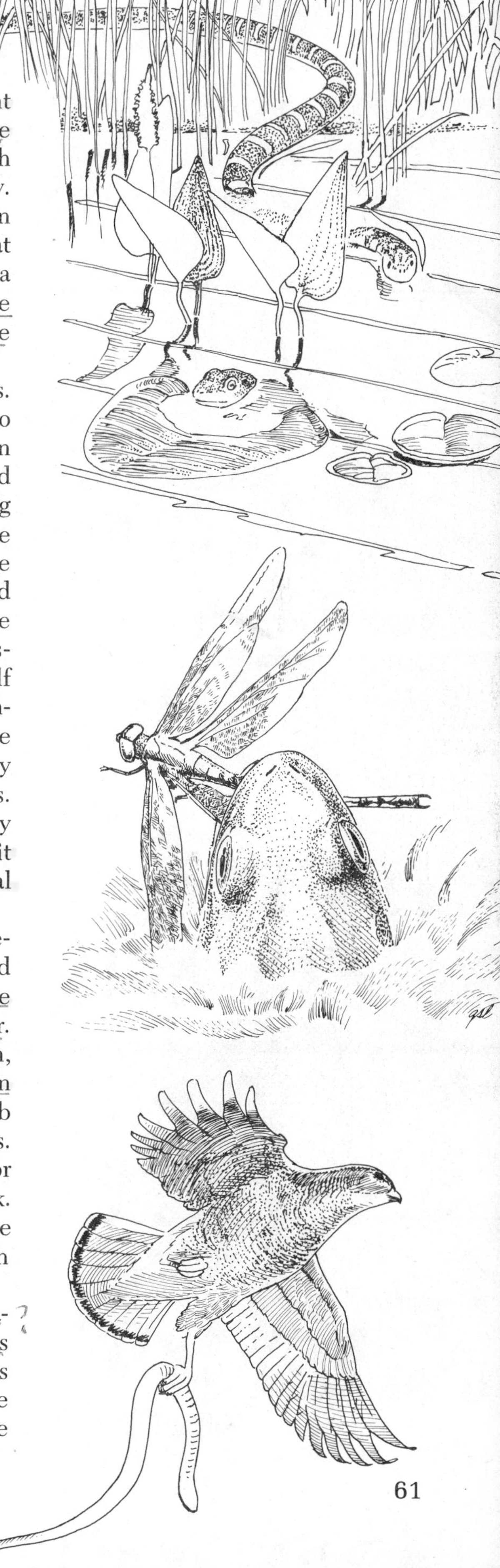

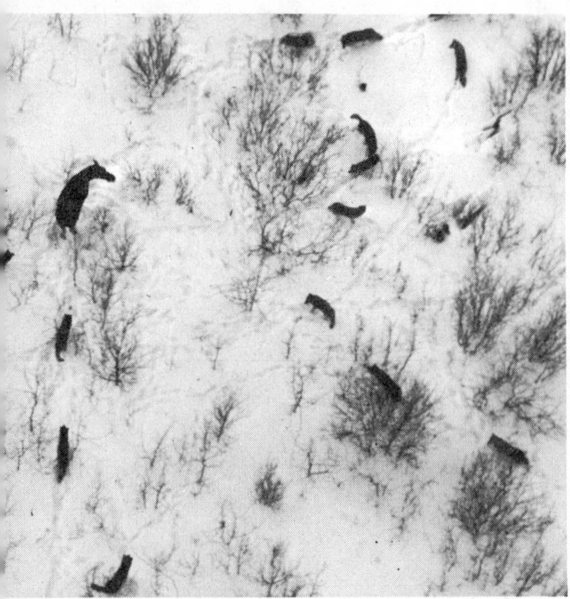

olves and moose on Isle Royale. (Courtesy of L. David ech.)

agle – a predator of small game. (Courtesy of H. Armstrong oberts.)

several predators in the community, one of which was a type of starfish. When the starfish were removed from an experimental area, the community was altered drastically. Interspecific competition for space and food became intense. Of the 15 original species, 7 were eliminated. Apparently, the niches of many of the organisms were quite similar. In spite of this similarity, however, competitive displacement did not occur in the undisturbed system because starfish are general rather than specific predators. They put the greatest hunting pressure on the most common species, thereby controlling overpopulation by any one type of organism.

Parasitism is a special case of predation in which the predator is much smaller than the victim and obtains nourishment by consuming the tissue or food supply of a living organism known as a **host.** Just as predator-prey interactions are balanced in healthy ecosystems, parasite-host relationships have also become part of the mechanism of homeostasis in nature. It must be stressed that this type of balance observed in old systems does not imply that a new parasite (or predator or grazer), artificially imported from another continent, will immediately establish itself as part of a stable system. On the contrary, a new species may find a new niche for itself and increase unchecked until, perhaps many years later, food supplies decline or another species migrates, is introduced, or evolves to control the rampant one.

Commensalism is a relationship in which one species benefits from an unaffected host. Several species of fish, clams, worms, and crabs live in the burrows of large sea worms and shrimp. They gain shelter and often eat their host's excess food or waste products, but do not seem to affect their benefactors.

A relationship that is favorable to both species is called **protocooperation.** Crabs often carry coelenterates on their backs, and move them from one rich feeding ground to another. In turn, the crabs benefit from the camouflage and protective stingers of their guests. (Not all crabs and coelenterates are mutually cooperative.)

Mutualism is an interaction beneficial and necessary to both parties. Lichen, which grows on bare rock, resembles an extremely thin layer of vegetation. Actually, the lichen is a mixture of a fungus and an alga. The fungus, which does not contain chlorophyll and thus cannot produce its own food by photosynthesis, obtains all of its food energy from the alga. In turn, the alga cannot retain water and, in some harsh environments, would dehydrate and die if it were not surrounded by fungus. Here the dependence is direct because the organisms must grow together in order to survive.

Another example of a mutualistic interaction can be found within our own bodies. Millions of bacteria live in the digestive tracts of every person. These organisms depend on their host for food but, in return, aid in the digestive process and are necessary to the survival of all of us.

2.6 COMMUNITY INTERACTIONS

(*Problems 26, 27, 28*)

Interactions within an ecological community are composed of a multitude of two-species encounters. The simplest type of community interaction is a chain reaction. Cats eat rats, rats attack beehives, bees pollinate flowers and produce honey. Thus, the popula-

Archer fish knocks ladybug off a leaf with accurately aimed drops of water. (From Roy Pinney – Globe Photos, Inc.)

Moose, elk, and American bison.

tion of wild flowers, and the price of honey, is partly dependent on the population of cats. Because every animal has a place in a food web, the removal, addition, or exploitation of any species will necessarily cause reverberations, large or small, beneficial or harmful, throughout the system. This rule holds for destruction of "pests" as well as for the introduction of new game species.

We have previously suggested that the plant life of an area is generally healthier if grazed than if left ungrazed. Upon closer examination, we find that plant communities thrive best if they are grazed by a community of grazer species instead of by a single species. A single selective grazer is apt to put pressure on a few species of plants, allowing others to dominate.

Interactions between diverse species tend to promote stable ecosystems. A good example is afforded by the role of three species of grazers—bison, elk, and moose—on the ecological balance of the Elk Island National Park in Canada. Moose eat saplings and small bushes. With brush growth held in check, grasses find room to grow. Bison eat grass. Elk eat either leaves or grasses. In this way, the community of grazers acts on the community of plants to ensure that an equilibrium *status quo* is maintained. Of course, there are many other interactions, such as the relationships between the community of predators and the community of grazers. The net result is a balanced, continuous, self-perpetuating ecosystem.

Diversity is also important in the plant kingdom. A single plant species does not constitute a stable system capable of buffering itself against changing weather. For example, both annual and perennial grasses grow in the prairie. The perennial grasses and some low bushes have deep roots, while the annual plants depend on much shorter and less extensive root systems. During dry years there is so little water that many annuals die. However, the perennials, which use water deep underground, are able to live; in doing so, they hold the soil and protect it from blowing away with the dry summer winds. In years of high rainfall, the annuals sprout quickly, fill in bare spots and, with their extensive surface root systems, prevent soil erosion from water runoff. Survival of both types of grasses is ensured by minimization of root competition because the different plants have root systems which reach different depths. In addition, not all species

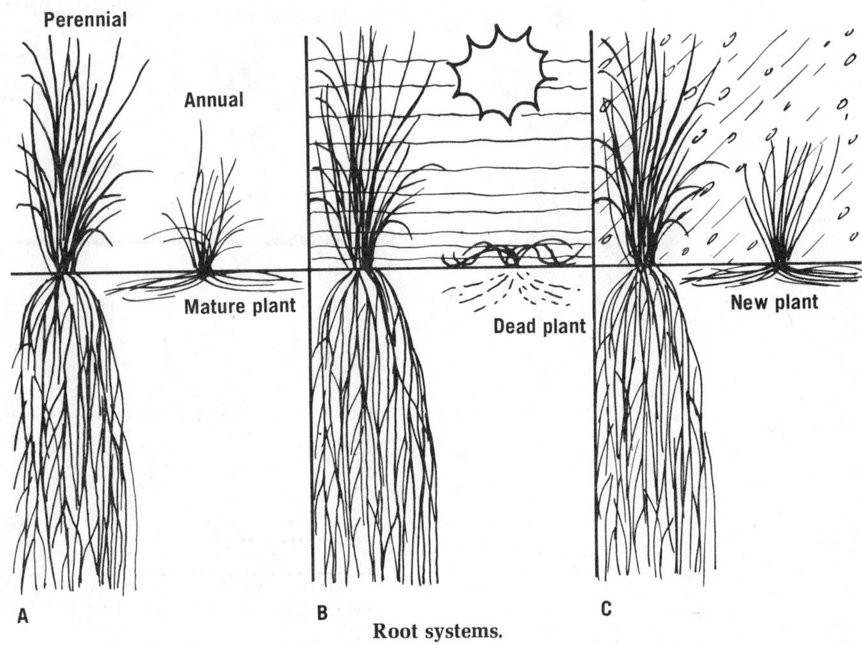

Root systems.

flower at the same time of the year, so that the seasons of maximum growth and consequent maximum water consumption differ.

In canyons in the Colorado Rockies, Ponderosa pine, juniper, and small cacti predominate on the dry sunny side, and the blue spruce, Douglas fir, and flowering plants predominate on the wetter, shady side. Although both sides of the canyon contain representatives of all species, the niche of the pine is sufficiently different from that of the spruce to preserve the separation. If a prolonged drought were to strike the canyon, there would probably be some shift in the plant population in favor of those species that are more effective in conserving water. Thus, changing conditions often change the order of dominance of species. The advantage of community diversity is that the changes that do occur are relatively mild and the ecosystem is not disrupted.

In numerous examples throughout this chapter, we have explained how in some instances species diversity engenders stability, and we have shown both that most stable systems are diverse and that most diverse natural systems are stable. It might, therefore, seem logical to think that *any* increase in diversity will automatically increase the stability of a system, or that diversity *causes* stability. However, this is not a valid conclusion. In Chapter 3 we will reexamine community interactions in such a way as to study the relationship between diversity and stability in more detail.

2.7 GLOBAL INTERACTIONS AMONG ECOSYSTEMS

(*Problem 29*)

We have emphasized that the various species within an ecosystem are interdependent. This dependence can be observed within a radius of a few inches as in the case of lichen, or within several square miles, as is the case for many predator-prey relationships. It is also apparent that entire ecosystems depend on other ecosystems which may be located thousands of miles away. The life around a large river depends on the yearly cycles of river flow. In turn, the river flow depends on the water balance at the tributaries. This water balance is controlled by the forest systems. Thus, the overall state of the forests on banks of a small creek has a direct effect on the life cycles of organisms at the mouth of the river.

In a larger sense, the life forms on the continental and oceanic masses of the earth are linked together into a single interdependent system. Southerly winds and warm ocean currents bring heat to the northern tundra, an area which does not receive enough direct solar energy to support much life. The global balance of carbon dioxide and oxygen in the atmosphere is the classic example of biospheric homeostasis. Before life evolved, the primary constituents of Earth's atmosphere were nitrogen, ammonia, hydrogen, carbon monoxide, methane, and water vapor. Oxygen was present only in trace quantities. Although some theorists believe that geological processes altered atmospheric composition, most scientists feel that the excess oxygen released by the first autotrophs built up slowly over the millennia until its concentration reached about 0.6 per cent of the atmosphere. Multicellular organisms could have evolved only at this point, because aerobic respiration is a prerequisite for their development. The emergence of various organisms about 600 million years ago triggered an accelerated biological production of oxygen. The present oxygen level of about 20 per cent of the atmosphere was reached about 450 million years ago. While there have been some more or less severe oscillations since that time, an overall oxygen balance has always been maintained.

Nitrogen, N≡N

Ammonia, N bonded to H, H, H

Hydrogen, H—H

Carbon monoxide, C≡O

Methane, C bonded to four H (H—C—H with H above and below)

Water, H—O—H (bent)

If the oxygen concentration in the atmosphere

were to increase even by a few per cent, fires would burn uncontrollably across the planet; if the carbon dioxide concentration were to rise by a small amount, plant production would increase drastically. Since these apocalyptic events have not occurred, the atmospheric oxygen must have been balanced to the needs of the biosphere during the long span of life on Earth. By what mechanism has this gaseous atmospheric balance been maintained? The answer appears to be that it is maintained by the living systems themselves. The existence of an effective homeostatic mechanism of the whole biosphere has led J. E. Lovelock to liken the biosphere to a living creature and call that creature *Gaia* (Greek for Earth). He believes that not only is the delicate oxygen-carbon dioxide balance biologically maintained but that the very presence of oxygen in large quantities in our atmosphere can be explained only by biological maintenance. If all life on Earth were to cease and the chemistry of our planet were to rely on abiological laws alone, oxygen would once again become a trace gas.

J. E. Lovelock: Gaia as seen through the atmosphere. Atmospheric Environment 6:579, August, 1972.

John E. Lovelock is an English chemist who has done important work on the analysis of trace gases, including atmospheric pollutants.

The concept of biological control over the physical environment warrants careful consideration. It says that, just as an organism is more than an independent collection of its organs and an ecosystem is more than an independent collection of its organisms, the biosphere is more than an independent collection of its ecosystems. The body chemistry of a human being can function only at or within a few degrees of 98.6°F, but we are normally not in mortal danger of very high or low body temperatures, because there are mechanisms by which our bodies maintain the proper temperature. Similarly, Lovelock believes that modern life can exist only in an atmosphere at or very close to 20 per cent oxygen, but we are not normally in mortal danger of conflagration or starvation because there are mechanisms by which we, the species of Earth, maintain the proper oxygen concentration.

An alternate theory claims that our physical environment has evolved through a series of inorganic reactions, and that biological and physical evolution were independent. The difference between these two beliefs is not trivial. If Lovelock is correct, then a large biological catastrophe such as the death of the oceans or the destruction of the rain forest in the Amazon Basin could cause reverberations throughout our physical world that might create an inhospitable environment for the rest of the biosphere. Alternatively, if the physical world did evolve independently of the biological and is presently controlled by inorganic processes, such a doomsday prediction concerning oxygen balance might be considered unnecessarily alarming.

2.8 NATURAL SUCCESSION

(*Problems 30, 31, 32, 33*)

We have discussed the mechanisms by which natural ecosystems maintain balance, but we must remember that these mechanisms do not prevent change; rather, they provide compensating processes that tend to drive the system back to its original state. These compensations do not always work.

Natural succession is defined as the sequence of changes through which an ecosystem passes as time goes on. The **climax** is the final stage, the stage that is "unchanging." Of course, the word "final" is used with reservation, because the slow process of evolution changes everything. The composition of the climax depends on temperature, altitude, seasonal changes, and patterns of rainfall and sunlight.

As succession continues, changes occur not only in the terrain and the types of species present, but also in the trophic structure of the system. The food web grows more complex, diversity generally increases, and organisms tend to develop highly specialized niches. In the final climax, photosynthesis and respiration have reached a balance. Much of the total respiration that does occur is initiated by decay organisms. Great quantities of organic matter are supported by a relatively low community energy consumption, and it is as if metabolism has slowed down in old age. Thus, a climax would be a poor agricultural system, for agriculture must produce. But the climax structure is the type of system that has traditionally produced things of beauty—the redwood forests, the great plains, and the majestic hardwood stands that used to grow along the Eastern seaboard.

To a large extent, our perception of change in a system depends on how coarse or fine a time scale we choose for our measurements. A small lake is often used as an example of a stable ecological system. Plants, large animals, and microorganisms all exist in balanced relationships. Temporary imbalances occur and are adjusted as in all natural systems. However,

A marsh evolving into a meadow. (From North American Reference Encyclopedia.)

for most lakes the incoming streams and rivers bring more mud into the environment than is removed by the outgoing streams. The effect is small; it may not be noticed in the lifetime of one man. A short-term study of the ecology of a lake would conclude that the ecosystem was in balance. But mass balance is disrupted by the steady addition of solid matter from incoming streams. In time, the lake will begin to fill up with mud. The vegetable and animal life will change. New plants that can root in the bottom mud and extend to the surface where light is available will appear. The trout give way to carp and catfish. If an ecologist studies the lake at this stage he might again say that it was a stable system. Again, minor imbalances and adjustments could be observed but the overall system appears stable. Yet mud continues to flow into the lake. In addition, since it is common for the plants in the lake at this stage to produce more food than is consumed by the herbivores, the bottom fills up with humus. Eventually the lake may become so shallow that marsh grass can grow.

The marsh system is characterized by a net overproduction of organic matter, as indicated by imbalance in the equations shown at right. The unequal lengths of the two arrows denote that more organic matter is produced than is oxidized by respiration. In biological terms, this means that there are not enough grazers to consume the large bloom of vegetation, resulting in a net accumulation of litter. Thus, the energy of photosynthesis is stored in plant tissue.

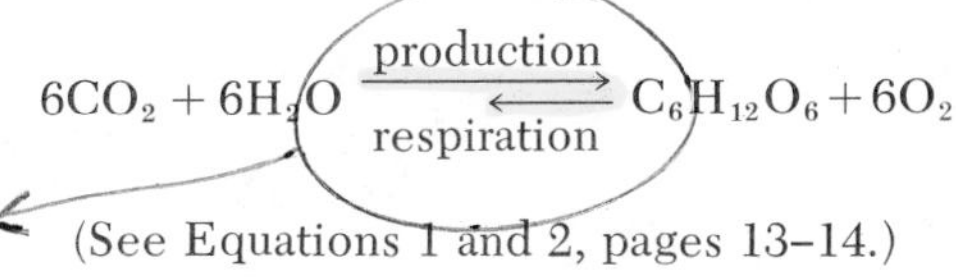

$$6CO_2 + 6H_2O \underset{\text{respiration}}{\overset{\text{production}}{\rightleftharpoons}} C_6H_{12}O_6 + 6O_2$$

(See Equations 1 and 2, pages 13–14.)

A meadow succeeding into a forest.

A "climax" originally referred to a gradually rising series, such as a gradation of more and more forceful expressions. More recently, the word has come to mean the highest point, or culmination, of such a series. It is in this latter sense that an ecological climax, such as a mature forest, means the end, or apex, of an ecological succession.

This tissue remains largely uneaten and fills the marsh.

A marsh is considered a stable system over a short range of observation, but in most cases it is gradually evolving into a meadow. Litter accumulates because decay in the marsh system is slow. Slow decay thus implies poor nutrient recycling. Such a system is inherently unstable, as is typical of any early successional stage. The accumulated litter gradually fills in the marsh which slowly becomes a meadow.

The meadow, too, may change. If the climate is right, trees can start to grow. First, shrubs appear, then quick-growing soft woods like birch, poplar, or aspen. The soft woods are replaced by pine; and finally, in what is called the climax, the pine is replaced by hardwoods.

During the succession of plant species, each species prepares the way for the next while contributing to its own extinction. Marsh grass could never grow without the rich soil of the partially decayed algae. However, by producing the rich soil, the algae helps to fill the lake and thereby destroy its own environment.

We cannot estimate the amount of time typically required to fill a lake, because that depends on many factors, such as the original volume of the lake and the net rate of accumulation of solids. For example, Lake Tahoe in California is so deep and clear that succession, if it occurs at all, would take geological time. The ecologist is generally not concerned with processes that are so slow. For all practical purposes, many lakes may be considered to be in a climax condition.

The time for progression from grassland to a climax forest has been measured. In the southwestern United States, grasslands give way to shrub thickets in one to 10 years, the shrubs become pine forests in 10 to 25 years, and the pine forests give way to hardwoods after about 100 years. We emphasize again that for an ecosystem to be indefinitely stable there must be a complete balance. This is never the case on Earth. Daily, seasonal, yearly, and long-term fluctuations occur in all natural systems, even in the most stable ones. The ultimate consideration of concern for the continuation of life on this planet is that the sum total of all the changes results in worldwide total balance.

The preceding discussion does not explain how a grassland, a marsh, or the tundra can be a climax ecosystem. We have presented situations which sometimes, but not always, represent reality. The great plains which once stretched unfenced from the eastern slope of the Rocky Mountains almost to the Mississippi River never became forests because the rainfall wasn't sufficient. Instead, the grasslands evolved into a climax system with all the trophic structure, energy balance, and stabilizing mechanisms characteristic of the most stable forest ecosystems.

Climax systems are often determined in large part by seasonal variations and geological cycles. A prairie plentiful with buffalo grass grew in Kansas because there wasn't enough yearly rain to support the forests, while the Everglades marsh has existed for a long time because the yearly rain comes in definite seasons. During the flood times, the Everglades behaves like an early successional marsh and is characterized by an overproduction of plant matter and a general silting of the streams and pools. If allowed to persist at this stage, the Everglades would soon follow the successional path toward a forest system. During the periodic droughts, the pools dry up and the litter decomposes, and sometimes burns rapidly. When the rainy season returns, the fallen leaves and logs have been recycled back to fertilizer, the pools have been cleared, and conditions are once again ripe for a bloom of marsh life.

The total cycle in the Everglades can't be classified neatly either as an early successional stage or as a climax system. During periods of bloom when net production greatly exceeds net consumption, we are reminded of a transient marsh, yet the overall stability and continuity of the system resembles that of a climax system.

The Everglades is the dominant biological community south of Lake Okeechobee in Florida. It includes extensive regions of sawgrass, cypress swamp, mangrove, and finally brackish waters leading to the Florida Bay and the Gulf of Mexico. It is a highly developed ecosystem with a large variety of plant and animal species and very complex biotic relationships, still not thoroughly understood.

The role played by fire in succession is also important. When the pioneers settled near the edge of the prairie in northern Wisconsin, plowed it, planted it, and controlled it, they discovered that any area left uncultivated soon became thick with new tree seedlings which sprouted and grew rapidly. A study of the area showed that the prairie that used to grow there had been periodically consumed by fire. Fast-growing grasses quickly regained control, while forests were never able to survive even though the soil and weather conditions were favorable. These grasslands exhibited a **fire climax,** a condition in which the continuance of a given system is maintained by fire. During growth after a fire, the ecology of the area resembled that of an early successional stage, for production was greater than consumption, and the standing quantity of organic matter was low. Eventually, evolution produced animals and plants which were able to accommodate to, or even depend on, the fires. Some grasses and trees developed seeds that sprouted only after being cracked open by fire. Those trees that did exist evolved thick, fire-resistant barks. Although fire returned fixed nitrogen to the atmosphere, this effect was counterbalanced because nitrogen-fixing legumes were common among the early successional plants.

It is interesting to consider what determines natural boundaries between climax systems. In many places, local weather patterns such as those observed around a large river or a mountain range result in rather rapidly changing climates with the resultant change in climax systems. In other places, soil structure or local factors such as human work will affect the change. Even under these conditions, one wouldn't expect an abrupt transition between ecosystems but rather some sort of border area which harbors representative species of each neighboring region. If you live near a place that is part farmland and part woods, take a walk in the country some day and direct your attention to the boundary between plowed fields and woodlots. In most places you will find a band, sometimes not more than 10 or 20 yards wide, in which birch, poplar, or aspen saplings mingle with various grasses and maybe a corn stalk or two. Flora from both the field and the woods will be present. Additionally, many types of bushes which aren't characteristic of either major ecosystem will be found. If you have the eye of a naturalist, or of a hunter, you will discover that more rabbits and upland game birds live here, where they can take advantage of the rich plant life, the good cover, and easy access to crops or woods. Such a border area is known as an **ecotone** and varies greatly in width. Ecotones generally support more life and numbers of species than either bordering climax area. Recall that an estuary is an example of an aquatic ecotone.

2.9 THE ROLE OF PEOPLE

People, too, occupy a position in the flow of energy through the biosphere and must necessarily interact with thousands of other species of plants and animals. Why, then, do we consider humans separately? The answer is that today people have unprecedented power at their disposal, and with it the ability to increase the productivity or destroy the ecosystems of the earth. Man has been so successful at reducing competition from other species that his numbers have risen precipitously in a time span that is insignificant on the evolutionary clock. Before man, species and ecosystems evolved together over long stretches of time, but man's social and technological changes are orders of magnitude faster than evolution.

Nowhere is the rapid effect of human technological advancement more pronounced than on the North American continent. Three hundred years ago, this land mass housed a set of diverse, balanced ecosystems where many types of plants and animals, as well as people, lived in a manner which had changed little since the last ice age. Today, some of the native ecosystems such as the tall-grass prairie and the Eastern hardwood forests are gone or altered almost beyond recognition. Those biomes that still exist, such as the Northern forest or the Rocky Mountain tundras are shrinking in the wake of progress, even as the original balances of species are being altered by predator control and hunting. In fact, every predator large enough to take a deer or a cow, with the exception of man himself, has been pushed into a few remote or bitter-cold areas. Even there the predators risk eradication. In place of the natural forests and ranges, millions of acres of intensely cultivated land produce large yields of foodstuffs to feed millions of non-farmers.

Such is our present condition. Moreover, it is completely unreasonable to believe that man will choose to return to the ways of the nomadic hunter or the Stone Age farmer. The problem for us is not to try to recapture the past but to make decisions for the future.

Many people believe that our present systems, controlled by man, are providing people with greater physical security and comfort than they have ever known. The major question of environmental policy is: what types of systems are viable and will continue to provide security and comfort far into the future?

Imagine a technology spectrum ranging from a Stone Age hunter to an automated city inside a huge environmentally controlled dome, where food is synthesized from sewage in large factories with the aid of sunlight. What levels of technology can survive for a long time, and of those that can, which are desirable? Before we attempt to find the answers to such grave questions, we must establish the criteria on which the answers must be based. If our environment is to be stable and hospitable to man for a long time, our decisions must be based on sound scientific reasoning by people whose goals go beyond personal gain or power, and whose interests extend beyond their own lifetimes. Public opinions on any given environmental problem, such as radioactive waste pollution, range from: "It doesn't present a problem," to "It is an annoyance but well worth the benefits gained," to "It's a debatable trade-off," to "It's an imminent threat to the existence of man." Of these, all statements must be given consideration except the first, for man-made pollution does indeed present problems.

Perhaps the main reason for the wide range of opinions from lay people as well as specialists is that the data can be interpreted in many different ways, depending on the individual's choice of priorities. We conclude this chapter on natural ecology with two examples of issues which are still to be decided. In both cases, the direct alteration of natural ecosystems is at stake.

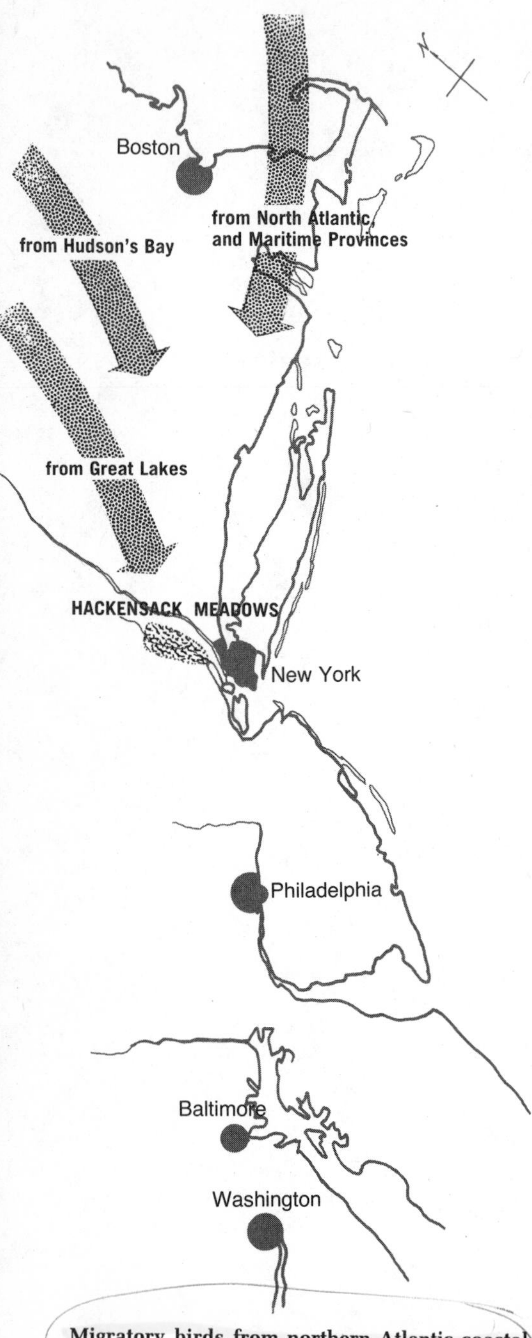

Migratory birds from northern Atlantic coastal and mid-west nesting areas use the Hackensack Meadows as a rest and feeding stop on their way south.

CASE I

Every spring, birds of many species must fly north to their summer breeding grounds, and every fall they must return, or many would starve and freeze to death. Along the Eastern seaboard of North America there is still room in the northlands and southlands for the migrants to nest, but the area in between, from metropolitan New York south to Washington, D.C., is becoming increasingly inhospitable. One great and ancient swamp does remain in northern New Jersey, right in the heart of one of the most industrialized areas of the world. This marsh houses many species of its own, and supports millions of migrants who must stop there for rest and food. Without this haven, many birds could not survive the next migration.

It has been proposed that the marsh be converted into a model urban community complete with environmentally controlled shopping centers, comfortable apartment complexes, and spacious parks and recreational areas. The construction of the new city would provide jobs for many workers and, when complete, many needed new living units for our expanding population. These living units will undoubtedly provide modern conveniences for the new inhabitants and relieve some pressures from the already overcrowded cities. Let us assume that the planners of the project will install the latest in sewage recycling systems, solid waste disposal units, public transportation systems, and other pollution-limiting features. But what is the environmental cost? This particular project wouldn't exterminate all the remaining migratory bird populations along the Eastern seaboard, but it would destroy many birds. The lives of many species would end if projects of this sort proliferated.

Some questions the proposed development presents are esthetic and moral. Does a goose have a right, as a member of the biosphere, to live out its life according to its ability and wits, and to eat or be eaten as its ancestors were? Or, from a person's point of view, is it part of one's birthright to stand on the doorstep on that first cold day in October and watch the geese grouping into as yet poorly formed V's and honking their way south? Will the residents of the new housing projects want to go to Vermont each fall to hunt? How important is natural beauty to man? Are we willing to destroy some of the natural quality of our lives in order to provide more technology?

Moreover, threatening the survival of the geese presents concrete problems. Lumber and paper are harvested from the geese's summer homes; salt-water fish grow to adulthood in the southern salt-water marshes where many geese winter. The removal of migrating birds would certainly affect the ecological relationships of both areas, but we are not sure of the effect on their ability to produce goods for man. Trees can be grown in monoculture farms, and perhaps the estuaries, without geese, could provide more food if we added more fertilizer, and perhaps both the forests and the estuaries could easily adjust to their new existence.

For the people who feel that nature must surrender to progress, the question remains: will monoculture forests and farmed estuaries be able to survive in time, or are we building our technology on a poor foundation? As technology improves, perhaps we will be able to construct plastic building materials and synthesize food from coal. But if the natural world is destroyed during the construction of a technological one, and the latter does not work, we will be able to re-create neither.

CASE II

Part of the shoreline of southern Long Island boasts some of the world's finest beaches. They provide recreation for millions of metropolitan New Yorkers. But the beaches are being eroded away by the sea. The wave action responsible for the sand removal isn't a factor initiated by some action of man. Rather, it is a natural geological process for oceans to form beaches and then displace them on land or move them out to sea. Shorelines and beaches have been moving and reshaping since the formation of the continents. Because of the recreational importance of this particular set of beaches, the Army Corps of Engineers has proposed to rebuild them at an estimated cost of 200 million dollars. The Corps plans to dredge some of the sand needed for the project from the inland bays and other estuary systems of the south shore. Conservationists argue that during the dredging process the ecological relationships in the estuaries would be severely disrupted. We know that many salt-water fish rely on estuaries as nursery grounds and that man relies on the fish as food. It is possible, then, that rebuilding the beaches would disrupt commercial offshore fishing.

Ultimately, a decision must be made: To what degree should man try to be the controller of his planet? Are humans wise enough to control the earth for their, and its, ultimate benefit or just powerful enough to change it for their own immediate gain?

PROBLEMS

1. **Definitions.** Define biotic potential, environmental resistance, and carrying capacity.

2. **Sigmoid growth function.** Examine the sigmoid curve shown in Figure 2.1 and mark the points where the rate of increase is largest and smallest. What is the smallest rate of increase on the curve?

3. **Unstable growth functions.** In some areas of the world, human populations increase rapidly during years of abundance, only to be faced with famine and starvation during years of drought. During these periods of famine, population growth slows down somewhat, and occasionally, the population decreases, but the long-range trend is toward increasing populations. Draw a hypothetical curve of human population size, illustrating these events. Show which points correspond to years of abundance and which to years of famine. How does your graph compare with the population graphs for yeasts, reindeer on St. Paul island, and sheep in Tasmania?

4. **Carrying capacity.** The worldwide human population has been increasing continuously for the past few hundred years. Can this trend continue indefinitely? Are human populations subject to the constraints of a worldwide carrying capacity? Explain.

5. **Time lag.** In Chapter 1, Section 1.5, the effects of drought on a grass-grazer-tiger system in a hypothetical valley in Nepal were discussed. Draw a graph of the imaginary population levels as a function of time. Do the population maxima and minima of the organisms in the three trophic levels coincide? Compare your imaginary graph with the oscillations in hare-lynx cycles. Briefly discuss time lag in population theory.

6. **Lemming cycles.** (a) The numbers of many of the lemming predators fluctuate much less dramatically than do the numbers of the lemmings themselves. Can you think of a reason for this? (*Hint:* See Section 2.6.) (b) The snowy owl is a predator which specializes in catching arctic rodents. Snowy owl populations fluctuate dramatically with lemming populations. Explain.

7. **Law of Limiting Factors.** Animal migrations can be considered to be an adaptation in response to the Law of Limiting Factors. Explain.

8. **Law of Limiting Factors.** Do you think that a human being living in a cold climate could survive as well on a severely limited diet as a person living in a moderate climate could? Explain.

9. **Ecotype.** Are there different ecotypes of human beings? Explain, give examples. Design a simple experiment that could be used to test your hypothesis.

10. **Ecotype.** Why do you think that northern jellyfish have a much broader temperature range than southern jellyfish?

11. **Environmental resistance.** Discuss density-dependent and density-independent factors. Give an example of each.

12. **Environmental resistance.** Explain why populations living in the tropics would be likely to be controlled by biological factors, whereas populations living in the Arctic would be likely to be controlled by physical factors.

13. **Environmental resistance.** The physical components of the environmental resistance are not always completely independent of the biological ones. Discuss this statement and give examples to support it.

14. **Niche.** Define ecological niche. Are you living in your optimum niche? Defend your answer.

15. **Niche.** Decay organisms living in flowing streams often attach themselves to rocks. One species may predominate on the lee side and another on the current side. What factors might be involved in establishing such a relationship?

16. **Niche.** Can two individuals of the same species occupy exactly the same niche? Explain.

17. **Niche.** Discuss some differences between the home territory of the nomadic hunters and the niche concept of animals.

18. **Two-species interactions.** List the eight major types of two-species interactions. Include a brief explanation of each.

19. **Diversity.** Explain how niche competition promotes diversity.

20. **Competition.** Would you expect competitive pressures to be density-dependent or density-independent? Explain.

21. **Competition.** Competition can be interspecific or intraspecific. Explain and give examples.

22. **Predation.** The ecologist Elton contends that predators live on capital, while parasites live on interest. Explain. Is this true from an individual or from a community viewpoint? Explain.

23. **Predation.** Would you expect a buffalo herd to be healthier after years of being hunted by men with bows and arrows or by men with guns? Explain.

24. **Predation.** After the removal of starfish from an intertidal ecosystem, the total number of species decreased. (See page 62.) Do you think that the total number of individuals also decreased? Defend your answer.

25. **Two-species interactions.** Refer to the eight types of two-species interactions and categorize each of the following examples: (a) mistletoe sucks the sap from the pine tree on which it grows; (b) a barnacle gains mobility by attaching itself to a whale, but does not consume any of the whale's tissue; (c) a small bird eats the meat caught between a crocodile's teeth—the bird gets fed and the crocodile has free dental care; (d) a paramecium eats a bacterium; (e) an elephant steps on an ant; (f) two trees growing side by side reach out for light.

26. **Community interaction.** Do you think that it might be economically profitable to raise deer along with cattle in some areas of North America? Explain.

27. **Ecosystem stability.** Which do you expect to be better able to survive a drought; a cornfield or a natural prairie? Explain.

28. **Natural balance.** Discuss the statement, "Natural systems are perfect because they are always in harmonious balance."

29. **Ecosystems.** Would you say that the Earth includes many ecosystems that are relatively independent of each other, or that it contains only one ecosystem that occupies the entire biosphere, or that both statements are true? Present arguments in favor of your position.

30. **Climax systems.** It has been observed that large, complex plants and animals are more characteristic of climax situations than of early successional stages (a) Name some organisms and their habitats that serve as examples of this observation. (b) To explain this observation, it has been suggested that species with long life cycles have evolved away from unstable environments. It has also been suggested that large, complex animal species cannot find proper sustenance from plants that grow in simple systems such as marshes. Repeat these hypotheses in your own words and argue for or against them.

31. **Succession.** Define natural succession. What factors bring about changes in an ecosystem? What is the climax of an ecosystem? Cite three examples of a climax ecosystem.

32. **Succession.** Imagine that a new island just arose in the South Pacific. Trace the succession that would be expected to occur. Estimate the time span required for the climax to be reached. Compare the energy balance of the early ecosystems of the island to the energy balance of a marsh. Compare the types of species present.

33. **Imbalance.** What do you think would happen to the Everglades if man built a set of dikes to ensure constant water levels all year round?

BIBLIOGRAPHY

The three basic textbooks cited in the bibliography of Chapter 1 all contain material germane to this chapter as well. Several books dealing specifically with population biology are:

Arthur S. Boughey: *Ecology of Populations.* 2nd Ed. New York, Macmillan Co., 1973. 182 pp.

Arthur S. Boughey: *Contemporary Readings in Ecology.* Belmont, Calif., Dickenson Publishing Co., 1969. 390 pp.

Kenneth E. F. Watt: *Ecology and Resource Management.* New York, McGraw-Hill, 1968. 450 pp.

Edward O. Wilson and William H. Bossert: *A Primer of Population Biology.* Stanford, Conn., Sinauer Assoc., 1971. 192 pp.

3

WHY ARE THERE THIS MANY SPECIES?

3.1 RIDDLES

Riddle 1: Why do trees in Africa look like this?
Answer: Because giraffes eat them like this.

Riddle 2: Why hasn't another species of tree browsers evolved with slightly shorter necks to eat the lower branches, and then why hasn't a third species evolved with longer necks to eat the higher branches? Why aren't there more species? Or, why have giraffes evolved at all? Why isn't there just one species of herbivore on the African Continent with a telescoping neck and the ability to eat grass and tree leaves at all heights? Why are there this many species—no more and no less?

Nonanswer: Of course, this second riddle is extremely difficult. Some scientific questions can be approached directly by experiment and observation. The answers to others must always remain hypothetical. How was the universe formed? Or, how did life first appear on the Earth? Since no one witnessed these events, our answers can never be certain; we strive to make them at least plausible. Riddle 2, which asks why life has evolved as it has, is such a question.

3.2 THE EVOLUTION OF SPECIES
(Problems 1, 2, 3, 4, 5, 6, 7)

> *Discrete* **means separate, detached, or individually distinct. Do not confuse it with discreet, which means judicious or cautious.**

Since the beginning of life on this planet, new species have continually been formed and existing species have continually been driven to extinction. The number of species on the Earth today represents the difference, the total number of new species which have evolved minus the total number of extinctions.

No two individual organisms are exactly alike; on the other hand, some groups of organisms are similar enough so that we recognize their common characteristics. Biologists who classify organisms into discrete groups recognize levels of similarity. Thus, they observe that any given plant is similar in many respects to all other plants and is basically different from animals, whereas all animals, in turn, have common characteristics. Subgroups within these two large groups also exist; for example, all animals with backbones form one such smaller unit. Within the larger subgroup of animals with backbones, all those that are fur-bearing and suckle their young form a distinct class called mammals. Of course there are many different orders of mammals, such as rodents, ungulates, and primates. As these groups become more numerous, the distinctions among them become finer and finer. A pertinent question is, When do we stop classifying? If we stopped too soon (for example, if we did not subclassify mammals), our distinctions would be too gross to be useful. On the other hand, if we stopped too late (for example, if we classified humans according to the curliness of their hair and the color of their eyes), our categories would be too numerous to be very useful.

Of course, not all trees in Africa are shaped like those shown on page 80; giraffes can bend down and eat lower branches. The picture is exaggerated to make the point. This riddle was suggested by Professor Robert M. May of Princeton University to dramatize the fact that the analysis of the diversity of species is a very complex problem.

Since individual animals or plants breed only with similar animals and plants, and not with dissimilar ones, it is convenient to classify organisms that breed together into discrete groups. A **species** is most satisfactorily defined as a group of animals or plants that exhibits reproductive isolation; that is, a group whose members form an interbreeding population but do *not* reproduce outside the group. The number of ways in which groups of organisms can isolate themselves reproductively from other groups is unexpectedly large. For example, potential mates may have different breeding seasons or spawning grounds and hence may never meet during the time that they are interested in mating. Characteristic behavior at the time of mating (for example, courtship and display behavior) may not be recognized. In some cases, mating may be physically impossible owing to incompatibility of reproductive organs. Mating may take place, but the egg may not be fertilized; or if it is fertilized, it may die before many cell divisions. An egg may even develop into an adult organisms, but one with greatly reduced fertility, as is the case with

Classification of Man

Kingdom	Animalia
Phylum	Chordata
Subphylum	Vertebrata
Class	Mammalia
Order	Primates
Family	Hominidae
Genus	*Homo*
Species	*sapiens*

the mule (which is the result of a cross between a horse and donkey). All these mechanisms operate in nature on various groups, singly or in combination, to ensure their identities as separate species.

The species is a fundamental unit of classification because parental characteristics are always passed on to members of succeeding generations. As members of these succeeding generations reproduce, the group characteristics become intermingled, but do not intermingle with characteristics outside the group.

Individual plants and animals face many pressures, such as intraspecies competition, interspecies competition at a given trophic level, interspecies encounter in the form of predation or parasitism, and climatic variation. Now, if we have a varied population subject to certain environmental stresses, some of the individuals may be more successful at reproducing than other members of the species. Obviously, individuals who produce the most viable progeny will make a greater contribution to the next generation than those who produce fewer offspring. Those individuals who enjoy a reproductive advantage over others are said to have greater *fitness.* This process by which environmental stresses give certain members of a species a reproductive advantage over others was termed **natural selection** by Charles Darwin, who was chiefly responsible for suggesting the overwhelming importance of this mechanism in evolution.

What determines which individuals have greater fitness and hence which will be preserved? This question is a reworded form of the ecological question: what niche adaptations enhance the survival of a species? There is no single answer to this question, for there are many ways in which different species adapt. Some individuals, like antelopes, may survive predation because they can run fast, and others, like lions, because they are good fighters and can win territorial battles. But strength and speed are only two of many types of attributes which are selected for. Recall the example of the two species of Scottish barnacles from Chapter 2. One species survived because it was somehow better able to compete for the underwater niche; the other species gained fitness by its ability to grow in spite of periodic desiccation. The nature of this interspecies competition is quite complex and goes beyond conventional concepts of strength and speed. Ability to survive is also manifested in disease and parasite resistance, which is a powerful evolutionary force. Some species compensate for weaknesses by various forms of cooperation, such as the mutual relationship between algae and fungi that form lichens, or the herd and troop defenses of many animals.

We should emphasize at this point two important features of the concept of fitness. In the first place, it makes no sense in most cases to talk about an abstract concept of fitness. A trait which is advantageous in the jungle could be disastrous in the desert. One must almost always speak of fitness with reference to a particular environment. (We add the "almost" to the preceding sentence because animals with some characteristics, such as a lack of reproductive organs, would have a fitness of zero, independent of environmental variations.)

Secondly, the term fitness refers solely to reproductive advantage. A particular lion may be the largest and strongest lion in Africa and may have won every territorial battle he ever fought, but his evolutionary fitness will be zero if for any reason he is incapable of reproduction.

An example of natural selection will serve to tie some of these ideas together. In the mid-nineteenth century in England, rare black variants of the common pepper moth were noted for the first time. By the end of the nineteenth century, however, this

FIGURE 3.1 Picture of moths having industrial melanism. (From Helena Curtis: ***Biology*****. New York, Worth, 1968. Courtesy of Dr. H. B. D. Kettlewell.)**

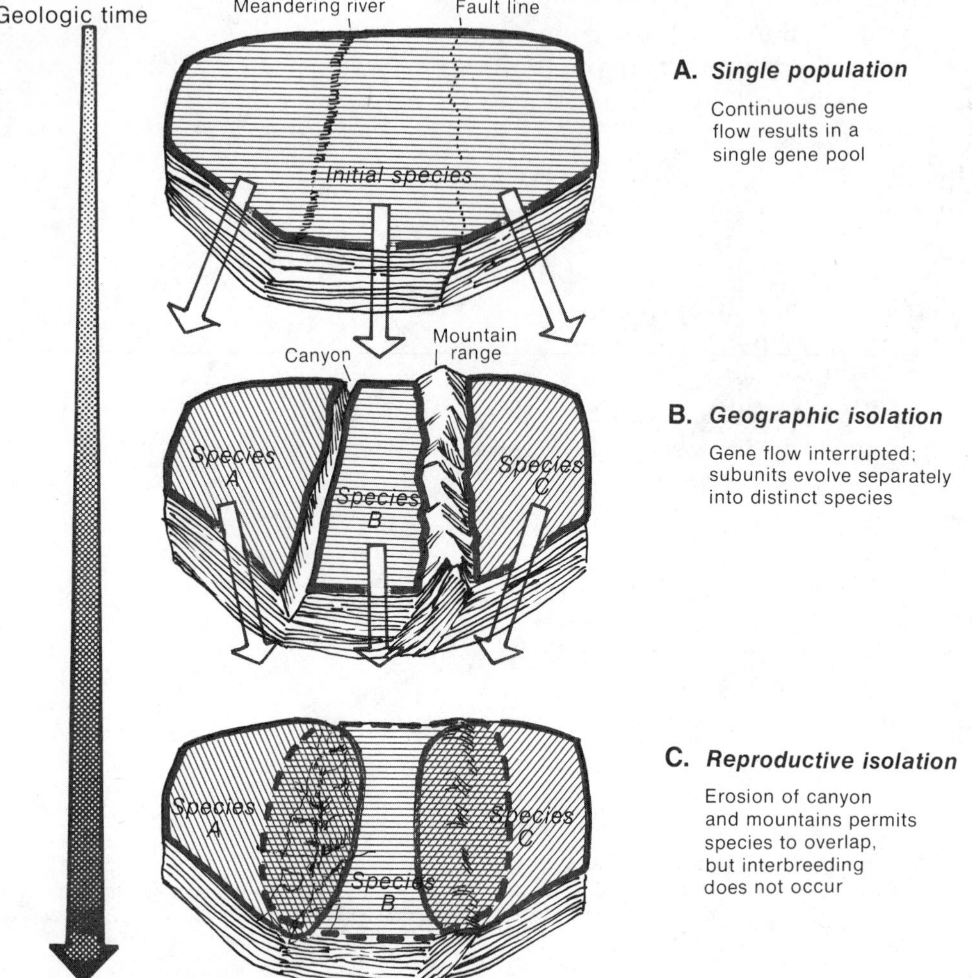

FIGURE 3.2 Theoretical stages in the evolution of new species from a single common ancestor.

Prolonged geographic isolation leads to permanent reproductive isolation. Interbreeding cannot occur, even if the new species again overlap. (From M. E. Clark: *Contemporary Biology*. **Philadelphia, W. B. Saunders Co., 1973, p. 570.)**

once-rare color variant had become the predominating variety in most of the industrial sections of Britain. Such a phenomenon has since been noted in many other moth species in Europe and North America as well (Fig. 3.1). The most likely explanation for these striking observations was that the dark color protects the black moths from their natural predators (birds) in environments darkened by industrial pollution. In heavily industrial areas, then, one might predict that any individual moth with black coloration will preferentially survive and thus have a reproductive advantage over more lightly colored individuals. Hence, these color traits will be enriched in subsequent generations.

Though this sounds like a reasonable explanation, it needs experimental verification, that is, a demonstration that color could be protective in different environments. This verification has been provided by experiments in which equal numbers of black and white moths were released together, both in woods whose trees were darkened by pollution and had no normal lichen on their bark, and in normal woods far from polluted areas. It turned out that there was great preferential survival of the light variants in the normal woods and of darkened variants in the polluted setting. This proves that the color of a moth may serve a protective function and strongly suggests a role for color in natural selection, at least for moths.

Natural selection explains how a species evolves but does not explain how a new species is formed. As long as there is sexual contact among members of a species, a new trait, if beneficial, is likely to be dispersed in the population. For two species to diverge, however, populations must be isolated reproductively from each other for a long period of time, say 2000 to 100,000 generations. During the course of this isolation, changes occur in each population independently, and since the environmental pressures are, in general, different on opposite sides of some natural barrier, eventually two distinct species may arise (Figs. 3.2 and 3.3). While only a large barrier, such as an ocean, a jungle, or a high mountain range, can separate groups of birds, different species of rodents have evolved even on opposite sides of a large river.

FIGURE 3.3 Different species of birds of paradise which evolved by isolation on New Guinea.

3.3 THE EXTINCTION OF SPECIES
(Problems 8, 9, 10)

Environments change. They change because weather patterns change, because the geology of a region changes, or because new organisms evolve or immigrate and alter the existing food webs. Some species adapt to certain environmental changes, but others cannot, and those that cannot become extinct.

When a component of the environmental resistance increases in a given region, the population of any particular species is likely to decline, but in most instances this decline does not result in global extinction. The species may be preserved by one or more of several mechanisms: (a) The population may adapt to the new situation as, for example, the pepper moths adapted to survival in polluted forests. (See page 83.) (b) When the population declines, the density-dependent component of the environmental resistance will decrease. For example, specific predators may alter their habits or may themselves become extinct. This relaxation relieves the added pressure, and the species may survive, although perhaps in reduced numbers. (c) Certain members of the species may migrate and find more favorable living conditions elsewhere. (d) Sometimes conditions are so severe that local communities will become extinct, but the species will survive in more favorable areas, as in the case of grizzly bears which are extinct in California, Arizona, New Mexico, and Nebraska, but remain in Montana, Wyoming, Idaho, Alaska, and Canada. The elimination of a species within a small area is a common occurrence; we will see in the next two sections that individual colonies living on oceanic islands or on mountain tops are frequently driven to extinction. These local extinctions are always reversible, for the "extinct" species may return when conditions become favorable. Global extinctions, on the other hand, are irreversible; a species cannot be replaced once all its members have died.

FIGURE 3.4 The passenger pigeon—a lesson learned too late.

Alternatively, the original pressures which led to the population decline may be so strong that the species does become extinct. For example, a certain species of crustacean was known to live in a few small Midwestern ponds. As these ponds were drained for farms and for construction, the crustacean populations declined. When the last pond containing that particular species of crustacean was drained, the species became extinct.

The crustacea in the previous example were driven to extinction by the same pressure that caused the initial population decline. In other instances the force driving a species to extinction has been less obvious. The populations of some species have been known to become extinct some time after the environmental pressures have been relieved.

Let us consider the case of the passenger pigeon (Fig. 3.4). In the late 1800's, approximately two billion birds flew over the North American continent in flocks that blackened the skies. Commercial hunters, shooting indiscriminately into the flocks, killed millions for food and many more for fun, for the species was thought to be indestructible. As hunting pressures increased, the pigeon populations naturally suffered, and by the early 1900's market hunting was no longer profitable. Yet thousands of pigeons survived. Then suddenly they all vanished. Poof! Ducks, geese, doves, and swans had all been hunted and they survived in reduced numbers; why did the passenger pigeons succumb?

The naturalist Aldo Leopold could offer none of the conventional ecological arguments and said of the pigeon,

> The pigeon was a biological storm. He was the lightning that played between two opposing potentials of intolerable intensity: the fat of the land and the oxygen of the air. Yearly the feathered tempest roared up, down, and across the continent, sucking up the laden fruits of forest and prairie, burning them in a traveling blast of life. Like any other chain reaction, the pigeon could survive no diminution of his own furious intensity. When pigeoners subtracted from his numbers, and the pioneers chopped gaps in the continuity of his fuel, his flame guttered out with hardley a sputter or even a wisp of smoke.

From Aldo Leopold: *A Sand County Almanac.* New York, Sierra Club/Ballantine Books, 1970.

In modern biological language, we say that the pigeon population was reduced below its **critical level,** and thus could not survive even though many mating pairs remained alive. The concept of the critical level is a general one and applies to nuclear reactors as well as to pigeons. However, it is often difficult to determine accurately the critical level for a given system. To understand how critical levels operate and why they differ quantitatively from species to species, the following aspects must be considered:

1. When animals face certain types of stress, many of them fail to reproduce normally. This phenomenon is fairly common in the animal kingdom; an

abnormally high percentage of American males became impotent during the Depression of the 1930's; female rabbits reabsorb unborn fetuses into the blood stream during drought; and many animals, such as the Javanese rhinoceros, have never bred successfully in captivity. We just don't know enough about behavior to predict how stress will affect fertility of a given animal, but we do know that in certain cases destruction of range and harassment by hunters alter the reproductive potential of the remaining stock. A sudden descrease in the reproductive rates in the face of mounting pressures would certainly lead to a rapid population decline.

2. Often the pressures which act on one species of an ecosystem do not affect the entire ecosystem equally, and severe imbalance can result which might lead to extinction. The original North American population of a few billion passenger pigeons must have supported a large population of various predators. When commercial hunters slaughtered a significant number of pigeons, they did not shoot a proportional number of predators, and it is possible that the pigeon was exterminated because the ratio of predator to prey was so unfavorable.

4. When the population of a species becomes low, that population is particularly subject to bad luck, and a few unfortunate events can lead to extinction. The situation is analogous to that of a gambler who is playing a game that is rigged so that the odds are in his favor. If he has a large cash reserve and can play for a long time, he will win, but if he starts off with a small stake, a few unlucky hands in the beginning can end the game. So it is with species. Any large population living in its own natural surroundings would be expected to survive. However, if only a small breeding stock exists, chance occurrences might lead to extinction.

An example is the Steller's Albatross. Despite decimation by hunters, a few viable flocks survived and breeded. Then, in 1933 as one flock was nesting peacefully on an offshore island, a volcano erupted, killing most of the adults and all of the young. Again, in 1941, another volcano erupted and destroyed a second nesting population. The species survived for another 20 years and reproductivity had just begun to accelerate when the last sizable flock was caught in a typhoon and destroyed. The future of the species is in question.

Sometimes, however, chance favors survival. The sea otter was extensively hunted for fur from about 1750 to 1900 (Fig. 3.5). By 1910 the otter was believed to be extinct. No one knows how many individuals actually survived the fur trade, or where they lived, but around 1930 this "extinct" animal reappeared and established itself in several locations. Obviously, a small breeding stock had survived and increased in spite of two decades of hard times.

Biological luck is more complicated than the presence or absence of a typhoon, volcano, landslide, avalanche, or other natural disaster. When breeding populations are very low, then genetic misfortunes can be disastrous. For example, appearance of a rare deleterious mutation in one offspring of a large population will have no effect on the survival of the population. But if the breeding population is tiny, such a mutation may threaten continuation of the species simply because the tiny community can ill afford to lose even one potentially reproductive member.

5. **Inbreeding** poses another danger to the existence of reduced populations. In general, unions between closely related animals, such as brother to sister and cousin to cousin, reduce the vitality of a population. Naturally, if a population is small, matings must occur between close relatives, and thus the vitality of future generations is endangered. For this reason, a species might not be able to survive an initial population decline even if favorable conditions return.

For all these reasons a species with a small population has an inherently lower chance of survival than a species with a large population.

One example which illustrates the concept of the critical level is the story of the introduction of the European starling into the United States in the early 1800's. The first flock to be released soon died out. The next two attempts were similarly unsuccessful. Finally, 80 birds were released in Central Park in New York City, and they began to breed successfully. By 1970 the starling had become a major bird species in the United States and Canada. Since the environment in the Western Hemisphere didn't change appreciably just before the successful starling introduction, we can only conclude that the early starling introductions failed because so few birds were imported that some series of misfortunes destroyed them all.

FIGURE 3.5 Sea otter surfacing with sea urchin (top) and abalone (bottom). (Courtesy of James A. Mattison, Jr., Salinas, California.)

3.4 ISLAND BIOGEOGRAPHY
(Problems 11, 12, 13)

In the last two sections we briefly discussed the mechanisms whereby new species are formed and are removed, but we still haven't explained the interrelationship between speciation and extinction. Does the evolution of a new species *cause* an extinction of an existing species? Or conversely, does an extinction vacate an empty niche leaving "room" for evolution? Are some ecosystems of the world saturated with species, and if so, which ones?

We will initiate our search for answers to these questions with a study of small oceanic islands. Several factors make islands ideally suited to this examination: (a) Many mobile species immigrate to coastal islands. Immigration is much faster than evolution, and therefore we can study the effect of the addition of a species to an ecosystem without waiting for evolutionary time spans. (b) If we wish to study how the disappearance of a species affects an ecosystem, it is again advantageous to examine the situation on an island, where emigration is likely to be more rapid than extinction. (c) Because islands are smaller than continents, many other aspects of the study, such as counting populations or estimating biomass, are easier to carry out.

There are nine islands near the coast of southern California (Fig. 3.6). The smallest of these has an area of about 1 square mile; the largest, Santa Cruz Island, is about 96 square miles. Since the closest island lies only 8 miles offshore and the farthest is 61 miles out to sea, we would expect birds to migrate easily to and from the islands. Almost 60 years ago a

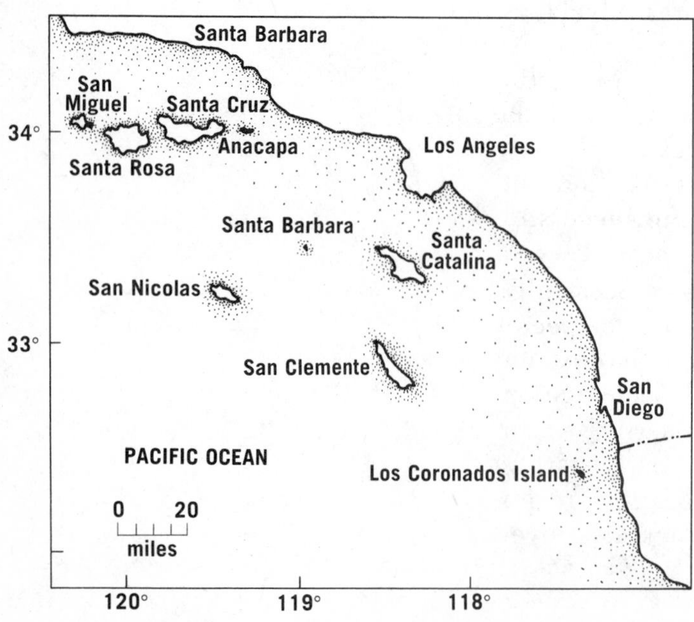

FIGURE 3.6 The Channel Islands.

scientist recorded the species of birds breeding on each island (Table 3.1). Recently a second survey was made and several significant observations were noted:

1. All the islands contained fewer nesting species than could be found in an area of equal size and of similar geography on the mainland.

2. The number of bird species on each island remained fairly constant during the time interval between the two studies.

3. Although the *number* of species remained fairly constant, the specific species changed. Many birds which were residents in 1917 became locally extinct by 1968, and likewise, many mainland birds immigrated to the island.

These data imply (but do not prove) that the number of species on each island is somehow an inherent property of the island. Perhaps some feature or combination of features such as the size, the distance from the mainland, the terrain, or the type of vegetation determines the maximum number of different species which can coexist on a given island.

The results from a second experiment support the observations made in the Channel Islands study. Four small islands in Florida Bay were observed, and all the resident species of animals (mostly insects, crustacea, and other arthropods) were recorded. The islands were then covered with a plastic canopy and sprayed to kill all the resident animals (but no plants). Next, the canopy was removed, and the islands were again observed. The results are shown in Figure 3.7. As you can see, immigration and recolonization were initially rapid, but later the number of species on the

The removal of all animal life is called defaunation.

TABLE 3.1 BIRD SPECIES ON THE CHANNEL ISLANDS*

	NUMBER OF SPECIES IN 1917	*NUMBER OF SPECIES IN 1968*	*EXTINCTIONS*	*ADDITIONS*
Los Coronados	11	11	4	4
San Nicolas	11	11	6	6
San Clemente	28	24	9	5
Santa Catalina	30	34	6	10
Santa Barbara	10	6	7	3
San Miguel	11	15	4	8
Santa Rosa	14	25	1	12
Santa Cruz	36	37	6	7
Anacapa	15	14	5	4

*Source: J. M. Diamond: Avifaunal equilibria and species turnover rates on the Channel Islands of California. Natl. Acad. Sci. Proc., *64*:57–63 (1969).

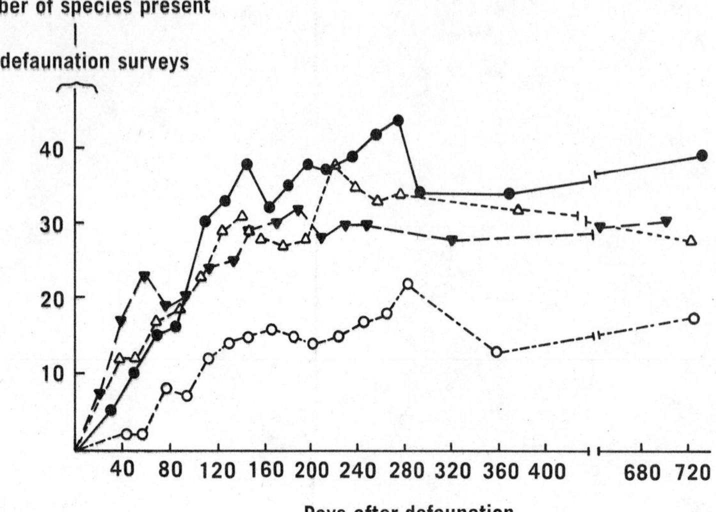

FIGURE 3.7 Recolonization of four islands in the Florida Keys. (From D. S. Simberloff and E. O. Wilson: Experimental zoogeography of islands. Ecology, *51*:936, (1970) Duke University Press.)

island stabilized. The number of species at equilibrium was about the same as the number of species that had lived on the island before the experiment was started. However, many of the new colonists were of different species than had existed previously on the island. As time went on, the number of species changed only slightly, but many immigrations and disappearances were observed.

Thus, it seems that after a certain number of species reside in a specific island habitat, a new addition is, in fact, balanced by a subtraction, or conversely, the removal of one species somehow "allows" a new successful one to appear.

We must understand that the maximum number of species does not necessarily correspond to the maximum number of individuals. Thus, one island might be able to support 20 species and 200 individuals, whereas another island might be able to support only 15 species but 300 individuals.

What characteristics of the environment determine the maximum number of species which can coexist? Recall from Chapter 2, page 56, that if two species of flour beetles are placed in a homogeneous jar of flour, one species always thrives, and the other becomes extinct. The results of this experiment and others like it have led us to believe that two species cannot share the same niche.

At equilibrium can the number of species in an ecosystem be correlated with the number of niches or habitats? We could try to count the number of available niches in the environment, then count the number of species, and see how closely the two numbers match; but niches aren't well defined cubbyholes on a shelf or squares on a checkerboard. No one can look into a forest and count the number of niches. A niche is defined as the combination of function and habitat of a species. We define a niche after an organism has occupied it, not before. Since no one knows how many niches there are, no one knows when they are all filled, or how many empty ones are available. A more productive line of reasoning is to assume that complex environments must have more niches than simple ones have and then to compare the number of species in simple and complex environments.

On islands as well as on the mainland there are almost always more species of birds nesting in a heterogeneous forest than in a homogeneous meadow, and there are more species of insects living in a heterogeneous natural meadow than in a homogeneous agricultural field. In a quantitative comparison of different forests, one examiner, R. H. MacArthur, made a measurement of the "foliage height diversity." A forest with dense foliage at many horizontal layers would have a high foliage height diversity, whereas a more homogeneous forest would have a low foliage height diversity. When the number of bird species was plotted against the foliage height diversity, the relationship shown in Figure 3.8 was observed; the taller and bushier forests housed more birds than did the short homogeneous forests.

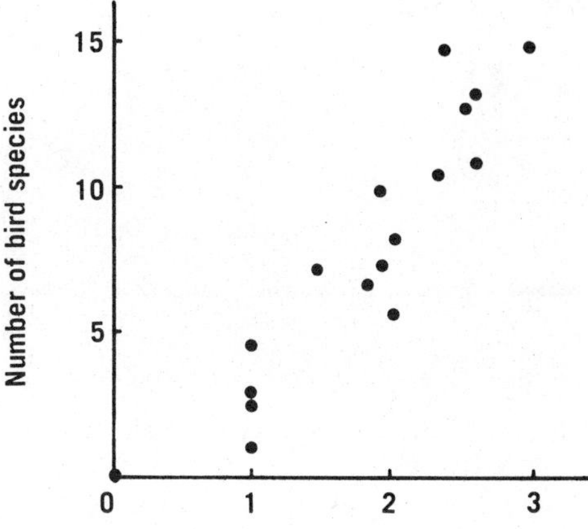

FIGURE 3.8 The number of bird species as a function of the complexity of the environment. [From Figure 7.3 (page 178) *Geographical Ecology* by Robert H. MacArthur (Harper & Row, 1972).]

> **It is often easier to rank two quantities than to count them. We understand enough about niches to believe that, for example, a rock with many cracks should be able to house more species than a smooth rock, but we don't know which cracks are habitable, or which ones are different enough to support different species.**

The correlation between environmental complexity and the number of species does not explain why an area on the mainland always houses more species than an environmentally similar area on a small island. For some reason, not all the available island habitats are occupied. The apparent paradox can be explained by examining chances for extinction. Let us imagine that a given species of bird is adapted for life in an oak grove near a grassy clearing. On the mainland there may be many oak grove-grassy clearing habitats, and birds move among them easily. On a small island there may be only one, and immigration is slow compared with exchange among habitats on the mainland. This one clearing may contain, perhaps, two nesting pairs. Four individuals, however, are a tiny population, and thus the island birds would be existing near or below the critical level at all times, and the probability of extinction would be high.

Forest with low foliage height diversity. Some areas contain little foliage.

If we compare topographically similar islands, the larger island almost always houses more species than the smaller island (Fig. 3.9). We explain this by hypothesizing that larger islands have a greater diversity of physical environments. Additionally, any particular type of environment, such as an oak grove-grassy clearing, is more likely to be found repetitively on a large island than on a small one. Thus, as would

Forest with high foliage height diversity. Foliage dense from ground up.

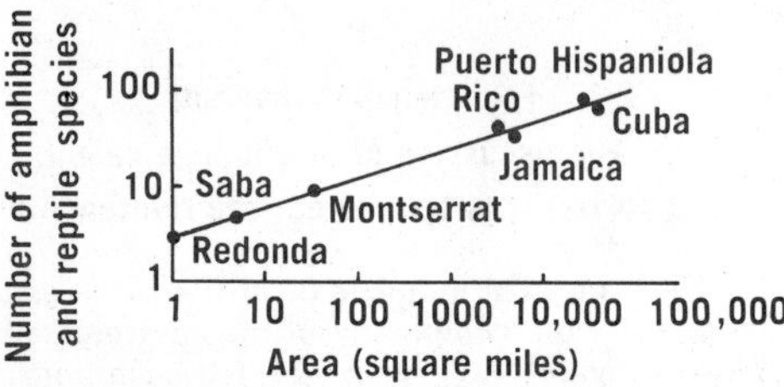

Figure 3.9 **The number of different amphibian and reptile species on some West Indian islands. [From** *The Theory of Island Biogeography,* **by Robert H. MacArthur and Edward O. Wilson (copyright © 1967 by Princeton University Press), No. 1, Monographs in Population Biology, Fig. 2, p. 8.]**

be expected, the number of species found on the largest islands approaches that found on the mainland.

The number of species on an island can also be correlated with the rate of immigration and, therefore, with the distance from the island to the mainland. Obviously, an island connected to the shore during low tide would be expected to house most of the organisms found on the nearby mainland, for movement to and from the mainland would be extremely facile. Conversely, on a small island in the central oceans the rate of immigration would be much slower. Therefore, if a species became extinct on an isolated island, replacement would be slow. This theory helps to explain the relative paucity of species on isolated islands.

Studies in island biogeography have illuminated two major points germane to our present discussion: (a) There does seem to be a maximum number of species in a given region (on islands at least), and once this maximum is reached future additions of new species are balanced by local extinctions. (b) The magnitude of this maximum depends on the climate, the terrain, and the size of the island, and on the distance between the island and the mainland.

3.5 THE NUMBER OF SPECIES ON LARGE CONTINENTS

(*Problems 14, 15, 16, 17, 18, 19*)

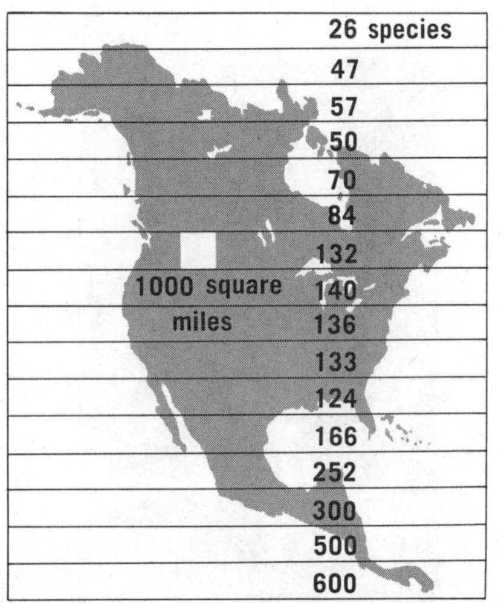

FIGURE 3.10 **Average number of bird species per 1000 sq. mi. of land area. Note that the number of species increases as we approach the equator.**

Some well defined areas within large continents are ecologically similar to oceanic islands. For instance, many species are adapted to living on mountain tops and have difficulty migrating across a valley or other lowland. Therefore, mountain tops can be considered to be **habitat islands,** and the diversity of species on a habitat island can be explained by use of the theories developed from studies of island biogeography.

The large biomes of the earth, however, are so massive, their ecology is so complex, and the interrelations between them are so varied that we cannot simply expand the generalizations learned from the study of island biogeography and apply this expanded theory to continental areas. For example, there are many more species of trees native to North America than to Europe. Yet the climates of these two regions are similar, and both large areas contain a great many favorable habitats such as mountains, valleys, coastal plains, rivers, and other similar topographical features. Although North America is clearly larger than

Europe, both continents are so large that favorable habitats are found repetitively.

Another perplexing phenomenon is that southern ecosystems typically contain more species than temperate or polar regions. Thus, 4 acres in a tropical rain forest may house roughly 225 different species of trees, whereas a 4 acre plot of a temperate deciduous forest is likely to contain only 10 to 15 species of trees. Animal communities are also more diverse in the tropics than they are in colder areas, as can be seen in Figures 3.10 and 3.11. Yet the tropical areas don't seem to be either geologically more complex or larger than colder areas; consequently, the number of species residing in an area must depend on other factors as well. Sometimes this explanation is offered: The polar regions are harsher than the tropics, so it is obvious that evolution should be easier in the tropics, and these regions should naturally be more varied. We reject this argument as insufficient because "harshness" cannot be adequately defined. The Congo River basin is certainly harsh to a polar bear. The bottoms of the central oceans have a great number of species, yet to our terrestrial way of thinking, those areas are harsh.

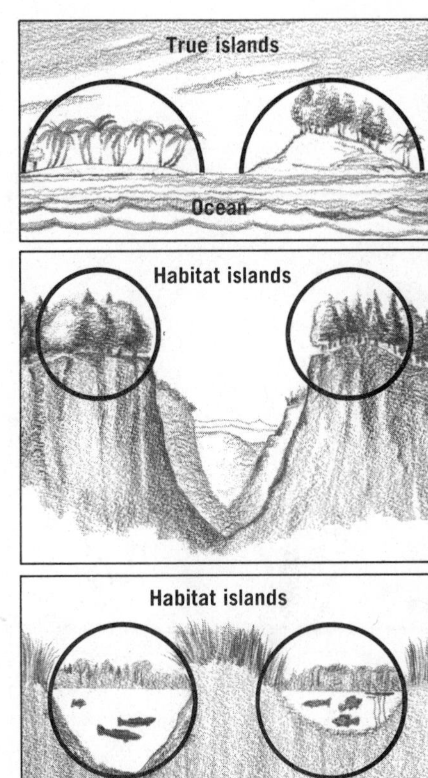

Several more attractive hypotheses have been rendered to explain the diversity of species in continental areas:

1. Historical Reasons. As we noted in Section 3.2, evolution and speciation are slow. Therefore, if some calamity had destroyed many species in a given area sometimes in the past, it is likely that the impoverishment would persist to some degree up to the present time. As mentioned previously, there are fewer species of trees in Europe than there are in North America. In North America the major mountain ranges run north-south, while in Europe the mountains form an east-west barrier (Fig. 3.12). When glaciers moved southward during the ice ages, the plant species in the Western Hemisphere could migrate southward toward warmer climates and then return to the north when the glaciers retreated. In Europe, on the other hand, plant species could not migrate readily over the Alps or the Pyrenees or across the Mediterranean Sea, and many extinctions occurred. Since the last Ice Age ended a mere 10,000

FIGURE 3.11 Number of ant species indigenous to selected areas of South America. Again, note that the number of species increases as we approach the equator.

to 15,000 years ago, there hasn't been sufficient time for diversity to be reestablished in Europe.

The tropics have experienced a more constant environment over many millennia than the polar regions have. Perhaps this historical circumstance, by itself, accounts for the observed north-south diversity gradient.

FIGURE 3.12 *A.* **Relief map of North America.** *B.* **Relief map of Europe.**

2. Climatic Stability. Seasonal temperature fluctuations in the tropics are less than those in northern regions. As a result, tropical plants and animals are generally active and productive for the entire year, and thus can produce more generations of young per year than do their northern counterparts. Since speciation requires 2000 to 100,000 generations (not years), evolution would be expected to occur faster in more constant climates.

A steady environment may also offer an inherent advantage over one with seasonal variations by its ability to support a large number of species. For example, in the tropics a fruit-eating bird can specialize in eating fruit and can maintain a constant diet for its entire life; similarly, an ant-eating bird can eat ants all year round. In northern latitudes, no animal could survive entirely on fruit or on ants, simply because the harvest season lasts only a few months of the year. A northern fruit-loving bird must supplement its diet part of the time with insects or with grain. Therefore, during certain seasons many temperate birds must compete for similar diets. Since direct competition often leads to extinction, temperate or polar habitats might be expected to be able to support fewer species than a geographically similar tropical habitat.

3. Increased Productivity. Imagine that one out of a thousand trees in a forest belongs to species A. If productivity is high and several thousand trees grow in a small river basin, several trees of species A will grow in the same ecosystem, and one tree is likely to fertilize its neighbor. On the other hand, if productivity is low and each basin houses only one tree of the rare species, cross-fertilization will be highly unlikely, and extinction may result (Fig. 3.13). Similarly, if a species of caterpillar specializes in eating the leaves of this rare tree, the survival probability for the insect will be higher if several trees exist within a small geographical area. Thus, areas with low productivity, like the polar regions, would be expected to be able to support fewer species than more productive tropical regions.

Although this hypothesis may correctly explain some patterns of diversity, there are also many instances in which the number of species decreases as the productivity increases. For example, polluted rivers support a much larger *number* of organisms than unpolluted ones, but the clean waterways support more *kinds* of organisms.

FIGURE 3.13A Productive areas are likely to support two or more rare plants in a single ecosystem.

4. Predation. In Section 2.6, page 63, we noted that nonspecific predators, such as the starfish on the Washington coast, tend to increase the diversity of a system. Since there are more predators in the tropics than there are in other regions, it follows that there should also be more of other kinds of species in the tropics.

Consider first, however, why there are more predators in the tropics than elsewhere. Many species of predators can coexist in the tropics for the reasons just discussed—a long history of evolution, a stable climate, and high productivity. Thus, these predators are likely to increase the diversity still further. For example, let us assume that the numbers of plant parasites and disease species are high in the tropics because the climate is stable. These organisms may then infect dominant tree species, thereby leaving room for the growth of rarer trees.

FIGURE 3.13B In less productive areas rare plants are isolated and cross fertilization probabilities are low.

3.6 THE RELATIONSHIP BETWEEN DIVERSITY AND STABILITY

(Problems *20, 21, 22*)

Imagine that we have an ecosystem with a total of n species, and we want to predict whether or not an $(n+1)^{st}$ species will be able to colonize or evolve there, or conversely, whether or not one species is likely to become extinct, leaving $(n-1)$ species. If the system were stable, population fluctuations would be small and extinction probabilities low. Furthermore, speciation or immigration in a stable system would be expected to be less favorable than in an unstable one. For example, if the soil were covered with a stable plant population, the seed of an immigrating plant might have difficulty displacing native species and establishing itself in the community. However, if the

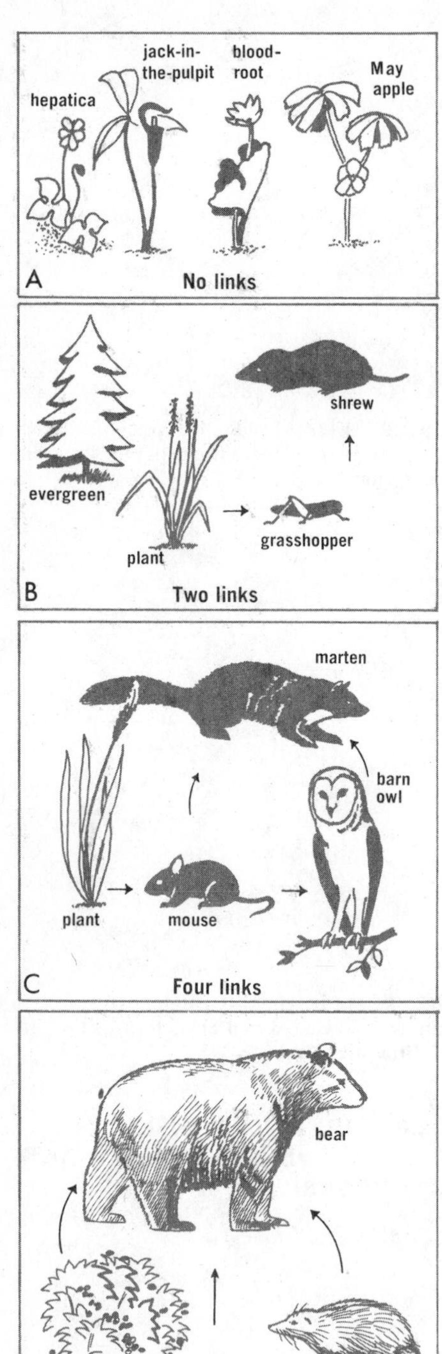

FIGURE 3.14 Four possible food webs for a four species system.

population levels tended to oscillate, then there would be periods when the plant populations would be low; at some time there would be bare spots on the ground. A new plant species could capitalize on this situation and establish itself in the ecosystem. Stable systems resist change, unstable systems invite change; these conclusions naturally follow from the definition of stability.

Therefore, it becomes obvious that the number of species in a system must be related to the stability of the entire system. Is there some optimum number of species which maximizes the stability of a system? Such a question is like asking, "Is there a relationship between the number of boards in a house and the strength of the structure?" It depends on how we put the boards together. Assume that there are four species in a simplified ecosystem. We could have one species of plant, one herbivore, one primary carnivore, and one higher-order carnivore. (We will omit decay organisms to simplify the discussion.) This system represents a simple food web (Fig. 3.14*C*). Alternatively, we could have four plant species and no heterotrophs (Fig. 3.14*A*), or a food web with plants, herbivores, carnivores, and omnivores (Fig. 3.14). Obviously, the interactions among the various trophic levels in the four different systems shown in Figure 3.14 are dissimilar even though each system contains the same number of species. Experiments and field studies both support the idea that the number of *links* in the food web is more closely related to stability than is the number of species. Therefore, we can rephrase the question posed at the beginning of the paragraph and ask, "Is there some optimum number of links in a food web which maximizes the stability of the system?"

Several observations imply that simple systems are unstable:

1. Many experimentalists have tried to construct ecological systems with just two, three, or four species, but most of these simple systems have been unstable.

2. When a farmer changes a natural ecosystem into an agricultural system (for example, when a forest is replaced with a cornfield), he usually replaces a complex ecosystem with a simple one. In general, cultivated ecosystems are less stable than natural ones. (See Chapter 7.)

3. Similarly, when man destroys the complex natural insect food webs by spraying farms with indiscriminate pesticides such as DDT, he often inadvertently causes instability, and local population explosions have occurred. (See Chapter 7.)

4. Small, ecologically simple islands are particularly vulnerable to invasion and are easily disrupted.

5. Mathematical models show that large population fluctuations are likely in simple systems.

If our reasoning is correct and simple systems are inherently unstable, then we know why one species of browsers with a telescoping neck hasn't evolved as the sole herbivore on the African continent.

Now let us discuss whether or not extremely complex systems are stable. One mathematical analysis of model systems concludes that both simple systems and highly complex systems are prone to be unstable compared with systems of intermediate complexity. This theory is difficult to test experimentally. We cannot simply introduce a species from one continent into an ecosystem on another and observe the stability of the system, because the species haven't co-evolved. We wouldn't expect a foreign invader to be previously adapted to a new environment. We do know that complex tropical rain forest ecosystems are highly stable, but we don't know if additional species would stabilize the system still further or would only bring about instability. If increased diversity would lead to an even greater stability, then we can surmise that there are fewer than the optimum number of species because there hasn't been enough time for the ecosystem to evolve to its most stable configuration. If, on the other hand, complex tropical systems were known to house the optimal number of species, then we could conclude that a system is destabilized by too many interactions in a food web. Unfortunately, we cannot distinguish between these two possibilities; we don't know the answer.

This model is discussed in great detail in Robert May's book, *Stability and Complexity in Model Ecosystems.* (See Bibliography.)

SUMMARY

1. Within a small isolated habitat, the number of species seems to be regulated by the size and the complexity of the environment.

2. In large continental areas the number of species might be linked to: (a) the evolutionary history of the area, (b) the stability of the climate, (c) the productivity, or (d) the degree of predation.

3. Extremely simple systems are generally unstable.

4. Perhaps extremely complex systems are also unstable, but we are not sure.

PROBLEMS

1. **Definitions.** Define species. Discuss the different ways that groups of organisms can isolate themselves reproductively.

2. **Evolution.** Outline briefly the types of environmental pressures to which an animal must adapt. Using this outline as a guide, discuss Darwin's concept of "survival of the fittest."

3. **Speciation.** Although intermarriage between ethnic groups is relatively rare in human societies, many such unions do occur. Are specific ethnic groups separate species? If your answer is no, are they likely to become separate species in the future? Explain.

4. **Fitness.** In primitive societies, where physical strength and stamina may have counted for more than they do today, physical fitness may have endowed its possessors with Darwinian fitness. Can you imagine a type of social organization in which the best physical specimens were actually at a reproductive *disadvantage* with respect to others? Describe such a hypothetical situation.

5. **Fitness.** A disease which infects individuals before they are old enough to reproduce or during the reproductive years will necessarily have a selective effect on a population. Suggest a way in which a disease which occurs principally in the *post*reproductive period could have a selective effect on a population.

6. **Adaptation.** Some anthropologists feel that the dark skin of the black races is adaptive for the warm climates and hot sun in which these people have spent thousands of years. What simple experiment might you design to test the hypothesis that blacks tolerate warm climates better than whites?

7. **Speciation.** Many zoos are finding it advantageous to breed specimens in captivity. If this practice continues, do you feel that there is a possibility that separate "zoo species" will evolve?

8. **Survival of species.** Discuss four mechanisms whereby a species may survive after the environmental resistance has increased.

9. **Critical level.** Define critical level. Give two examples.

10. **Critical level.** Discuss how the critical level might differ greatly from species to species. In your answer, briefly explain how each of the following factors could affect different species differently: (a) reproductive failure, (b) ecosystem imbalance, (c) ability of males to find females, (d) luck, and (e) inbreeding.

11. **Colonization.** The first colonists on an uninhabited island may survive and thrive even if they are not particularly well suited to that environment. Later, as more well suited species arrive, the early colonists often become extinct. Explain.

12. **Colonization.** The word "niche" can be used to describe economic as well as ecological habitats and functions. Imagine that a new town were started near an oil drilling site on the north slope of Alaska. Ap-

plying your knowledge of island colonization, draw a graph illustrating your predictions of the number of businesses in this small town as a function of time. Discuss what happens when a new business moves in; when an old business goes bankrupt. Would some types of businesses be likely to be more successful than other types? Explain with examples.

13. **Heterogeneous environments.** There are more species of insects in a heterogeneous natural meadow than in a homogeneous cornfield. The cornfield is more susceptible to population blooms of a single pest species. Explain.

14. **Habitat islands.** Mountain tops are habitat islands for many resident species, but not for all. Explain, and give examples of the latter category.

15. **Habitat islands.** State whether each of these areas is or is not a habitat island, and defend your answer: (a) the Great Plains, (b) a small glacial lake, (c) an oasis in the desert, (d) the Mediterranean Sea, (e) the Amazon Basin, (f) an island in the center of the Mississippi River.

16. **Speciation on the continents.** Discuss briefly five reasons why there are more species per unit area in the tropics than in polar regions.

17. **Climatic stability.** Unstable environments are generally harsh, but harsh environments are not necessarily unstable. Explain. Discuss reasons why instability is a better criterion than harshness by which to judge ecosystems.

18. **Climatic stability.** Many arctic animals migrate southward during the winter. Discuss how these seasonal migrations might affect the number of species in the forests south of the tundra.

19. **Climatic stability.** The number of species of lizards in the desert can be correlated with the quantity of brush growing in different regions. Since productive deserts are also heterogeneous, we need more information before we can determine whether the diversity of lizards is responsive to increased productivity or to the increased heterogeneity of the productive areas. Design an experiment to distinguish between these two possibilities.

20. **Species diversity.** Draw six different food webs with four species in each.

21. **Ecosystem stability.** List five observations that imply that simple systems are unstable. Does this *prove* that simple systems are unstable? Defend your answer.

22. **Importance of basic research.** In recent years government funding of basic research has often been reduced because many scientific endeavors have been

considered to be irrelevant to the plight of modern man. A bird watcher in southern Illinois, noting the decrease in the number of his sightings over the years, learns something about the relationship between environmental complexity and the stability of ecosystems. A farmer in Iowa finds that his insect pest problem has increased after he has destroyed bushes and shrubbery near his fields. (a) Explain how these two findings may be related to each other. (b) Suggest how the knowledge gained from bird watching research could be relevant to the price of food in the supermarket.

BIBLIOGRAPHY

A good summary of modern genetics is given in:

R. P. Levine: *Genetics.* New York, Holt, Rinehart and Winston, 1962.

Several excellent works dealing with various aspects of evolution are:

Theodosius Dobzhansky: *Genetics and the Evolutionary Process.* New York, Columbia University Press, 1970.

———: *Mankind Evolving.* New Haven, Yale University Press, 1962.

Ernst Mayr: *Populations, Species, and Evolution.* Cambridge, Mass., Harvard University Press, 1970.

George Gaylord Simpson: *The Meaning of Evolution.* New York, Bantam Books, 1971.

A summary of the research on industrial melanism is given in:

H. B. D. Kettlewell: The phenomenon of industrial melanism in the Lepidoptera. Ann. Rev. Entomol. 6:245, 1961.

Two comprehensive books dealing with the distribution of species on islands and continents are:

Robert H. MacArthur: *Geographical Ecology—Patterns in the Distribution of Species.* New York, Harper & Row, 1972. 269 pp.

Robert H. MacArthur and E. O. Wilson: *The Theory of Island Biogeography.* Princeton, N.J., Princeton University Press, 1967.

An excellent book which discusses mathematical models of stability in natural ecosystems is:

Robert M. May: *Stability and Complexity in Model Ecosystems.* Princeton, N.J., Princeton University Press, 1973. 235 pp.

4

THE EXTINCTION OF SPECIES

4.1 INTRODUCTION

In recent years the number of species has declined dramatically in a great many ecosystems, and many species have become totally extinct. Therefore, as a sequel to the question posed in the last chapter, "Why are there this many species?" we must now ask, "Why, all of a sudden, are so many of them disappearing?"

We are living in a period of rapid ecological change. Entire biomes have been disrupted in just decades, and small ecosystems are altered practically overnight. Examples appear in many places and can be observed frequently if we are but attentive to them. For example, if you lived near Nooseneck, Rhode Island before 1965, you would have seen many quail frequenting the nearby meadows. A section of Interstate Highway 95 was then constructed nearby, and the next fall the birds could not be found. Perhaps the barrier interrupted small migrations for food; it could be that the noise disturbed the birds, and they moved away; perhaps the construction workers poached the delicious quail during the summer months; or perhaps the new road provided easy access for licensed hunters who then put a severe pressure on the population; we do not know. We do know that this example is not an isolated one; myriads of small ecosystems across the globe are being disturbed by local roads, farms, housing projects, and other activities of man. Larger ecosystems are also being affected. For example, the Aswan Dam, recently built across the Nile River, has changed the old annual cycle of spring floods and summer droughts. The change disrupted the lives of all organisms which were adapted to this cyclic pattern. Some species of the Nile basin have become extinct, while others have capitalized on the new situation and their populations have expanded.

Let us view the situation on a larger scale: Two hundred years ago the Great Plains of North America encompassed a stable grass-bison-predator system. Now the bison no longer roam the plains, nor do the wolves, the bears, or the Indians, and the sturdy prairie grasses survive mostly in neglected cemeteries, old railroad rights-of-way, and other isolated areas. In their place grow corn, wheat, barley, and oats which are harvested by machine and trucked to food processing plants or to beef cattle herded in feedlots.

Change in itself is not necessarily beneficial or harmful. In this and subsequent chapters, we shall study the nature of some of the changes that are occurring in modern times and consider how these changes may affect the ecosystems of the world.

4.2 THE EXTINCTION OF SPECIES IN PREHISTORIC TIMES

(Problems 1, 2)

Throughout geological history, certain species have developed gradually, flourished, and then suddenly vanished, leaving behind them an empty niche. For example, approximately 500 million years ago, enormous numbers and varieties of primitive sponges populated the seas. Although for 30 or 40 million years these sponges dominated the seas from pole to pole, most species later disappeared, and different ecosystems developed. In more recent times, the rise and fall of the dinosaurs was certainly one of the most outstanding events in the history of the earth. Small dinosaurs first evolved some 225 to 250 million years ago. For perhaps 25 to 50 million years they slowly evolved and grew in numbers and variety until they dominated the earth. The dinosaurs reigned for 100 million years and then they all died, whereupon the established ecosystems and the balances of the food web collapsed. Dinosaurs have been often belittled in popular literature as being stupid and clumsy. But in fact the great reptiles were outstanding evolutionary successes; they dominated the earth longer than any other order has. The extinction of the dinosaurs reminds us that seemingly stable biological systems may someday collapse.

Of course, the period following the extinction of dinosaurs was not devoid of life, and during this time, the previously slow evolution of mammals accelerated. Many species, including man, arose. The next significant wave of extinctions occurred about 10,000 years ago at the end of the Pleistocene Age. Before this time the mammalian fauna of the world were significantly more varied than at present. Mammoths, mastodons, camels, horses, wild pigs, giant ground sloths, giant long-horned bison, woodland musk oxen, tapirs, bear-sized beavers, and may different kinds of now-extinct deer roamed the North American continent (Fig. 4.1). They were hunted in part by sabre-toothed tigers, giant jaguars, and dire wolves. All these animals have become extinct, for the most part without ecological replacement. In fact, estimates are that 95 per cent of all large animal species in North America were lost, and mass extinctions occurred simultaneously in what are now South America, Northern Asia, Australia, and Africa.

What caused these extinctions? There are many uncertainties. Perhaps climatic changes had an effect. We know that rapid alterations in weather patterns occurred at this time, but we also know that the same animal species that suffered extinction had already lived through several advances and retreats of the glaciers. Also, few small animals followed their larger cousins into oblivion, and it is hard to explain why climatic variation affected large animals differently from small ones. It seems certain that the hunting activities of primitive people played some role in the drama. We know, for example, that all the herbivores that are now extinct were pursued by early humans. However, no major extinctions preceded the development of advanced hunting techniques such as throwing spears, shooting arrows, and setting fires to drive animals over a cliff or into a trap (Fig. 4.2). If large herbivores were hunted to extinction, the specialized carnivores like the sabre-toothed tiger must have perished soon after.

But now we come to an important question. How could man, the primitive hunter, have been even a

"You're being recalled—He's going to try mammals."

FIGURE 4.1 Extinct North American animals.

FIGURE 4.2 How the Tule Springs site, Nevada, may have looked 10,000 to 12,000 years ago, with scattered sagebrush (*Artemisia*) and saltbush (*Atriplex*), and springs with ash trees, cattails, and other mesic species. (Courtesy of John Hackmaster.)

TABLE 4.1 PAST AND PRESENT EXTINCTION RATES

YEAR SPAN	*AVERAGE NUMBER OF EXTINCT SPECIES OF MAMMALS PER YEAR*
1–1800	.02
1801–1850	.04
1851–1900	.62
1901–1950	.93

small factor in the destruction of some 200 genera of animals across the face of the earth, and even if he were, how could he have survived after his prey had perished? One reasonable explanation lies in people's powerful ability to adapt culture to new situations. Imagine the relative plights of two great predators in post-glacial North America: the human being and the sabre-toothed tiger. Both depended at first on the mammals of the plains for their food. Of course, their hunting techniques differed. The tigers had evolved a specialized body structure to kill their prey. On the other hand, humans though physically no match for their competitors, were extremely adaptive. When big game became scarce, people turned to other foods: freshwater mussels, insects, fruits, and berries. They could hunt larger prey only when it was convenient. But the tigers could not change their habits quickly—their hind legs were not structured for speed, their teeth and claws were not well adapted for opening mussel shells, their digestive system was not capable of utilizing berries—and they perished. The exact details of the extinction of the mammalian species in Pleistocene times will never be fully explained, but the fact is that man and a few big game animals adjusted and thrived.

For a period of five to six millennia very few additional species became extinct, and then suddenly, in recent times, a new age of species destruction has begun. Not only have extinction rates been increasing

FIGURE 4.1 *Continued.*

rapidly in the last half century (see Table 4.1), but a great many more animal and plant species are seriously endangered today. Unless current trends reverse, extinction rates will continue to accelerate in the near future.

4.3 FACTORS LEADING TO THE DESTRUCTION OF SPECIES IN MODERN TIMES

(*Problems 3, 4, 5, 6, 7, 8, 9, 13*)

We know that climate has been relatively constant in recent years; certainly grizzly bears weren't expelled from California, or passenger pigeons from Wisconsin, because the weather changed. We also know that devastating disease epidemics have not recently occurred in wild animal populations, and that vegetation hasn't changed. We know, in short, that people are the major agents in the extinction of species today. Some of the mechanisms are discussed in the following paragraphs.

Destruction of Habitat

The relationship between human beings and nature is no longer a simple predator-in-the-forest system. Perhaps the greatest single cause for species extinction in the civilized world is man's destruction of natural ecosystems. This can occur in several ways: (1) The prodigious growth of human populations creates demands for living space. Accordingly, cities, suburbs, and roads displace animal and plant communities. (2) Agriculture preempts enormous areas of land; farmers have systematically destroyed climax

FIGURE 4.3 *A.* **Trailing arbutus. (Photo by Alvin E. Steffan, National Audubon Society.)** *B.* **Giant white pine tree—a reminder of the past. (From the book** *The Great American Forest* **by Rutherford Platt. © 1965 by Rutherford Platt. Published by Prentice-Hall, Inc., Englewood Cliffs, New Jersey.)**

ecosystems and have replaced them with croplands. (See Chapter 7.) (3) Fuel consumption and energy utilization are multifaceted sources of environmental disruption, and the mining, transportation, and combustion of fuels (see Chapter 6) often destroy natural habitats. (4) Insecticides and herbicides have at times made water, air, and food supplies unhealthy for wild animals and plants.

When we think of the destruction of species, we usually visualize animals, but many plant species, too, are rapidly facing extermination, mostly from destruction of habitat. The prairie that once covered the midwestern United States and Canada was a beautiful sight, but it is gone and may never return. Small holdouts of true prairie plants still exist in a few old cemeteries and along old roadways, but grain fields and domestic grasses have displaced most of the old flora. An untold number of prairie plant species have become, or are in danger of becoming, extinct. But prairie plants are not the only species to suffer; forest plants, such as the trailing arbutus (Fig. 4.3*A*), can thrive only in the soils of climax systems, and as climax forests are being converted into stands of second-growth timber, the trailing arbutus and many other plants are becoming increasingly rare (Fig. 4.3*B*).

The Introduction of Foreign Species

A few species are able to disperse themselves around the world. One small crustacean, *Calanus finmarchicus*, which lives predominately in the cold waters of the North Atlantic near Newfoundland, has been able to migrate across the warm equatorial seas to Madagascar because cold ocean currents sink below the surface of the northern ocean and travel south as an underwater river. The current passes under the warm waters, which would kill this particular species of crustacean, and rises again in another hospitable environment. Plant species are also able to cross inhospitable barriers. For example, coconuts can float, and some isolated Pacific islands are covered with coconut palms which grew from nuts which had been washed ashore after a lengthy ocean voyage.

Madagascar is an island near the southwest coast of the African continent. It is now an independent nation called the Malagasy Republic and is slightly larger than France. The distance between Newfoundland and Madagascar is about 8500 miles.

Most organisms, however, cannot migrate across large natural barriers. Oceans or high mountain

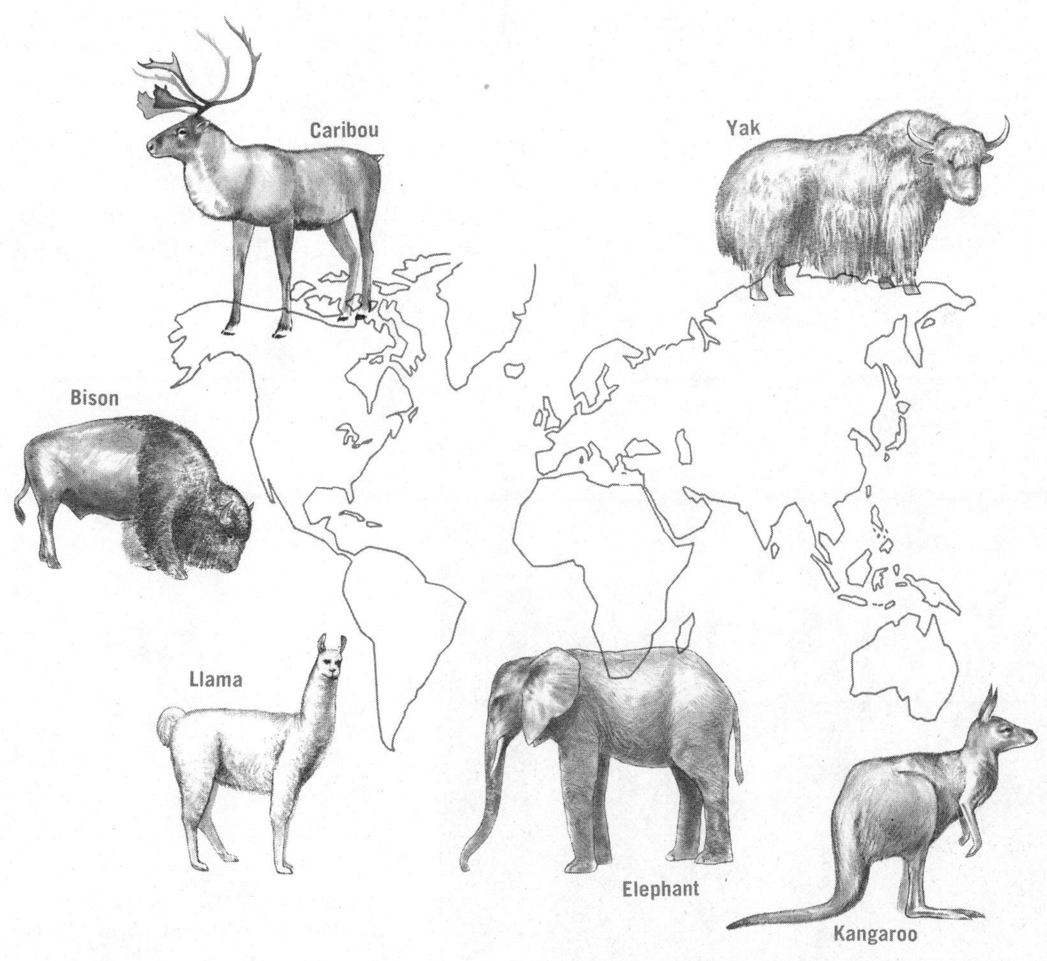

ranges are impenetrable to a wide range of plants, fish, birds, and terrestrial animals. Consequently, evolution and speciation have occurred independently in various areas of the world, and each continent has given birth to its own native species. The results of this independent evolution are dramatically shown by the contrasts among herbivores in each of various sections of the globe: the bison in North America, the llama in South America, the yak in Himalayan regions, the elephant in Africa, and the kangaroo in Australia. Many plants also originated in specific areas: corn in North America, wheat in the Middle Eastern region, and rice in Eastern Asia.

Geological changes have aided migrations in past ages. Thus, the rise of the Isthmus of Panama from the ocean permitted movement of organisms between North and South America. Such geological change, however, is slow. Formation of a land bridge may take thousands, or even millions, of years. In modern times plants and animals no longer need land bridges or mountain passes to facilitate their movement. Thousands of species have migrated with unprecedented ease as passengers on the ground, air, and ocean vehicles of man. To name but a very few examples, African mosquitoes have been transported to Brazil, European starlings to the United States, Chinese mitten crabs to Europe, sea lampreys to

FIGURE 4.4 Chestnut blight. (From Odum: *Fundamentals of Ecology*. **3rd Ed. Philadelphia: W. B. Saunders Co., 1971.)**

Lake Michigan, the European spruce sawfly to North America, and the American vine aphid to France. At one extreme, many organisms have been unable to compete in foreign lands, and colonizations have been unsuccessful. At the other extreme, some organisms have succeeded so well in their new environments that they have disrupted the ecological balance and have endangered other species.

Some efficient invading predators and parasites have upset local ecosystems by decimating populations of prey. For example, the sea lamprey is a predator-parasite that attaches itself to a living fish and sucks out the fluids and flesh until the prey dies. Lampreys have long been a part of aquatic ecosystems along the St. Lawrence River and the North Atlantic Ocean, and have been controlled by natural homeostatic mechanisms. In 1829, a ship canal was built to bypass Niagara Falls, thus allowing ships (and lampreys) access from the sea to the Great Lakes. For some reason, lampreys failed to traverse the canal for 100 years, but when they finally did, they gained access to Lake Huron and Lake Michigan, where they multiplied rapidly and, in a 10-year period almost exterminated their major prey, the lake trout. This incident shows us that predators do have the potential to destroy a prey species, and may realize this potential if they find themselves in a new and hospitable environment.

The American chestnut tree used to forest much of the middle and southern Atlantic coastal areas of the United States, but now the chestnuts are seriously endangered because a parasitic tree fungus imported from China was lethal to the American trees (Fig. 4.4). Both American and Chinese trees had developed resistance to the parasites that grew in their own habitats but not to foreign parasites. As a result, all large American chestnut trees on the east coast died within 50 years of the arrival of the Chinese fungus. Perhaps some young trees will survive and provide a new breeding stock for the species, but the outcome is uncertain.

Other invaders compete so successfully that indigenous animals cannot survive. For example, the thylacine (Fig. 4.5) is a doglike carnivore which is native to Australia. Sometime in the distant past Australia became geographically separated from Southeast Asia by an ocean barrier. Wild dogs thrived in

FIGURE 4.5 *A.* Thylacine. *B.* Dingo.

Thylacines, also known as Tasmanian wolves, survived and thrived in Australia until the last century. When white men settled the thylacine's range, this carnivore preyed on man's cattle. As a result, the settlers initiated a campaign to eliminate the thylacines. The species is now bordering on extinction. A few animals may remain alive in some remote highlands.

Asia, and thylacines thrived in Oceana. Although wild dogs and thylacines are similar in size and weight, and both practice similar hunting techniques, they were both able to survive because they lived in separate ecosystems and could not compete. When a species of wild dog, called dingos, was brought to Australia by the aboringines about 6000 years ago, these predators proved to be so efficient that they displaced the thylacines and drove them to extinction on the Australian continent.

Successful ecological invasions also present direct problems to man. In many cases, invaders quickly become pests and are responsible for large financial losses. The sea lamprey (Fig. 4.6) virtually eliminated a lake trout industry of 8.5 million pounds per year; the Japanese beetle, imported from the Orient, feeds on many crops, such as soybeans, clover, apples, and peaches; the American vine aphid, imported to France from the United States, was responsible for destroying three million acres of French vineyards. In fact, over half of the major insect pests in many areas are imports.

In addition to such direct economic disturbances or to the extinction of some native species, the invasions of plants and animals are responsible for more subtle perturbations. A few hundred years ago, individual continents possessed their own unique types of ecosystems. Thus, for example, temperate

FIGURE 4.6 Young sea lampreys, *Petromyzon marinus*, attacking brook trout in an aquarium. (From Lennon, R. E.: "Feeding Mechanism of the Sea Lamprey and its Effect on Host Fishes." Fish Bulletin; U.S. Fish and Wildlife Service 56 #98; 247–1293, 1954.)

grasslands in America, Asia, and Australia all contain different groups of organisms. With the rapid immigration of some species and the destruction of others, the differences between ecosystems are becoming smaller, and the biological world is becoming simpler.

Hunting for Sport or Fashion

Perhaps one of the saddest and least excusable types of environmental disruption caused by man in the world today is the wholesale killing of endangered species to satisfy the sport or fashion demands of a few individuals.

FIGURE 4.7 Giant sable antelope. (Courtesy of Richard D. Estes.)

Conservationists estimate that between 500 and 700 wild giant sable antelopes live today in west central Africa (Fig. 4.7). These animals are endowed with a set of magnificent curved horns, sometimes five feet long, which may easily be the animals's downfall, for hunters come from many parts of the world to shoot a big bull and bring home a trophy. Additionally, many hunters and tourists who are unable or unwilling to shoot a giant sable purchase their trophy from native hunters and return home with an expensive wall decoration. The species is endangered but will probably survive in small protected populations.

The snow leopard is one of the few large predators living in the high country of the Himalayas. Leopard skins are considered to be elegant and snow leopard skins are warm as well. As a result, there are fewer than 500 snow leopards left alive. Snow leopard coats can still be purchased, even though some major furriers no longer sell them.

These two examples were chosen out of several hundred cases where sport and fashion hunters are responsible for large-scale killing of animals and the destruction of natural systems. Other well-known examples include Nile crocodiles and American alligators for handbags and shoes, many jungle cats for furs, and several species of birds for their elaborate feathers.

Hunting for Food

After the wave of Pleistocene extinctions, primitive hunting societies seemed to have existed in ecological balance with their prey. Yet today, many systems are so seriously disturbed that once again some species are being eliminated by food gatherers.

The pigmy hippopotamus has coexisted with tribal hunting societies in central Africa for a long time, but as industrialization reduces the animal's range and brings hunger to many people in the area, the hippo is being hunted more systematically today, and extinction may result.

Predator Extermination

Man has always shared his planet with other predators, and during the millennia of co-evolution, delicate relationships have developed. The most striking aspect of the interactions between man and his carnivorous competitors is that the large terrestrial predators of the earth habitually avoid attacking man. This does not mean that people are never killed by other predators, but just that confrontation has been minimal. For example, the grizzly bear is one of the largest and strongest carnivorous animals alive today (Fig. 4.8). These giants weigh from 450 to 900 pounds when mature and measure up to four feet high at the shoulder. They can run at about 25 to 30 miles per hour on flat ground and are strong enough to crush the skull of a large cow with one blow. Man's fear of this formidable animal is reflected by its scientific name, *Ursus horribilis.* Yet the bear's record is far less fearsome than its name. One of the last significant ranges of the grizzly south of Canada and Alaska is in the Rocky Mountains of northern Montana in and near Glacier National Park. In the past 50 years millions of hikers have toured the park and adjacent wilderness areas, thousands of unarmed people have sighted grizzlies, thousands of armed people have hunted and sometimes killed the bears, and on only two occasions have the giant animals killed persons in the park.

FIGURE 4.8 Alaskan brown bear (grizzly) with salmon. (Photo by Leonard Lee Rue, National Audubon Society.)

The record of the relationship between man and wolves is even more remarkable. European literature is filled with horror stories of wolves, and English-speaking children learn the lesson early when they listen to the tale of Little Red Riding Hood. Yet, there have been *no* documented cases of wolves attacking man in North America during the past 50 years.

The list could continue on: American alligators do not attack unless cornered, lions generally avoid man, and cheetahs have *never* been known to attack man unless pursued. Other species, like the Bengal tiger, the polar bear, and the Nile crocodile attack man more often, but even for these animals, man is only a rare prey. Yet, though they do not constitute any real threat to life, predators have been hunted, sometimes vigorously, in the name of securing safety for humans (Fig. 4.9).

FIGURE 4.9 Mountain lion cornered by bounty hunters in Colorado. (Photo by Carl Iwasaki, *LIFE* Magazine.)

A second major source of conflict between predators and man is competition for food. Wild carnivores do kill game and domestic stock, and consequently hunters and farmers have often advocated the extermination of predators. But elimination of all predators does not mean more food for man. As an example, let us examine the relationships among ranchers, grass, sheep, and coyotes in the western United States and Canada. A coyote's diet includes many different items, among which are field mice, rabbits, moles, pack rats, sage hens, carrion of lambs that have died of natural causes, and freshly killed lambs. If the coyotes were exterminated, then the rodents would flourish, and since rodents eat grass, the quantity of forage available for the sheep would decrease. On the other hand, a spiraling coyote population would prey on range lambs and inflict financial losses on the ranchers. Thus, though coyote control is defensible, extermination is ecologically unsound. Even limited control programs have been questioned by some study groups, who contend that the cost of the anti-predator campaigns does not warrant the meager livestock savings that are realized. Unfortunately, mass poisoning programs are being advocated by many ranchers who do not appreciate the true complexities of ecosystems.

The coyote is not now an endangered species, so why use this particular example here? We cite this case because it demonstrates the value of predation in a natural system and points to the foolishness of attempts to exterminate a generally useful and beneficial animal.

The Zoo and Pet Trade

Zoos and pet dealers have been castigated for their role in species extinctions, for the populations of many species cannot support the pressures caused by trapping. Orangutans, for instance, have been severely affected by this trade (Fig. 4.10). After orangutan populations were severely reduced by the intrusion of agriculture, their already small community was invaded by hunters and trappers. Since orangutans are too smart to be trapped efficiently, and since adults are difficult to ship anyway, hunters often shoot a nursing mother and take the infant. As the survival rate of infants without maternal care is only about 15 per cent, for each orangutan that reaches the market in western cities about a dozen have died.

FIGURE 4.10 Mother orangutan and young. (From *LIFE*, March 28, 1969, p. 83. Photographer, C. Rentmeester.)

Zookeepers around the world have recognized the danger and in recent years have taken positive steps to reverse the trend toward extinction of many species. Reputable zoos will no longer purchase animals belonging to endangered species except in rare cases when an animal that is being overhunted can be saved by maintaining a viable breeding stock in captivity. Pet trading in endangered species, however, continues even in some areas in which it is prohibited by law.

Miscellaneous Mechanisms of Extinction

Some paths to extinction are strange. Powdered horns of the Javanese rhinoceros (Fig. 4.11) are re-

FIGURE 4.11 Indian rhino—close cousin of the Javan rhino. (From *LIFE*, Feb. 7, 1969, p. 68. Photographer, Eliot Elisofon.)

puted to be powerful aphrodisiacs by many Asians, and a single horn is worth $2,000 in Bangkok. The medicine may or may not work, but the species will probably not survive another decade.

The Florida Key deer was almost eliminated just after World War II. Since then the tiny population of 25 individuals has increased to 250, and perhaps the species will be saved. But a new and serious danger to the deer has arisen. They seem to have developed a taste for the tobacco in the cigarette butts thrown from passing cars. Unfortunately, roadways are hazardous places to graze, and many Key deer are killed in automobile accidents.

4.4 TWO CASE HISTORIES OF ENDANGERED SPECIES

The road to extinction is different for each individual species, and no single factor is responsible for the extinction of any species. Two case histories will be given here to illustrate how several factors combine to endanger a species.

The largest flying bird alive today is the California condor. Spreading their nine-foot wings and gliding along in search of food, these big birds still grace the mountains of southern California, but their days appear to be numbered. They are scavengers that have traditionally fed on the carcasses of elk, deer, bear, or other inland game, and on fish and an occasional whale washed ashore by the sea. The

FIGURE 4.12 Andean condor. (From *LIFE*, June 13, 1969, p. 74. Photographer, John Dominis.)

development of west coast lands for ranching and farming excluded game animals from these areas and thus reduced food available for condors. Their habitat preempted, they were forced to move further back into the mountains. At this point it is interesting to compare the condor with another avian scavenger, the magpie. Magpies have fared well in the twentieth century and are often seen along highways feeding on dead dogs, cats, and rodents. A condor could easily displace a flock of magpies, and the carrion on the California plains and beaches could support a sizable condor population. But the big birds do not come. Perhaps they are frightened away by the noise of cars on the nearby highways; we do not know.

But perhaps it is just as well that condors do not share space with man because they have been, and continue to be, hunted extensively. Many ranchers believe, quite incorrectly, that condors are predators and kill live lambs or chickens. Some hunters find sport in killing these magnificent birds. Shooting condors has been illegal since 1880, but in nearly 100 years only once (in 1908) has someone been punished for breaking the law. Condors are also inadvertently trapped or poisoned by programs to exterminate coyotes.

There are perhaps 50 to 75 birds still alive. Rapid population expansion is impossible because the condor's reproductive rate as compared to that of other birds is very low. A female does not breed until she is five or six years old. Then she lays a single egg, which she incubates for 42 days. The newborn chick is totally dependent on *both* parents for seven months, and partially dependent on them for another seven. Only about eight condor eggs are laid per year in the entire world, and since pressures that caused the original decline in condor populations are still operative, death rates appear to be higher than survival rates.

FIGURE 4.13 A bandicoot. (Photograph courtesy of Anthony Robinson.)

Apparently, condors are endangered because they are large, magnificent, unwieldy, unadaptive, and slow to reproduce. But don't be too quick to draw conclusions. If instead we shift our attention to the bandicoots of Australia, New Zealand, and Tasmania, we find that they are small, rodent-like, quick, and relatively fecund, yet four species have become extinct in recent times and three more are seriously endangered (Fig. 4.13). What has happened? At some early stage in evolutionary history, the islands of Oceania separated from the rest of the world and contact was not reestablished until modern times. The bandicoots that evolved on their islands were well adjusted to the life there; they faced and survived predation from monitor lizards and aborigines and did well eating insects, grubs, and some vegetables. When Europeans arrived, they brought a whole arsenal of new enemies. Cats, dogs, and rats preyed upon the bandicoots and proved to be effective hunters in this previously isolated world. Then man himself began hunting them, for valuable furs and tasty meat could be harvested with relative ease. Finally, after introducing the rabbit to Australia, man then decided that the rabbit was undesirable, and initiated a massive trapping, poisoning, and shooting campaign. Many bandicoots were inadvertently killed during the rabbit slaughters, and this type of predation added enough extra pressure to existing populations to exterminate some species and endanger others.

4.5 CURRENT TRENDS IN SPECIES EXTINCTION

(*Problems 10, 11, 12*)

What characteristics do the endangered species have in common? Highly specialized or immobile animals and plants are particularly vulnerable to pressures from loss of habitat. As an example, the ivory-billed woodpecker feeds only on insects that live in the decaying wood of standing virgin timber in the southeastern United States and nests only in the upper regions of large trees. It does not eat the small parasites of second-growth stands, nor does it approach the ground to search for food in fallen debris. As the large trees in the bird's range have fallen before loggers' saws, the ivory-billed woodpecker population has shrunk and the animal is now believed to be extinct. Disappearance of this magnificent bird has attracted much attention, but it is probably only one of many species that failed to survive the destruction of the climax forests. Undoubtedly, many insects and mites were also specialized to live in the large old trees of the southern forests, and uncounted numbers of these, too, are probably lost.

It is difficult to know how many invertebrate species are becoming extinct, but the number must be large. As an example, a species of freshwater crustacean was discovered about 30 years ago living only in small ponds in Ohio. Biologists, wishing to restudy this animal in recent years, found the original ponds drained, and the crustacean gone.

Among more mobile and adaptable creatures, large animals seem to be particularly at risk for several reasons. First, their habitat is most easily destroyed. Compare two major grazers of the Great Plains of North America—the buffalo and the field mouse. The bison population that now exists in Wyoming and Montana is not allowed to expand because the pastures and wheatfields of the West are not compatible with wild herds. Because it is easy to fence them out, the buffalo have few places left to roam free. In contrast, mice which cannot be barred by conventional fences, still range throughout the plains.

But destruction of habitat is often more subtle than is illustrated by this example. The white-tailed deer population has increased in the last quarter century in the United States. This increase has been hailed as a great achievement of conservation programs. Game management statistics often fail to mention that at the same time the larger grazers, such as moose, elk, and buffalo populations, have been dramatically reduced. Deer have flourished partly because they can thrive in small woodlots or woodlot-pasture ecotones. In contrast, the larger animals need more land to support themselves, and each one needs a sizable range.

Secondly, hunting pressures are generally more severe against large animals than small, for large game animals are more desirable trophies. Also, relatively small grazers like deer usually flee from their nonhuman predators, while a mature moose or bison is strong enough to stand his ground against wolves or mountain lions. Therefore, when men with modern firearms reached North America, they found that large game animals were far easier prey than the elusive deer. Consequently, the strong have fallen in proportionally greater numbers than the quick.

Voles include various genera of rodents related to rats and mice. Voles are generally stouter, have smaller eyes, limbs, and tails, and move less briskly than rats and mice.

Thirdly, small animals like rodents and insects usually have a higher reproductive rate than larger animals. For example, a female bank vole is independent of her parents at 2½ weeks, reaches sexual maturity in 4 to 5 weeks, and in one year can give birth to five litters of five infants each. Insects lay several thousand eggs a year. These species have evolved amidst powerful predators and appear to adapt well to the addition of one more. By contrast, the mighty creatures have traditionally bred more slowly and maintained population growth through maternal care or defense by the herd. These protective measures are not effective against the repeating rifle.

Many *predator* species are seriously endangered for reasons which should be clear by now. Large predators have traditionally known almost no external enemies and are even more poorly adapted to predation than are large herbivores. In addition, their niches are on the top of the food chain and they often need a large hunting range. Thus, as destruction of their habitat constricts their natural range so that the wild lands can no longer support them, the hunting animals often roam toward farmers' fields and prey on cattle. As a result, they are hunted extensively in extermination programs.

4.6 THE NATURE OF THE LOSS

(Problem 3)

What will happen to the biosphere if these current trends continue? Ecosystems and food webs will become simplified, but the pattern of primary, secondary, and tertiary consumption will continue. If the deer becomes the largest grazer and the coyote becomes the largest non-human predator, grazing and predation will continue, so why worry about extinction?

There are several reasons to worry. In some ways the most compelling is the aesthetic and religious argument. Different individuals may express their feelings in different ways. To some, species and the wilderness should be preserved simply because they exist. Others might reformulate this in saying that man has no right to exterminate what God has created. To still others, an unobtrusive passage through an untouched wilderness area is a source of enormous aesthetic gratification, as valid and moving an experience as a great play or string quartet. As many of the greatest and noblest creatures of the earth fall prey to man's thoughtless acts, so too, the richness, variety, and fascination of life on this planet diminish with their passing.

A second reason is concerned with developments in medicine and the biological sciences, which have always been dependent on various plants and animals as experimental subjects. The range of species used in different experiments has traditionally been very wide. For example, when Thomas Hunt Morgan initi-

Thomas Hunt Morgan (1866–1945) was an American geneticist whose experiments formed the basis for his theory that paired elements or "genes" within the chromosome are responsible for the inherited characteristics of the individual. He won the Nobel prize in medicine in 1933.

ated his famous studies of genetics, he needed an animal that was easy to breed in large numbers and had few, large, and accessible chromosomes. He chose the fruit fly, *drosophila,* not because the life style of these insects was of particular interest in itself, but because the animals were easy to study and the lessons learned from them could be generalized and applied to studies of other creatures. Another serious scientific loss in this category is that of wild populations of plants and animals that have traditionally been used in breeding hybrid species for agriculture. For example, domestic corns are often susceptible to various diseases. Geneticists have periodically crossed high-yielding susceptible corns with hardy wild maize in an effort to develop high-yielding resistant corn. However, natural maize growing along fence lines and roadways is often considered to be a weed and has been combated across North America with herbicides. The loss of these species would be a severe blow to modern agriculture.

A third argument is related to ecosystem stability. Chapter 3 showed that simple systems are unstable and are not well buffered against fluctuations of populations. Man has simplified many natural ecosystems by displacing or destroying some of the native species. Many of these simplified systems have been plagued by frequent outbreaks of pests which then require the application of technical control methods. Perhaps man's labors would be more fruitful if more care were taken to preserve the natural control mechanisms which have operated for millennia. Consider, for example, what is occurring among the herbivores on the African savannas. A simplified picture of the savanna ecosystem, with its migrating species was discussed in Chapter 2, page 59. In many places natural grazers are being displaced to make way for cattle ranches. But cows are unable to utilize grass as efficiently as the heterogeneous African herds. Domestic stocks tend to overgraze certain species of plants selectively, thus upsetting the water table, and in addition they are susceptible to many tropical diseases, especially sleeping sickness. The result is that an untouched savanna is capable of an annual production of 24 to 37 tons of meat per square kilometer in the form of wild animals, while the best pasture-cattle systems in Africa can yield only eight tons of beef per square kilometer per year. Yet in the

Impalas in Africa. (In T. A. Vaughan: *Mammalogy,* 1972, W. B. Saunders Co., p. 249. Courtesy of W. Leslie Robinette.)

name of agricultural progress, many ungulates are being threatened with extinction, and other herd sizes are being substantially reduced.

The final reason concerns the possible long-term effects of worldwide ecosystem alteration, which were mentioned on page 00. When 95 per cent of all large mammals died off in North America 8000 to 10,000 years ago, climax plant systems and the human race survived intact. Does this mean that another wave of extinctions could occur without catastrophic side effects?

If the study of ecology and the environment teaches us only one lesson, it is that perturbations at one end of a delicately balanced web of interrelationships may shake the whole structure. Our understanding of ecosystems and their stability is still elementary. Therefore, it is notoriously difficult to predict with any assurance what will happen to a given ecosystem when one or more of its species are extinguished; the system is far too complex and poorly understood. But it is a foolish person indeed who participates in, supports, condones, or ignores the needless destruction of species while maintaining steadfastly that no ill effects will supervene because none is obvious at present. For just as events at the cellular level generally require short times to become observable or measurable, events at the level of the ecosystem may take years, centuries, or millennia.

What we have inherited from the evolutionary process is surely likely to be far more stable and rich than what we are likely to produce with aimless, hit-or-miss destruction.

PROBLEMS

1. **Pleistocene extinctions.** Discuss the role of man in the Pleistocene extinctions. Do you feel that man, alone, caused these extinctions? Defend your answer.

2. **Pleistocene extinctions.** If the sabre-toothed tiger could be reintroduced into North America today, do you think that it would survive? Defend your answer.

3. **Habitat destruction.** List the major factors which lead to habitat destruction in the modern world. What factors would you list as being most disruptive to wild animals? Explain.

4. **Introduction of foreign species.** Explain why invading species are able to disrupt ecosystems if they are efficient predators or efficient competitors in their new environment.

5. **Introduction of foreign species.** During the millions of years that North and South America were separated by ocean, evolution and speciation occurred independently on each land mass. About 13 million years ago, the land rose and created an isthmus across which many species could migrate. What effect do you think this event had on the number of species on either continent? Did some species become extinct? Explain.

6. **Introduction of foreign species.** Cattle have successfully displaced bison on the North American plains. Has this displacement occurred through direct competition? What other factors have been involved?

7. **Predators and man.** Design an experiment to show whether or not coyote extermination programs are economical. Would you need a lot of land to conduct this experiment, or would, for instance, 10 to 20 acres suffice? You learned in Chapter 2 that populations of wild animals naturally oscillate. Would population fluctuations affect your experiment? How would you compensate for these fluctuations? Could a definitive experiment be conducted in one year?

8. **Predators and man.** Considering the grizzly bears' record, do you feel that it is safe to feed park bears despite regulations forbidding it?

9. **Extinction of species.** *Homo sapiens* occupies a much broader niche than any other species. Therefore, man competes with many different organisms simultaneously. Show how diversity in man has made it possible for him to endanger the existence of many different types of species. Give examples.

10. **Survival in the twentieth century.** Female polar bears give birth to one or two infants every three years. Discuss the survival potential of this low reproductive rate in past ages and during the present.

11. **Survival in the twentieth century.** Consider the following fictitious species, and comment on the survival potential of each in the twentieth century: (a) A mouse-sized rodent that gives birth to 40 young per year and cares for them well. This creature burrows deeply and eats the roots of mature hickory trees as its staple food. (b) An omnivore about the size of a pinhead. Females lay 100,000 eggs per year. This animal had evolved in a certain tropical area and can survive only in air temperatures ranging from 60° to 100° F. (15.5 to 38° C). (c) A herbivore about twice the size of a cow adapted to northern temperate climates. This animal can either graze in open fields or browse in forests. It is a powerful jumper and can clear a 15-foot fence. Females give birth to twins every spring.

12. **Justification for the preservation of species.** Discuss the importance of wild grasses in the modern world.

The following problem requires arithmetical computation:

13. **Zoo and pet trade.** On page 116 it is stated that if orangutans are obtained for zoos by shooting nursing mothers and taking the infants, of which only 15 per cent survive, then a dozen orangutans have died for each one that reaches the market. Check this arithmetic.

BIBLIOGRAPHY

Four books which deal specifically with the destruction of animal species are listed below:

Roger A. Caras: *Last Chance on Earth.* New York, Schocken Books, 1972, 207 pp.
Kai Curry-Lindahl: *Let Them Live.* New York, William Morrow & Co., 1972. 394 pp.
H. R. H. Prince Philip, Duke of Edinburgh, and James Fisher: *Wildlife Crisis.* Chicago, Cowles Book Company, 1970. 256 pp.
Vinzenz Ziswiler: *Extinct and Vanishing Animals.* New York, Springer-Verlag, 1967.

Two valuable books which deal more specifically with ecosystem alterations are:

David W. Ehrenfeld: *Biological Conservation.* New York, Holt, Rinehart & Winston, 1970. 226 pp.
Charles Elton: *The Ecology of Invasions by Animals and Plants.* New York, John Wiley & Sons, 1958. 181 pp.

A detailed, fascinating, and advanced discussion of the Pleistocene extinctions is given in:

Paul S. Martin and H. E. Wright, Jr., eds.: *Pleistocene Extinctions.* New Haven, Yale University Press, 1967. 453 pp.

5

THE GROWTH OF HUMAN POPULATIONS

5.1 INTRODUCTION

(Problems 3, 4)

The 1970's promise to be a decade of unprecedented population growth. Not only will each year bring an increase in the total number of people on Earth, and not only is the rate of growth expected to become higher each year, but even the rate of the rate, the acceleration, will probably continue its current rapid ascent. Our forefathers did not live in times of such rapid change. Estimates of world populations before the twentieth century are very approximate. (Indeed, even current population data are sketchy for most of Africa, for all of China, and for various other areas.) Anthropological evidence suggests that modern man evolved one hundred thousand years ago. During prehistoric times, the total human population of the Earth must have fluctuated widely. In some years there were more deaths than births, causing human population to decrease temporarily. By the first century A.D., however, population had already established its present pattern of almost uninterrupted growth. Rates of growth were then very slow and extremely variable. The Black Plague of the fourteenth century is believed to have caused a temporary decrease in world population, while the European Renaissance marked the beginning of a rapid rise in world population. At the time of the discovery of America, there were about one-quarter billion people alive on Earth. In 1650, about a century and a half later, world population had doubled to one-half billion. In another 300 years, world population multiplied fivefold to 2.5 billion persons. During the 1950's, the population increased almost another one-half billion and, by 1970, world population was approximately 3.5 billion persons. In other words, the *increase* in world population from 1950 to 1970 was about twice the *size* of world population in 1650. Or consider another comparison. Today there are more people in China than there were people on Earth in 1650. Or yet another: two-thirds of all people born since 1500 are alive today.

Figure 5.1 presents a schematic graph of world population size since the emergence of modern man. Note that the smoothness of the curve reflects ignorance of details of population data rather than regularity of population increase. It is clear that the curve of growth is becoming steeper and steeper in time. In fact, a glance at Figure 5.1 may well cause you to panic. If the population continues to grow ever more steeply, or even if it continues to grow at its current rate, very soon there will be too many people for Earth to support. If there are poverty and starvation now, how can economic and agricultural development be expected to keep pace with an exploding population? Destruction of land, depletion of natural resources, production of waste, and pollution of the Earth can all be expected to increase with increased population. You have probably read dire predictions

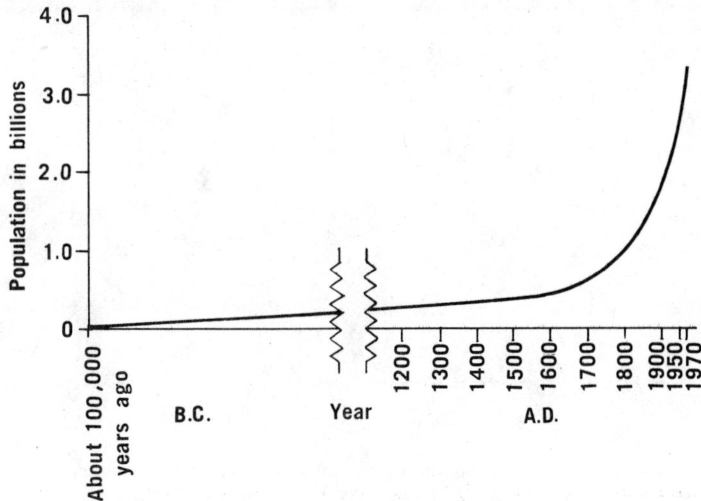

FIGURE 5.1 World population size from emergence of *Homo sapiens* **to 1970.**

based on extrapolation of the growth curve of Figure 5.1. One projection which has gained a certain currency in both lay magazines and scientific journals says that if the present rate of world population increase were to continue, there would be one person for every square foot of the Earth's surface in less than 700 years. If we accept the premise, the conclusion is necessary. But we know that the conclusion must be false. It is impossible for men to stand elbow to elbow on this planet. One square foot of Earth's surface cannot feed, clothe, and shelter a person. Thus the premise must be untenable. In other words, human population cannot continue to grow at current rates indefinitely. Indeed, similar reasoning should convince you that *population size cannot*

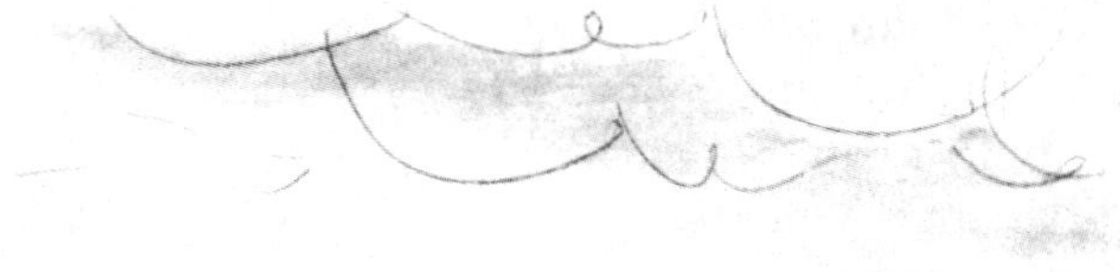

"Excuse me, sir I am prepared to make you a rather attractive offer for your square."
(Drawing by Weber; © 1971 The New Yorker Magazine, Inc.)

grow forever, even very slowly. It will be checked by such factors as limits of space and food, by explicit decisions of families and nations, by famine and diseases, and by complicated interrelated social forces.

Clearly, though, it is of great importance to know where the curve in Figure 5.1 is going. One reason we need such estimates is to plan for the future. How much food must be produced during the next decades? How many schools should be built? Where should roads be placed? Parks? Power plants? Without some methods for projecting future population growth, planners would have insurmountable difficulties. Lest these questions appear to imply that population growth acts as an inexorable and independent force, we must emphasize that social and economic factors help determine population size just as population size is one determinant of social and economic situations. For example, a well-educated and well-fed society typically tends to grow slowly, for people are both aware of population control and motivated to practice it. Conversely, a society that grows slowly is more likely to be well fed. These are not firm rules. After World War II, for instance, American population growth was quite rapid.

Thomas Robert Malthus (1766–1834) was an English economist best known for his *Essay on Population*, published originally in 1798 and republished in 1803, considerably revised and enlarged. In his first edition, Malthus asserted that war, famine, pestilence, misery, and vice prevent population from increasing beyond the limits of subsistence. In the second and later editions of his work, he suggested that "moral restraint" (postponement of marriage and strict sexual continence) acts as a further check on population growth. Throughout his life, Malthus remained pessimistic in his assessment of the future progress of mankind.

A second reason for knowing how to predict population growth is of a political nature. If we can arrive at a reasonably accurate estimate of population size at some future date, and we can show it to be too large to be consistent with societal well being, we have a numerical weapon with which to fight for the implementation of population control measures. In order to evaluate for yourself the validity of statements, predictions, and proposals concerning population size, you must understand the mechanism of population growth and the terms used to express it.

5.2 EXTRAPOLATION OF POPULATION GROWTH CURVES

(Problems 2, 5, 6)

We noted in the previous section that the most obvious method for predicting population growth is to construct a graph that plots population size against time and to guess how the curve will continue. Guessing points on a curve outside the range of observation is called **extrapolation.** Extrapolation is a subtle art. Look at Figure 5.1. If you didn't know the labels of the axes and were asked to continue the curve, what would you do? Such an exercise is frivolous. One person might think the curve will con-

TABLE 5.1 COMPARISON BETWEEN GEOMETRIC AND ARITHMETIC GROWTH

YEAR	*GEOMETRIC GROWTH*
0	10 people
1	$2 \times 10 = 20$
2	$2^2 \times 10 = 40$
3	$2^3 \times 10 = 80$
4	$2^4 \times 10 = 160$
.	.
.	.
.	.

YEAR	*ARITHMETIC GROWTH*
0	100 pounds
1	$(2 \times 1) + 100 = 102$
2	$(2 \times 2) + 100 = 104$
3	$(2 \times 3) + 100 = 106$
4	$(2 \times 4) + 100 = 108$
.	.
.	.
.	.

tinue to go up indefinitely; another might try to draw a curve that peaks and then falls below zero. Still another might finish it by continuing up for a while and then leveling off. And someone else might think that the curve will become wiggly and erratic. If, however, you knew you were graphing population, your historical knowledge would allow you to exclude several kinds of curves. Infinite and negative populations would be impossible, a zero population highly improbable for many years, and wide fluctuations unlikely. Exclusions, however, wouldn't construct your curve. On the other hand, if you had a theory of population growth you would have a basis for extrapolation.

A model for a mechanism of population growth was introduced in 1798 by the Reverend Thomas Robert Malthus. He noted that population, when unchecked, grew at a **geometric rate** of increase. For example, if there were x people in year 0 and ax in year 1 (where a is greater than 1), there would be $a(ax)$, or a^2x, in year 2, a^3x in year 3, and a^nx in year n. However, Malthus said that food supplies increase at an **arithmetic rate:** that is, if there were y pounds of food in year 0, there would be $y + a$ in year 1, $y + 2a$ in year 2, and $y + na$ in year n. Table 5.1 presents examples of both types of growth. In Figure 5.2A both types of growth curve are depicted graphically. Geometric growth is eventually much faster than arithmetic. Malthus therefore predicted that any uncontrolled population would eventually grow too large for its food supply. He continued:

> By that law of our nature which makes food necessary to the life of man, the effects of these two unequal powers must be kept equal.
>
> This implies a strong and constantly operating check on population from the difficulty of subsistence. This difficulty must fall somewhere and must necessarily be severely felt by a large portion of mankind.
>
> . . . The race of plants and the race of animals shrink under this great restrictive law. And the race of man cannot, by any efforts of reason, escape from it. Among plants and animals its effects are waste of seed, sickness, and premature death. Among mankind, misery and vice. The former, misery, is an absolutely necessary consequence of it. Vice is a highly probable consequence, and we therefore see it abundantly prevail, but it ought not, perhaps, to be called an absolutely necessary consequence. The ordeal of virtue is to resist all temptation to evil.

Thomas Robert Malthus: *An Essay on The Principle of Population.* Published originally in 1798. Republished in 1970 by Penguin Books Ltd., Harmondsworth, Middlesex, England, pp. 71–72.

The premises on which Malthusian theory is

based have not proved true historically. First, because of rapid improvements in agricultural techniques, man's food supply has increased much faster than arithmetically; second, geometric growth curves do not adequately describe human population increases. However, the core of the Malthusian argument cannot be ignored, for there are limits to the number of persons Earth can support, and unless growth is checked rationally, something akin to the Malthusian's "misery and vice" *will* afflict mankind.

Recall from Chapter 2, page 39, that population curves for some species are sigmoid, while the curves for other species may oscillate, or even decline to zero. We don't know what the shape of the human population curve will be in the years to come. Hopefully, man will act with foresight and will gradually control his own population to coincide with the carrying capacity of the biosphere. If the population curve for man eventually becomes sigmoid, examination of Figure 5.1 gives no clue as to when the top of the S will begin to form. Indeed if we had looked at the curve depicting growth from the beginning of man until 1600, we would have had no idea when the population would start to increase rapidly. A much more reliable method of predicting population is to look not at changes in total population size with time but at patterns of changes in *rates of growth*. For this purpose, the reader needs to understand what rates of growth are and how birth and death rates affect the size of populations. He must be able to see the relationship between current and future growth and decline. These concepts are introduced in the following sections.

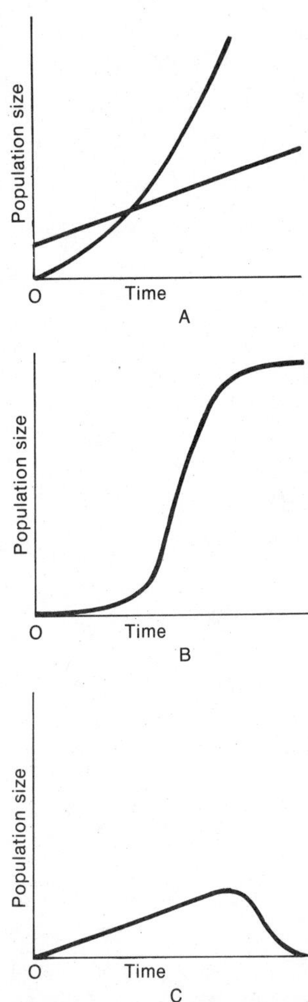

FIGURE 5.2 **Schematic growth curves.** *A.* **Arithmetic (straight-line) and geometric (curved-line) patterns of growth.** *B.* **Sigmoid curve of growth.** *C.* **Growth curve of a population that becomes extinct.**

5.3 AN INTRODUCTION TO DEMOGRAPHY

(Problems 1, 7, 8, 9, 10)

In order to investigate the population dynamics, or the **population ecology,** of a species of animal or plant, the ecologist studies the geography of the region of interest, the food supplies, climatic fluctuations, predation by other species, and competition between and within species. For human populations, however, social and economic forces are more important determinants of population growth. Man has no important predators except, of course, other men, for intraspecies competition in the form of exploitation, subjugation, and war is at least as old as recorded history. If an animal species grows too plentiful for its habitat, its members often starve; human overpopulation and consequent food shortages, however, more frequently result in emigration, or importation of food, or even the development of more effective modes of agriculture. Because of the uniqueness of human patterns of growth, we shall consider it in demographic, rather than ecological, terms.

Demography is that branch of sociology or anthropology which deals with the statistical characteristics of human populations, with reference to total size, density, number of deaths, diseases and migrations, and so forth. The demographer attempts to construct a numerical profile of the population he is studying. He views populations as groups of people, but is not professionally concerned with what happens to any given individual. He wants to know facts concerning the size and composition of populations. For instance, he may want to know the number of males in a population or the number of infants born in a given year.

The subject may sound terribly dry to those of you who are uncomfortable with numbers and computations, but we emphasize that demographic data often reflect in fascinating ways the history of the country studied, the trends in medical care, and the occurrence of social changes.

In addition to studying the composition of populations, the demographer is interested in how populations change in time. He studies changes by counting the number of **vital events**—the births, deaths, marriages, and migrations. If he knew the composition of a population at any given time and the number of vital events occurring between that time and another, he would know the composition of the population at the end of the period. For example, suppose that in 1924 the population of some village were 732. In the next two years, if there were 28 births and 15 deaths and if four people moved in and one moved out, there would be $732 + 28 - 15 + 4 - 1 = 748$ people at the end of 1926.

The word "vital" is from the Latin *vita*, meaning life. Vital means "pertaining to life" or "necessary for life," hence, "basic" or, even more loosely, "very important." When the term "vital" is used in "vital rates" or "vital statistics," the purpose is not to emphasize that the data are important, but rather that they pertain to lives.

Demographers often study **vital rates**, the number of vital events occurring to a population during a specified period of time divided by the size of the population. For example, the 1968 birth rate for the United States is the number of live births in 1968 divided by the midyear population in 1968. (Since the population itself changes during a given year, the population size is usually defined to be the number of people alive at midyear, June 30.)

Before we consider demographic rates, we must understand how a population grows. If we have $100.00 in the bank accumulating 3 per cent annual interest, we know at the end of the year there will be $103.00. We can think of the 3 per cent operating on each dollar within the population of dollars in the bank. Therefore, each individual dollar grows to $1.03, or we can say that each dollar is growing at an annual rate of 3 per cent. Taking one dollar out of the bank will not affect the rate of growth of the remainder.

But what does it mean to say that a population is growing at 3 per cent? Certainly, if we know that 100 people were alive on January 1 and 103 were alive the following December 31, the annual growth rate has been:

$$\frac{(103-100)}{100} \times 100\% = 3\%.$$

However, if we thought of the 3 percent operating on each person, we would have to think of each person becoming 1.03 people. What does that mean? A 3 per cent growth rate means only that for every 100 people there were three more births and immigrations than deaths and emigrations. Even in very simple cases, removal of one person from the population will affect the rate of growth of the remainder. Suppose we have 100 people on January 1, and during the year there are no migrations, five births, and two deaths of people over 65. There would be 103 people on December 31, a 3 per cent increase. Now, suppose we looked at only 99 people, omitting one infant. There would still be five births and two deaths, because births do not occur to infants. So there would be $99+5-2=102$ persons at the end of the year. The annual rate of growth would be:

$$\frac{102-99}{99} \times 100\% = 3.03\%.$$

On the other hand, suppose we looked at the rate of growth of the population excluding, instead of an infant, one of the women who had a child. Now we would have four births occurring to 99 people and the rate of growth would be:

$$\frac{101 - 99}{99} \times 100\% = 2.02\%.$$

Finally, omission of one death would have produced a population of size 99 + 5 − 1 = 103, with a rate of growth of:

$$\frac{103 - 99}{100} \times 100\% = 4.04\%.$$

This very simple example has pointed out some of the difficulties confronting the student of population size, but it also leads to three important insights that are necessary for an effective approach to the investigation of growth:

1. An overall "rate of growth" is really the difference between a rate of addition (by birth or immigration) and a rate of subtraction (by death or emigration). The rate of growth is positive only when there are more additions than subtractions.
2. The probability of dying or of giving birth within any given year varies with age and sex.
3. The age-sex composition, or **distribution,** of the population has a profound effect upon a country's birth rate, its death rate, and hence its growth rate.

Three very basic measures of population growth are the **crude birth rate,** the **crude death rate,** and the **rate of natural increase.** For any geographical area, or ethnic group, being studied, we define:

$$\text{(a) Crude birth rate per thousand during year } X = \frac{\text{Number of live children born during year } X}{\text{Midyear population in year } X} \times 1000$$

and

$$\text{(b) Crude death rate per thousand during year } X = \frac{\text{Number of deaths during year } X}{\text{Midyear population in year } X} \times 1000.$$

Since the denominators of the birth and death rates are identical, we can define:

$$\text{(c) Rate of natural increase per thousand during year } X = \text{Crude birth rate per thousand} - \text{Crude death rate per thousand}$$

$$= \frac{\text{Number of live children born during year } X - \text{Number of deaths in year } X}{\text{Midyear population in year } X} \times 1000.$$

TABLE 5.2 LONG-TERM INTERNATIONAL MIGRATION DATA FROM SELECTED COUNTRIES*

CONTINENT AND COUNTRY	YEAR	TOTAL NUMBER OF IMMIGRANTS† IN THOUSANDS	RATIO OF IMMIGRANTS TO BIRTHS	TOTAL NUMBER OF EMIGRANTS‡ IN THOUSANDS	RATIO OF EMIGRANTS TO DEATHS
America, North					
Canada	1969	162	0.4	?	?
Mexico	1969	84	0.04	63	0.1
Panama	1969	0.9	0.02	?	?
United States	1971	370	0.10	?	?
America, South					
Colombia	1968	3	0.004	5	0.03
Venezuela	1967	0.3	0.0008	2	0.03
Asia					
Israel	1969	33	0.5	7	0.4
Japan	1969	756	0.4	747	1.1
Europe					
Bulgaria	1969	0	0	0.2	0.002
Czechoslovakia	1967	4	0.02	14	0.1
France	1968	93	0.1	198	0.4
Germany (West)	1969	701	0.8	440	0.6
Netherlands	1969	67	0.3	45	0.4
United Kingdom	1969	206	0.2	293	0.4
Oceania					
Australia	1969	249	1.0	108	1.0

*Data from *U.N. Demographic Yearbook* (1970) and *Statistical Abstract of the United States* (1972).

†Long-term immigrants are defined as persons entering a country with the intention of remaining at least one year.

‡Long-term emigrants are persons leaving a country with the intention of not returning for at least one year.

Another name for the rate of natural increase is the **crude reproductive rate.**

For simplicity, we shall assume that migration rates are negligible and therefore we shall not discuss data from Israel, Australia, or other countries where the assumption is patently invalid. Table 5.2 presents some data on international migration.

Patterns of migration do not greatly affect growth except when a large proportion of the population migrates, especially when the migrants are predominantly of one sex. For example, in the last century many Irish young men emigrated, leaving a surplus of women behind. Consequently there were many childless women. Conversely, at the beginning of this century, there were many more men than women in the United States. The differential can be attributed in large part to the fact that the heavy immigration of the period was male-dominated. In general, groups of migrants are usually quite different with respect to age, sex, and vital rates from the inhabitants of the country either from which they leave or to which they go. In particular, migrants are often healthy males of reproductive age.

5.4 MEASUREMENTS OF MORTALITY

(Problems 11, 12, 13, 14, 33, 37)

For long-term assessment of historical trends, the three crude rates introduced in the previous section are concise, useful, and graphic. Figure 5.3 plots birth and death rates for the developed and the underdeveloped parts of the world. The United Nations defines the developed regions as Europe, the United States, the Soviet Union, Japan, temperate South America, Australia, and New Zealand. The rest of the world is called underdeveloped. (See the map on pages 138 to 139.)

Overall trends in birth and death rates, and the marked changes in times of war are clearly depicted in Figure 5.3. Yet for current data, analyzed by individual countries, these rates can be very deceptive. For example, in Taiwan, in 1970, the crude death rate was 5.1; in Japan, which is much more economically ad-

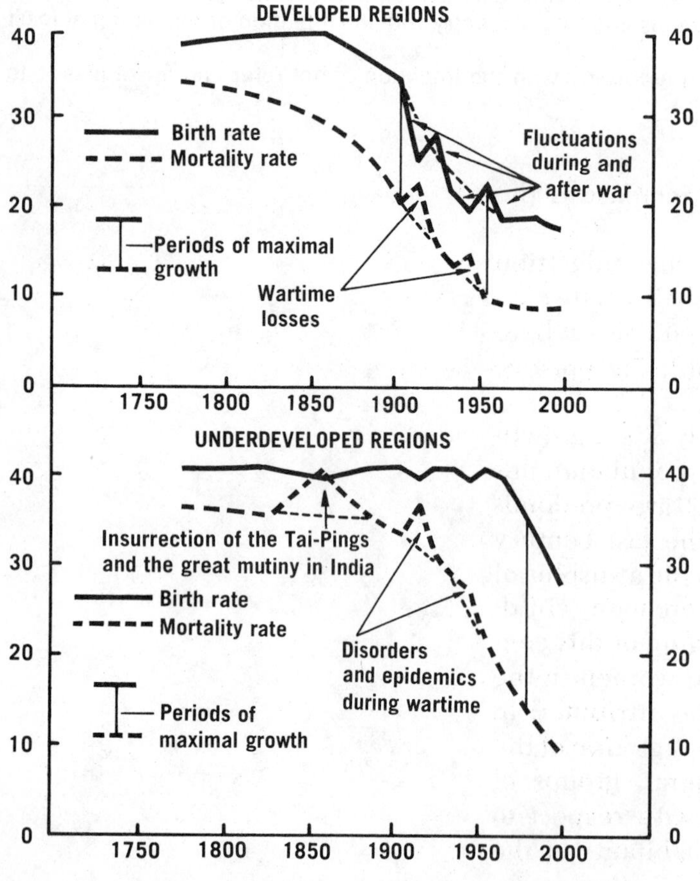

FIGURE 5.3 Estimated and predicted crude birth rates and crude death rates for the period 1750–2000 in the developed and underdeveloped regions of the world. (Adapted from *The Demographic Situation of the World in 1970*, United Nations, 1971.)

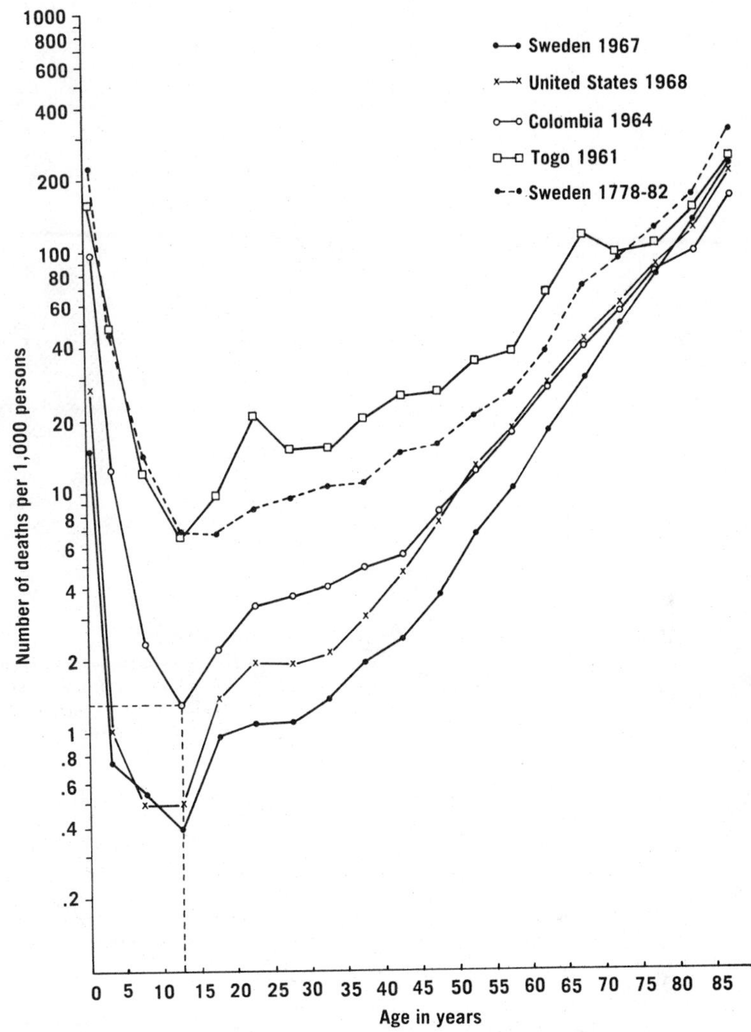

FIGURE 5.4 Age-specific death rates of males in Sweden (1967), the United States (1968), Columbia (1964), and Togo (1961). Example: The point shown at the intersection of the dashed lines tells us that in Colombia in 1964, the male death rate for 10- to 14-yea-olds was 13 per thousand. (Data from Keyfitz and Flieger: *World Population*, Chicago, University of Chicago Press, 1968; and from *U.N. Demographic Yearbook.*)

vanced, the rate was 6.9. In the Canal Zone, populated by Panamanians and Americans, the crude death rate in 1970 was 1.7; the rate in Panama itself was about 8 and in the United States 9.4. Sweden, with one of the world's best systems for delivering health care, had a crude death rate of 9.9 in 1970, and in Scotland, where free medical care is available to all, the rate in 1970 was 12.3. Why are there such apparent anomalies? The answer is that crude death rates are responsive not only to probabilities of dying, but also to the age distribution of the population. The populations of Sweden and Scotland are, on the average, much older than the population of Japan, which in turn is older than that of Taiwan. The Canal Zone is primarily a military installation, populated in large part by young Americans, and its very low death rate is simply a re-

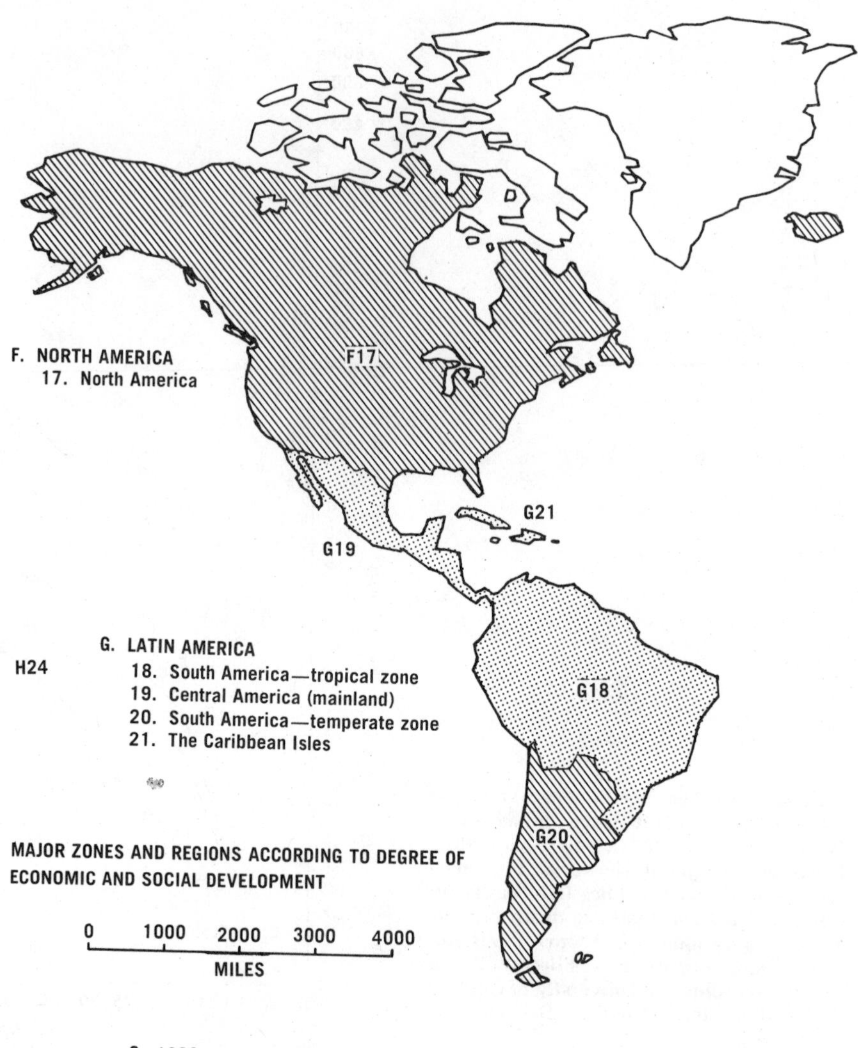

Developed and developing areas of the world.

flection of the fact that not many young men die each year.

For these reasons, serious studies of population growth use not crude death rates, but age- and sex-specific rates, that is, the number of people of a given age and sex who die in a given year divided by the total number of people of that age and sex. In general, these rates are high in infancy, reach their lowest values at about ten years, and then increase slowly. They become higher than infant mortality rates at some time after age 60. Particular societal conditions or events will cause changes in this overall pattern of mortality rate. For instance, maternal mortality is very high in primitive societies. And young men in war-torn coun-

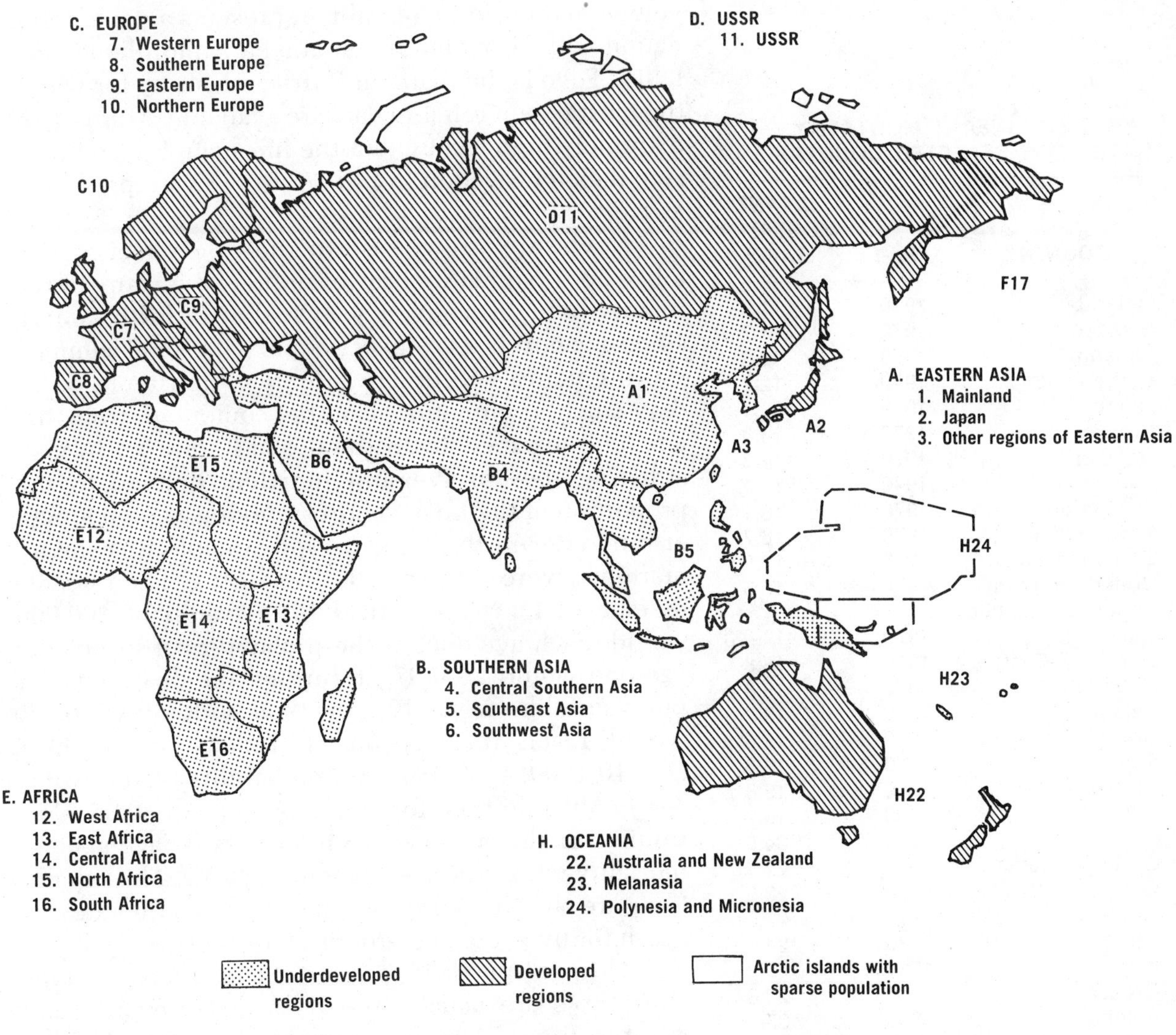

tries have abnormally high death rates. Figure 5.4 presents some recent mortality data for males in four countries: Sweden, the United States, Colombia, and Togo. Sweden, a country in which fine medical care is available to all segments of the population, enjoys low mortality throughout most of the life span, and the high crude mortality rate is simply a function of the age distribution of the population. In the United States, by contrast, large numbers of people have poor medical care. As a result, the **infant mortality rate** in the United States is higher than in Canada, Taiwan, Japan, Australia, and much of Western Europe (see Table 5.3). Indeed, mortality for much of the life span is higher in the United States than in many other de-

Infant mortality rate is usually defined as:

$$\frac{\text{Number of deaths of infants less than one year old in a given time period}}{\text{Number of live births in the same given time period}} \times 1000.$$

The definition of "live birth" varies from country to country, rendering international comparisons of infant mortality rates rather tricky. In particular, in some countries, infants who die within the first 28 days of life are not considered to be live births. Since the United States includes all live births in its definitions of infant mortality, the high infant mortality in the U.S., as compared with that of many other nations, may be in part an artifact of data collection and not an indictment of U.S. systems of delivering medical care. Even if this is the case, however, the differences in rates between whites and non-whites in the U.S. seem largely due to differences in medical care.

veloped countries. Colombia represents a developing nation with very high infant and early childhood mortality. Togo is the mainland African country for which the most recent reliable data are available. Mortality is extremely high throughout the life span. For comparison, data for Sweden from 1778 to 1782 are presented as well.

Now, suppose we wish to know the effect of mortality on a population; for example, how many of the infants born in a given year (called a **birth cohort,** since a cohort is a group of people with some common starting point) will live to celebrate their tenth birthday. Such information can be computed from the age- and sex-specific death rates of Figure 5.4 by constructing the **survivorship function,** defined as the proportion of the birth cohort surviving a given length of time after birth. Suppose, for example, that infant mortality were 30 per 1000 live births. Then 97,000 members of a hypothetical birth cohort of 100,000 would reach age one, or the probability of survival to age one would be 0.97. If, further, the death rate for one-year-olds were 10 per thousand, only 96,030 would reach their second birthday, for 97,000 × (1 − 10/1000) = 96,030. The probability of survival to age two would therefore be approximately 0.96. Continuing in this manner with successive age-specific mortality rates produces the entire curve. The calculations are slightly more complicated when the rates are given for five-year age groups instead of singly.

The age-specific rates used to construct a survivorship curve are usually the rates that prevail at any point in time. They are examples of so-called **period measures,** and do not reflect the mortality experience of any current or historical cohort. For instance, in 1970, the age-specific death rate for 70-year-olds is the death rate for people born in 1900, while the infant mortality rate refers to people born in 1970. Period measures present a profile of many cohorts at a point in time. A true cohort survivorship cannot be completed until all members of the cohort have died, but even incomplete cohort data may provide insights often lacking in period measures. Unless otherwise stated, reported demographic indices are period measures.

The survivorship functions for Sweden, the United States, Colombia, and Togo are plotted in Figure 5.5. Notice how smooth the curves are. Figure

TABLE 5.3 INFANT MORTALITY RATES FOR SELECTED COUNTRIES*

COUNTRY	*YEAR*	*INFANT MORTALITY RATE†*
Iceland	1970	11.7
Sweden	1970	11.7
Finland	1970	12.5
Netherlands	1970	12.7
Japan	1970	13.1
Norway	1970	13.8
Denmark	1970	14.8
France	1970	15.1
Switzerland	1970	15.1
New Zealand	1970	16.7
China (Taiwan)	1970	17.5
United Kingdom	1970	18.3
People's Republic of China	1970	17–20
Germany (East)	1970	18.8
Canada	1970	18.8
Ireland	1970	19.2
Hong Kong	1970	19.6
Singapore	1970	19.8
United States	1970	19.8
White	1968	19.2
Other	1968	34.5
Czechoslovakia	1970	22.1
Germany (West)	1970	23.6
Israel	1970	22.9
USSR	1970	24.4
Italy	1970	29.2
Greece	1970	29.3
Poland	1970	33.2
Yugoslavia	1970	55.2
Argentina	1967	58.3
Costa Rica	1970	67.1
Mexico	1970	68.5
Ecuador	1970	76.6
Guatemala	1970	88.4
Egypt	1969	119.0

*Data from *UN Demographic Yearbook* (1971) and *Statistical Abstract of the United States* (1972). Figure for People's Republic of China from *Studies in Family Planning,* Vol. 3, Number 7. Supplement.

†Rates are number of deaths of infants under one year of age per corresponding 1000 live births. These rates are difficult to compare because the definitions of live births and the effectiveness of birth and death registration vary from country to country.

TABLE 5.4 EXPECTATION OF LIFE FOR SEVERAL SELECTED COUNTRIES

COUNTRY	*LATEST AVAILABLE YEAR*	*CRUDE DEATH RATE PER 1000*	*EXPECTATION OF LIFE*	
			Male	*Female*
Denmark	1967–68	9.8	70.6	75.4
Iceland*	1962	6.8	71.4	76.5
Netherlands	1968	8.3	71.0	76.4
Sweden	1967	10.1	71.9	76.5
Australia*	1965	8.8	67.7	74.1
Bulgaria	1965–67	8.5	68.8	72.7
Canada	1965–67	7.5	68.8	75.2
Czechoslovakia	1966	10.0	67.3	75.6
England and Wales	1967–69	11.5	68.7	74.9
France	1968	11.1	68.0	75.5
Germany (East)	1965–66	13.0	68.7	74.9
Germany (West)	1966–68	11.5	67.6	73.6
Israel	1969	7.0	69.2	72.8
Japan	1968	6.7	69.0	74.3
New Zealand*	1965	8.7	68.2	74.2
Austria	1969	13.4	66.5	73.3
Hong Kong	1968	5.0	66.7	73.3
Poland	1965–66	7.3	66.8	72.8
United States	1968	9.7	66.6	74.0
Soviet Union	1967–68	7.6	65.0	74.0
Taiwan	1965	5.5	65.8	70.4
Uruguay	1963–64	10	65.5	71.6
Yugoslavia	1966–67	8.5	64.7	69.0
Mexico	1965–67	9.5	61.0	63.7
Guatemala	1963–65	16	48.3	49.7
Pakistan	1962	20	53.7	48.8
India	1951–60	20	41.9	40.6
Burundi	1965	25	35.0	38.5
Chad	1963–64	25	29.0	35.0
Nigeria	1965–66	25	37.2	36.7
Togo*	1961	29	33.4	40.2

*Data without asterisks from *U.N. Demographic Yearbook* (1970).
Data with asterisks from Nathan Keyfitz and Wilhelm Flieger: *World Population, 1968.*

5.6 plots male and female survivorship curves for Sweden in 1965. The women have lower infant mortality rates and generally higher survivorship. (Where is infant mortality plotted on the survivorship functions?) Later we shall see how very useful the age-specific death rates are for projections of population size. They are, however, very cumbersome for a static picture of the general health of a population as reflected by its mortality experience. For this, we use what is called **expectation of life.** The expectation of

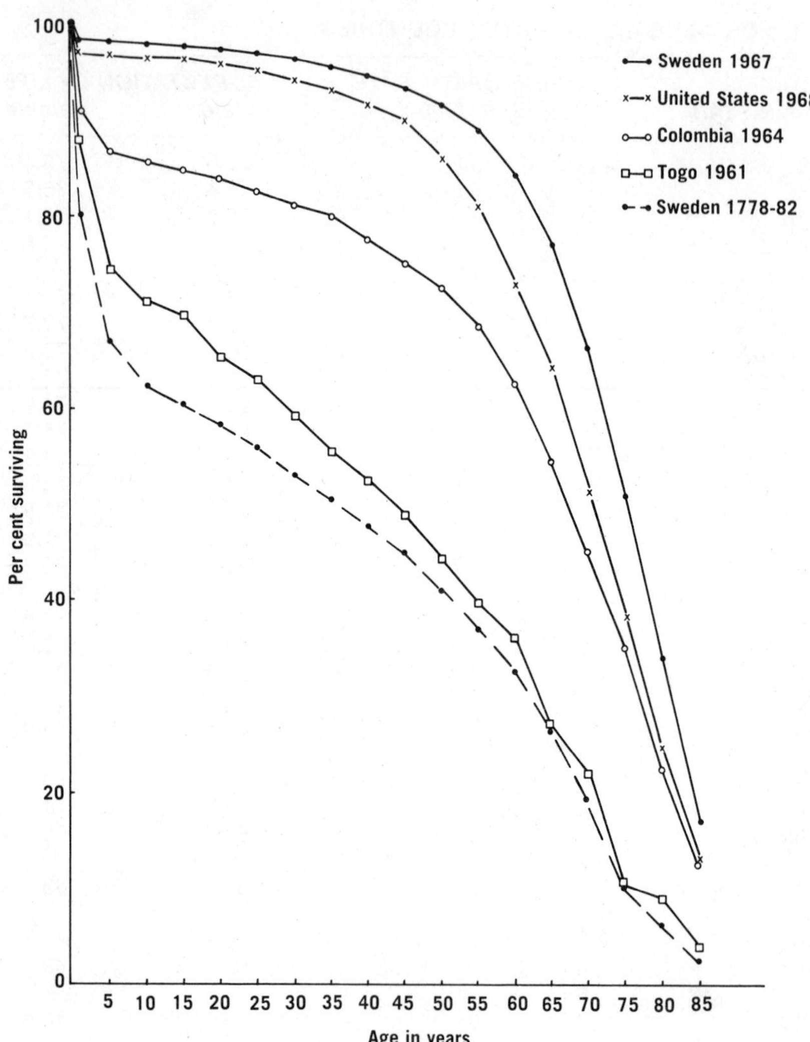

FIGURE 5.5 Male survivorship curves for several selected countries. (Data from Keyfitz and Flieger: *World Population*. Chicago, University of Chicago Press, 1968.)

Life table methodology is used to obtain average life span from the survivorship function. See any introductory text in demography.

Country	*Year*	*Expectation of life* *Male*	*Female*
Sweden	1967	71.9	76.5
United States	1968	66.6	74.0
Colombia	1964	58.2	61.6
Togo	1961	33.4	40.2
Sweden	1778–82	36.0	38.5

life is the average number of years a newborn infant will live given a set of age-specific rates. The measure depends only on age-specific death rates and not on the prevailing age-sex distribution of the country. Therefore, expectation of life is much more useful than crude death rate in comparing mortality experience between nations. For the countries represented in Figure 5.5, the expectations are shown in the margin.

Now consider a hypothetical population with fixed age-specific vital rates. Further, pretend that we may double the number who live through any one year of age we choose. (This assumes a population with death rates no lower than 50 per cent in the year of our choice. See Table 5.5 for a demonstration that doubling survivors is not the same as halving the death rate.) What age should we choose to double if we desire the greatest increase in population? Clearly, we will be most effective if we reduce infant mortality (deaths from age 0 to age 1), for if in the long run there were no other changes in rates, two times as many viable infants would lead, in just a century, to a doubling of population size. On the other hand, if we were to double the number of surviving 80-year-olds, the population size would hardly change. Why? First, few people reach 80; second, even healthy 80-year-olds cannot survive very much longer; finally, the aged do not produce babies.

To gain insights into the numerical effects of re-

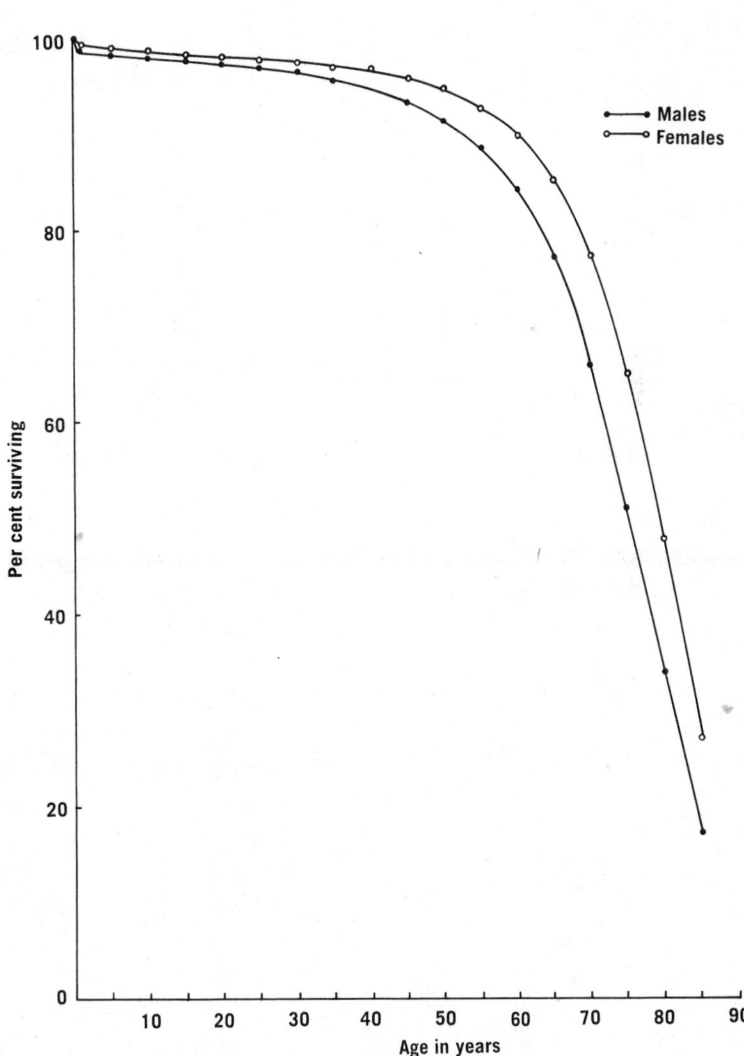

FIGURE 5.6 Male and female survivorship curves in Sweden (1965).

TABLE 5.5 WHY IS HALVING THE DEATH RATE NOT THE SAME AS DOUBLING THE SURVIVAL RATE?

EXAMPLE 1.* HALVING THE DEATH RATE

(1) Death Rate/1000	*(2) Survival Rate/1000*
1000	0
500	500
250	750
125	875
↓	↓
Successive halving	Decreasing change
↓	↓
8	992
4	996
2	998
1	999

EXAMPLE 2. DOUBLING THE SURVIVAL RATE

Survival Rate/1000	*Death Rate/1000*
1	999
2	998
4	996
8	992
↓	↓
Successive doubling	Increasing change
↓	↓
100	900
200	800
400	600
800	200

*In both examples, column (2) = 1000 − column (1).

duction of infant mortality, we shall compare Chile in 1934 to Sweden in 1970. In Chile in 1934 the infant mortality rate was extremely high—262 deaths for every 1000 live births! By contrast, the infant mortality rate in Sweden in 1970 was about 13 deaths per 1000 live births. What would have happened to the population of Chile after 1934 if all birth rates and death rates had remained constant but the infant mortality had suddenly dropped to levels approaching Sweden's? It is easy to see that the sudden reduction in infant mortality with no concomitant change in birth rate would have quickly led to a rapid increase in total population. In 1934, the probability that any infant who was born in Chile would reach the age of 1 was about 75%. If he reached 1 year, the probability that he would reach 15 years was about 90%. Thus, for every 1000 babies born, only about 650 reached reproductive age. On the other hand, if infant mortality had been reduced to about 13 per thousand per year, and if all other mortality rates had remained unchanged, about 900 of every 1000 infants born would have reached 15 years of age. This would have been about a 30% increase in the number of people who would have reached the reproductive age. Furthermore, if mortality rates in the entire pre-reproductive age had been reduced to the level of Sweden's, 950 out of every 1000 people would have reached 15, causing a small further increase in population growth. Thus, lowering rates of death in the pre-reproductive years, without changing birth rates, can induce a very rapid population increase.

5.5 MEASUREMENTS OF BIRTH RATES

Measurements of **natality** are data concerning the number of births or the rate of birth in an area and time period of interest. The simplest measure of natality is the previously mentioned crude birth rate, the number of live births in a calendar year divided by the total population of the area being studied. The rate is easily calculated, but, like the crude death rate, it reflects the current population distribution. Since the probability of a woman's giving birth is a function of her age, the most useful indications are age-specific birth rates which are the ratios of births to women in specified age groups. Figure 5.7 plots some age-specific birth rates for several selected countries.

The set of age-specific birth rates may be used to compute simple indices of birth rates independent of

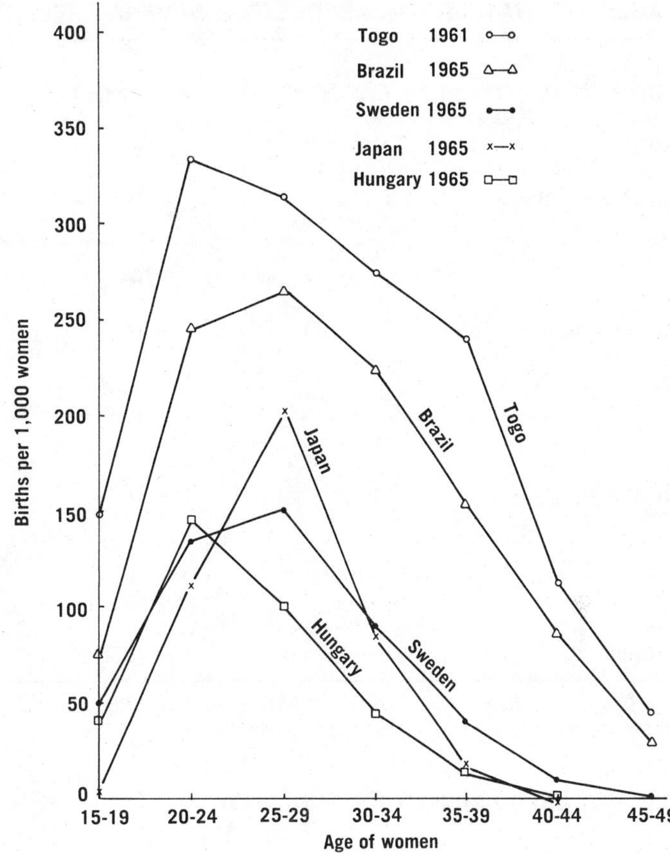

FIGURE 5.7 Age-specific birth rates in five selected countries.

population distributions. The most widely used of these is the **total fertility rate,** or TFR, which can be interpreted as the average number of infants a woman would bear if she lived through age 49 and if the age-specific birth rates were to remain constant for her generation. To compute the TFR for Sweden in 1965, for instance, we need the seven age-specific rates shown in the margin. Adding the seven, multiplying by five (because the age intervals are five-year spans and the age-specific birth rates are expressed per year), and dividing by 1000, the total fertility rate was 2.4. (The comparable rate in the U.S. was 3.0; in Canada 3.2; in Hungary, 1.8; in Costa Rica, 6.8.)

Because the TFR is based only on age-specific rates and not on population distributions, it is useful for international comparisons of child-bearing. Table 5.6 present crude birth rates and TFR's for several countries. In the developed countries, a TFR of roughly 2.1 is the so-called **replacement level**—that value of the TFR which corresponds to a population exactly replacing itself.

Age-Specific Birth Rate in Sweden, 1965

Age	Age-specific birth rate per thousand
15–19	49.3
20–24	136.4
25–29	152.1
30–34	90.0
35–39	39.6
40–44	10.1
45–49	0.7

TABLE 5.6 NATALITY MEASURES FOR SEVERAL SELECTED COUNTRIES*

CONTINENT	*COUNTRY*	*YEAR*	*CRUDE BIRTH RATE/1000*	*TOTAL FERTILITY RATE*
Africa	Togo	1961	54.5	7.1
North America	Canada	1965	24.3	3.2
	Costa Rica	1965	42.3	6.8
	Mexico	1965	44.2	6.7
	United States	1965	19.6	3.0
South America	Colombia	1965	36.8	6.6
Asia	Japan	1965	18.6	2.1
	Pakistan	1963–65	49.0	7.6
	Taiwan	1965	32.7	4.8
Europe	France	1965	17.6	2.8
	Hungary	1965	13.1	1.8
	Italy	1965	19.2	2.6
	Netherlands	1965	19.9	3.0
	Sweden	1965	15.9	2.4
Eurasia	Soviet Union	1966–67	18.2	2.4

*Data from Keyfitz and Flieger: *World Population,* 1968 and from *Interim Reports of Conditions and Trends of Fertility in the World,* 1960–1965, United Nations, New York, 1972.

5.6 MEASUREMENTS OF GROWTH

(Problems 20, 21, 34, 35, 36)

Why do we want to measure the population growth of a country? Our purpose will in large part dictate the kind of measure we use. To learn how many more people were inhabitants at one time than at some other time, we need only subtract the two populations. For example, the population of India in 1950 was 358 million persons. By 1970, the population had reached 550 million. The total growth of population in the 20-year span from 1950 to 1970 was 192 million people, or slightly less than the current population of the United States. By contrast, in the United States, there were 152 million people in 1950 and 205 million in 1970, a difference of 53 million. Thus, the United States, with many times the wealth of India, and about three times the land area, had about one quarter the population increase.

It should be clear that these population differences give little feeling about how fast a country is growing relative to other countries. More useful are rates of growth, defined in Section 5.3. For the data from India, the average annual rate of growth from 1950 through 1970 was:

$$\text{Average annual growth rate (India, 1950-1970)} = \frac{(550 \times 10^6 \text{ persons} - 358 \times 10^6 \text{ persons}) \times 100\%}{358 \times 10^6 \text{ persons} \times 20 \text{ years}} = 2.4\%/\text{year}.$$

For the United States in the same period the average growth rate was:

$$\text{Average annual growth rate (U.S., 1950-1970)} = \frac{(205 \times 10^6 \text{ persons} - 152 \times 10^6 \text{ persons}) \times 100\%}{152 \times 10^6 \text{ persons} \times 20 \text{ years}} = 1.7\%/\text{year}.$$

These rates are averages, and do not imply, for example, that the population of India in fact grew by 2.4 per cent per year. For if it had, that percentage would be compounded. Thus, for the second year, the average growth would be 2.4% × (1 + .024) and so on.

A common popular measure of population growth is the **doubling time** (t_d) of the population. The doubling time is defined as the length of time a population would take to double if its annual growth rate (r) were to remain constant. The relation between doubling time and growth rate is:

$$t_d = 0.693/r.$$

To compute doubling time from rate of growth, think of an analogy with compound interest at the bank. If a rate of growth (interest rate) is applied once a year to a population of size P_o (capital in the bank), the population (capital) at the end of one year is:

$$P_1 = P_o + P_o r = P_o (1 + r).$$

More generally, if population growth is compounded n times each year, then:

$$P_t = P (1 + r/n)^{nt}.$$

It is reasonable to suppose that populations grow continuously, or that $n \rightarrow \infty$. From elementary calculus, $\lim_{n\rightarrow\infty} (1 + r/n)^{nt} = e^{rt}$. The doubling time is then the solution of the equation:

$$2P_o = P_o e^{rt_d}.$$

Or, taking logarithms on both sides of the equation:

$$t_d = 0.693/r.$$

Overall, such growth rates are interesting measures of long-term historical trends, but as our study of birth and death rates has shown, population composition affects overall rates too severely for them to be useful for reliable prediction. Note that although the population size and composition are irrelevant to calculating a theoretical doubling time, they are not irrelevant to the time a real population will take to double. Therefore, beware of conclusions made on the basis of projecting doubling times. Remember that although the term "doubling time" sounds like a projection, the measure uses only the crudest available population data.

To project future population size more accurately, the simplest effective approach is to apply current age-sex specific birth and death rates to the present population and calculate future size. However, even this type of prediction is subject to large error, for it assumes constant age-specific vital rates. Figure 5.8 which shows birth and death rates for women aged 20 to 24 in France, demonstrates that these rates are in fact not always constant. (Certainly, the birth rates have been much more stable than the death rates.) Accurate forecasting of population size is impossible unless we know something about how vital rates are changing.

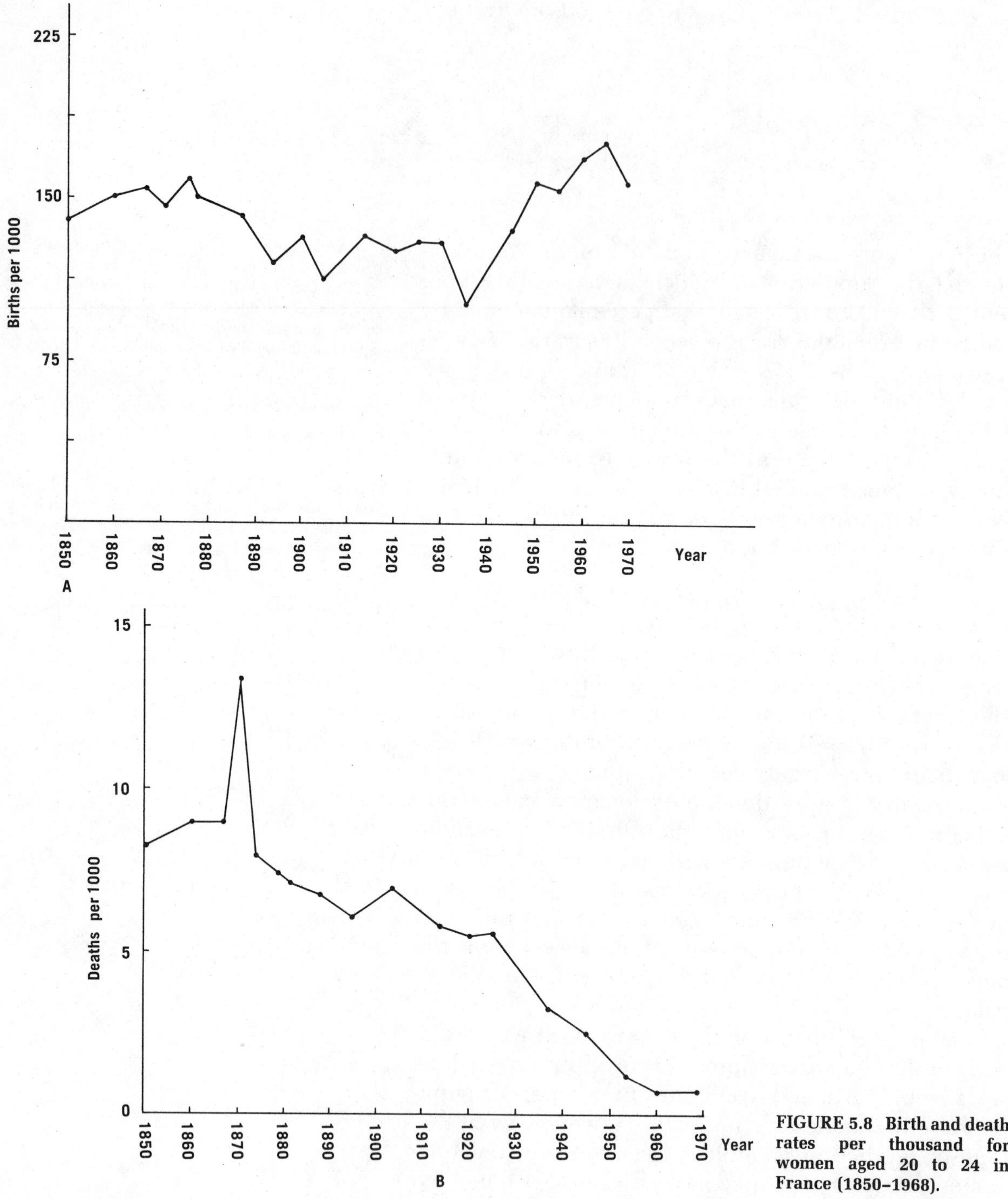

FIGURE 5.8 Birth and death rates per thousand for women aged 20 to 24 in France (1850–1968).

The current age-sex distribution of an area should be used in predicting future population sizes. In the hypothetical age-sex distribution of Figure 5.9, each age group has the same number of males as females. In particular, there are 500 boys and 500 girls under 10, and 50 men and 50 women between 90 and 100

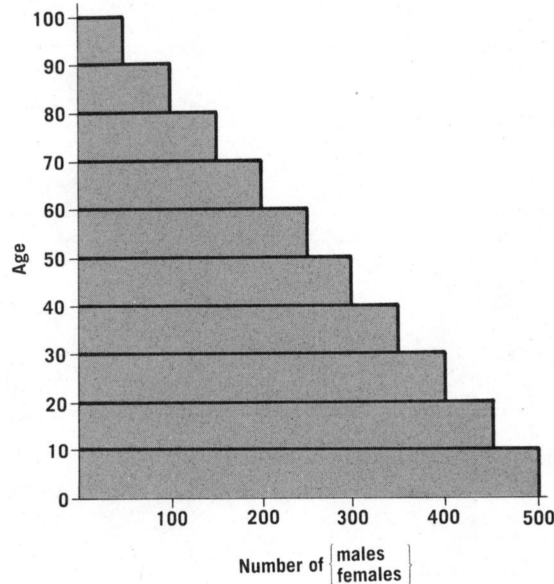

FIGURE 5.9 Age-sex distribution of an ideal population.

years of age. Furthermore, there are exactly 50 fewer men and 50 fewer women at each succeeding age decade. Thus, while there are 300 of each sex between the ages of 40 and 50, there are only 250 of each sex between 50 and 60.

Do human age-sex distributions look like Figure 5.9? Not at all! Figure 5.9 would represent a population in which (a) boys and girls were born with equal frequency; (b) the same number of persons were born every year for over a century; (c) everyone died by the age of 100; and (d) any person, at birth, had an equal chance of dying throughout each year of his life span. However, in real human populations, about 106 boys are born for every 100 girls. Nor is the probability of dying constant throughout man's life span. Instead, as we have discussed in Section 5.4, a relatively large proportion of people die when they are very young, comparatively few die between the ages of 10 and 50, and the proportion of people dying each year after 50 increases rapidly. In addition, there are marked sex differences in mortality (the number of deaths which occur in a given period). Women have a higher probability of surviving from one year to the next throughout the life span except, in some societies, during the childbearing years.

This figure represents a worldwide average. There is considerable geographic variation.

Consider the effects of realistic patterns of vital events on a birth cohort. The greater survivorship of women over men means that even though more boys are born than girls, the ratio of women to men increases as the cohort grows older. By the time the cohort is elderly, there are considerably more women than men. Also, data are usually collected in such a way that we know only the total population of each sex over 70, 80, or 85. Therefore, the graphs can be only approximate for the very old age groups. Figure 5.10 presents an age-sex distribution which more nearly reflects these demographic characteristics.

In addition to these reasonably predictable phenomena, many less predictable changes can occur. Population growth is affected by such events as war, famine, medical advances, and changes in social custom. For example, Figure 5.11 presents the age-sex distribution of France in 1965. This is a very bumpy curve. To interpret it we need knowledge from many fields. For example, the graph shows that in 1965 only 38 per cent of the population of France between 70 and 74 were males. Why so few? We have already learned that men die earlier than women, but the observed discrepancy is much larger than can be explained solely by natural differences in rates. History provides an answer. This cohort was born between 1890 and 1894. At the outbreak of World War I, in 1914, the cohort born in this five-year period was between 19 and 23 years of age. Its members, then, included many of the men who fought World War I. Therefore, much of the difference in numbers of men and women in 1965 reflects the mortality of French

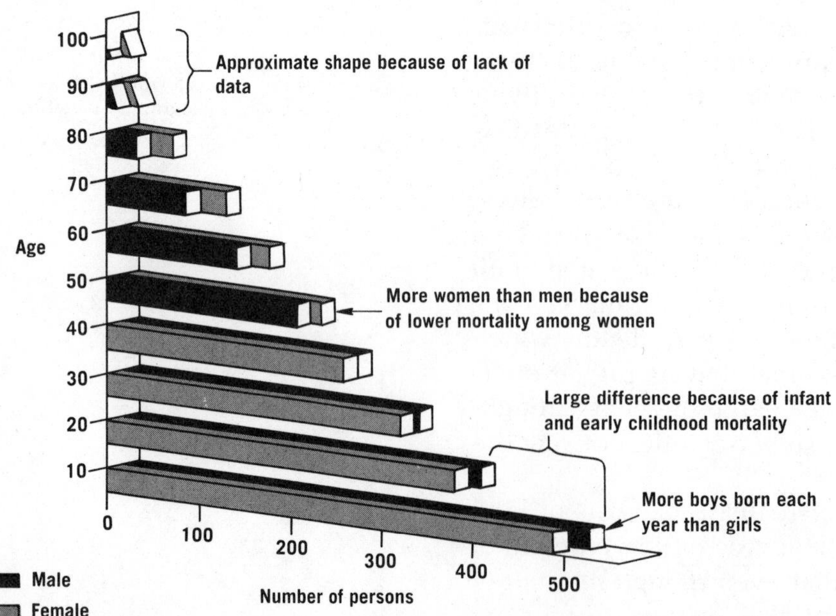

FIGURE 5.10 Typical age-sex distribution.

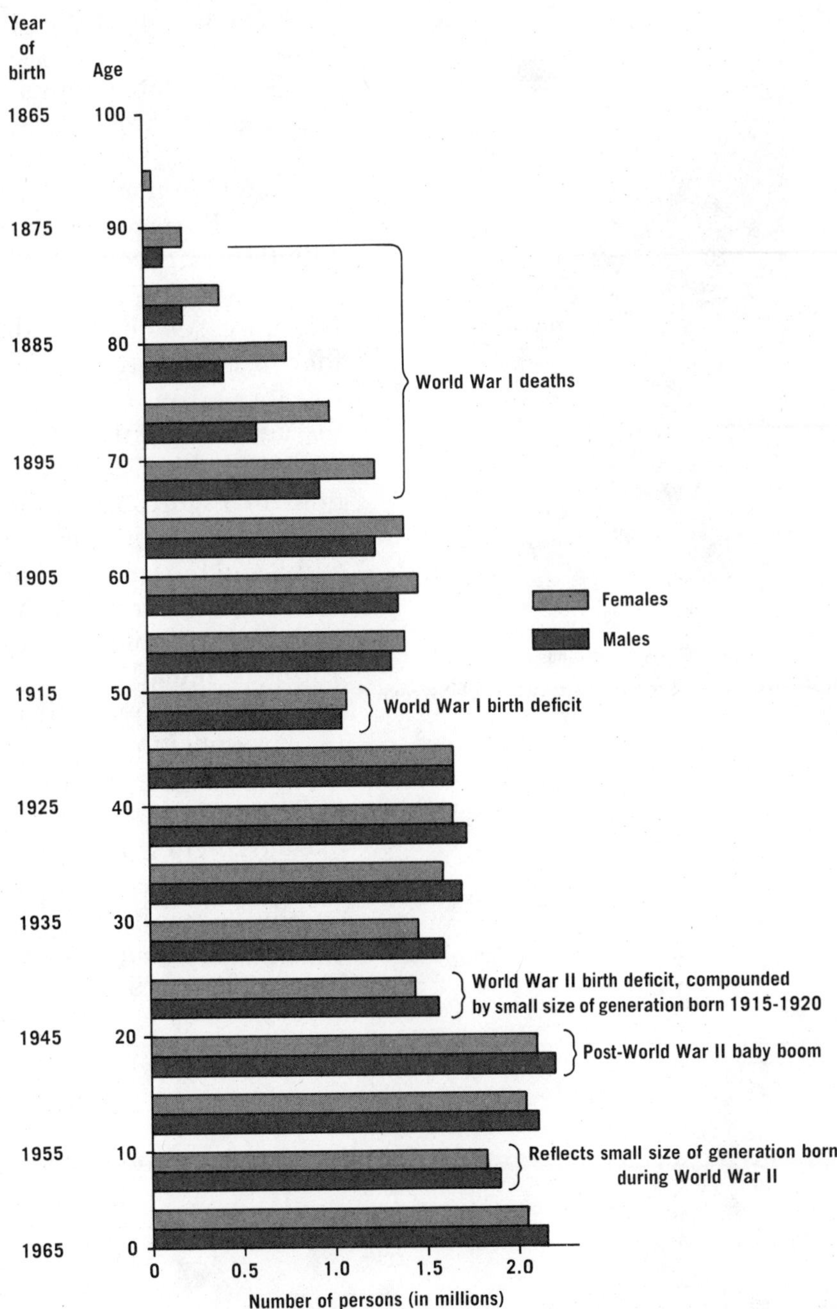

FIGURE 5.11 Population of France in 1965.

men during World War I. We can immediately see how the distribution curves can sometimes aid in predicting future population growth. We would expect that since the males of the 1890–1894 cohort were away from home during some of their reproductive years and since so many were killed, there should occur some very small cohort about one generation later. Indeed, the birth cohort of 1915–1920, the people

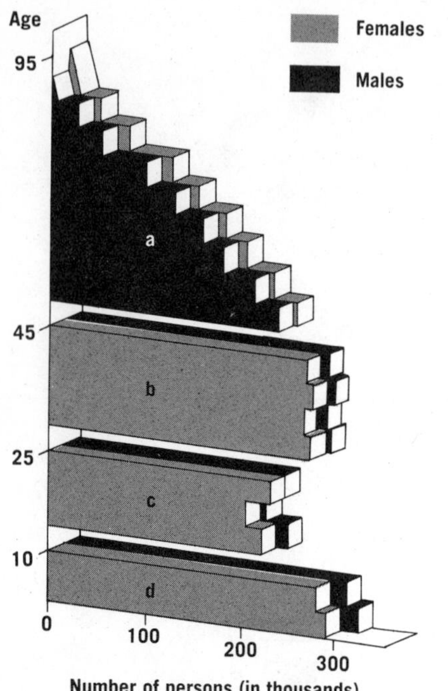

FIGURE 5.12 Age-sex distribution of Sweden in 1950.

born for the most part during World War I, is very small.

To illustrate another example of predicting population growth from age distributions, we examine the population distribution of Sweden in 1950 (Fig. 5.12). The total population in 1950 was nearly seven million persons. First, note that the male and female age distributions are very similar. Therefore, we need consider only one sex to examine population growth. Next, we see that the distribution is separated into four distinct parts. Since the distribution is very narrow for persons of reproductive age, we would expect the age distribution for women 10 years later to look something like Figure 5.13*A*. Each of the distinct sections of Figure 5.12 would move up the age axis 10 years, with each decreasing in size because of mortality, and the small reproductive class would produce a small number of births. In fact, the shape of the actual distribution curve in 1960, shown in Figure 5.13*B* is remarkably similar to Figure 5.13*A*. The male age distribution is nearly the same shape.

A population curve need not be bumpy to be a useful predictor of growth. The age distribution for women in India in 1951, shown in Figure 5.14, looks like a triangle with a very wide base. From this graph, we would expect that such a population would be rapidly increasing. Hence, we would guess that the base of the triangle described by the age distribution 10 years later (1961) would be very wide. Figure 5.14

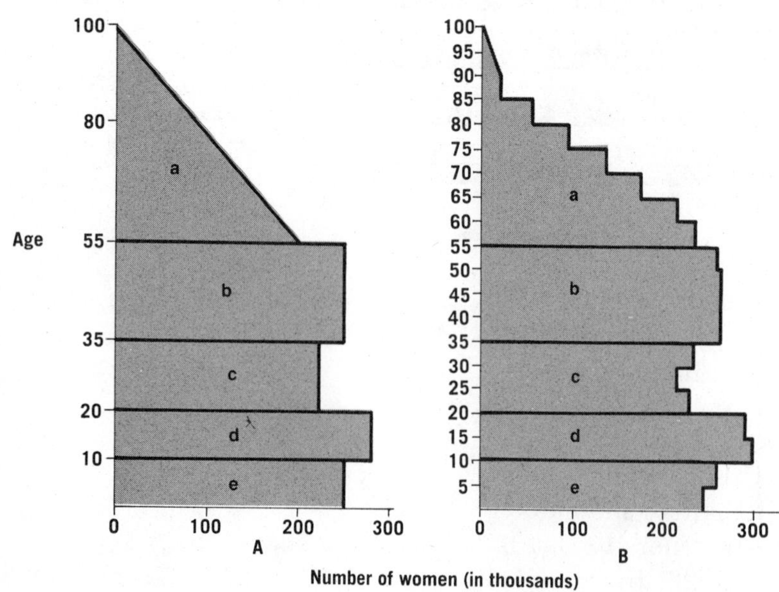

FIGURE 5.13 Female age distribution of Sweden in 1960. *A*. Distribution guessed from age distribution of 1950. *B*. Distribution based on Swedish government population data (1960).

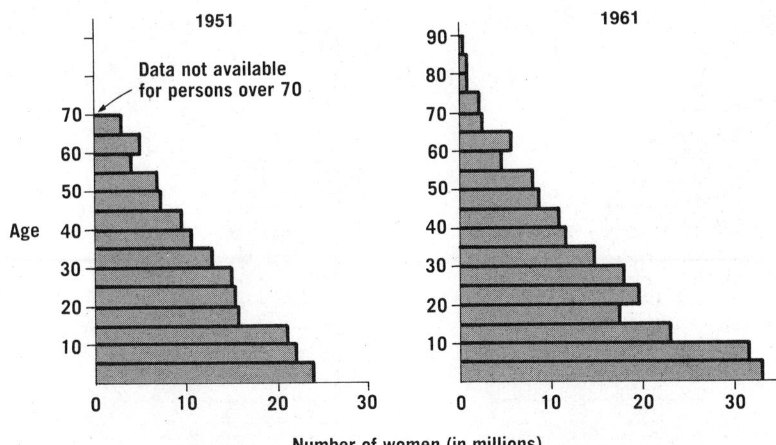

FIGURE 5.14 Female age distributions of India in 1951 and 1961.

shows that in 1961 the base had widened considerably, that the total population had increased by about 25 per cent, and that the proportion of the population under 10 had increased from about 25 per cent to about 30 per cent. The population distribution of 1961 predicts an even more rapidly increasing population during the next decades.

The deficit of 15- to 20-year-olds could not have been predicted from examining the 1951 distribution alone.

The shapes of smooth population curves tell much about the growth of a population. Rapidly growing populations have very large bases; populations that are remaining constant have relatively narrow bases, and declining populations have pinched bases. (See Figure 5.15.)

Predictions about future population growth from the age of the distribution alone are not sufficient if there is considerable change in mortality and fertility. For example, Figure 5.16 shows that the distribution for Sweden looked like a triangle in 1910. Not knowing anything about changes in vital rates, we would predict a triangle in 1930. However, in 1930 the base of the age distribution was pinched. Reexamination of Figure 5.11 shows that the pinched base persisted through the birth cohort of 1940. A demographer

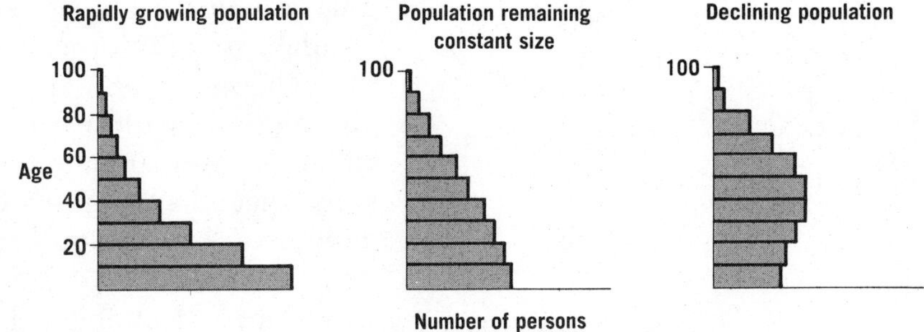

FIGURE 5.15 Schematic population age distributions.

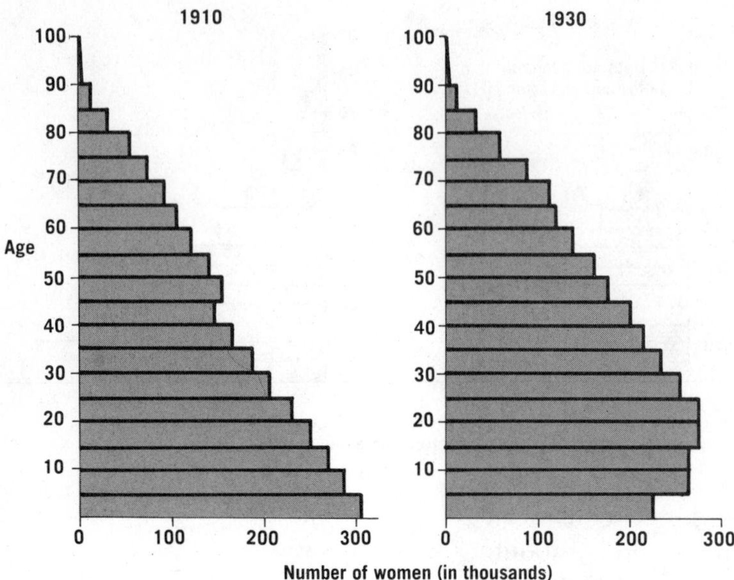

FIGURE 5.16 Female age distributions of Sweden in 1910 and 1930.

would have had to predict World War I and its profound economic and social effects on all of Europe to have projected accurately the population of Sweden from 1910 to 1930.

5.7 ACCURACY OF THE DATA

(Problems 17, 18)

Some readers may have found the choice of examples inexplicable. Why cite data for Togo, an African country with roughly one and a half million persons, and totally ignore Nigeria with close to 60 million persons? Why use Taiwan as an Asian example and not the People's Republic of China? Why, for South America, no mention of Brazil? And why not use 1970 data for all countries? The reason for the choices is practical. We have used Togo as an example of a continental African country not because it necessarily represents social, political, and demographic trends in Africa, but because Togo is the only continental African country with reliable data. We cannot use mainland China, or Brazil, for data are unavailable. The choice of years was dictated by availability of information.

The more refined the measure we wish to report, the more difficult it is to collect the data. Knowledge of the current population size of a country requires only that one number. Crude birth rate needs two pieces of information—total population and number of births within a year.

Data collection for the purpose of computing age-specific rates is much more complicated. Registration systems would require that the person reporting a birth know the mother's age, while a person reporting a death would have to know the deceased person's birthday. Many opportunities for error arise. The reported information may be incorrect, and the procedures used for tabulating the data may introduce serious inaccuracies. Data-gathering techniques other than registration systems, such as sample surveys, are expensive and require considerable expertise.

Even when data appear reliable, the cautious demographer must question their validity. Figure 5.17 depicts the age distribution by five-year age groups for Mexico in 1960. The data were collected in the decennial (tenth annual) **census,** or count, of the population. Each person was asked to report his age. The graph appears odd, especially between the ages of 30 and 60. There are more 36- to 40-year-olds than 31- to 35-year-olds; more 46- to 50-year-olds than 41- to 45-year-olds; more 56- to 60-year-olds than 51- to 55-year-olds. The population dips cannot be attributed to suc-

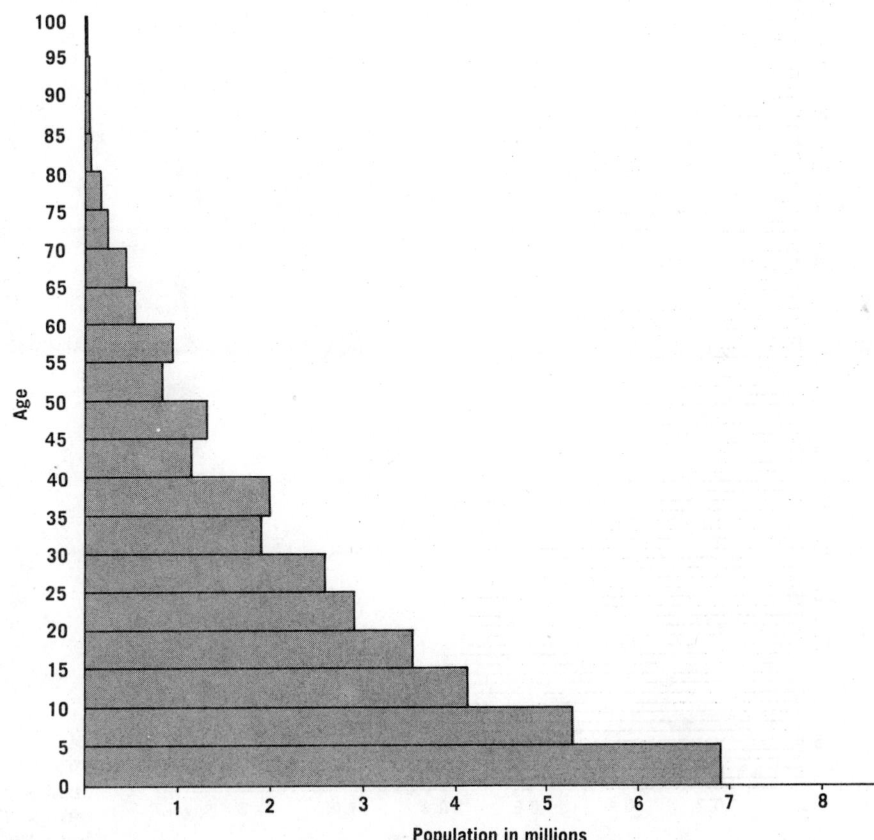

FIGURE 5.17 Population distribution of Mexico by age categories in the 1960 census. (Data from *U.N. Demographic Year-Book,* **1962.)**

cessive birth deficits because they are too close together. If we replot the data by single year of age (Fig. 5.18), the results are shocking. According to the 1960 census of Mexico, people were much more likely to report an age divisible by 10 than to report the next year. If we accept the reported data as true, we would have to believe that there were three times as many 30-year-olds as 31-year-olds; five and a half times as many 40-year-olds as 41-year-olds; six times as many 50-year-olds as 51-year-olds; 10 times as many 60-, 70-, 80-, and 90-year-olds as 61-, 71-, 81-, and 91-year-olds, respectively. We would have to postulate a demographic process which could explain the over-abundance of people whose age ended in five, and the lack of people whose age ended in seven. The answer, of course, is that we are dealing here with a tendency of people to report their ages in some round form, for instance, to the nearest five, or to some even number close to the true age. Such marked **digit preference** is

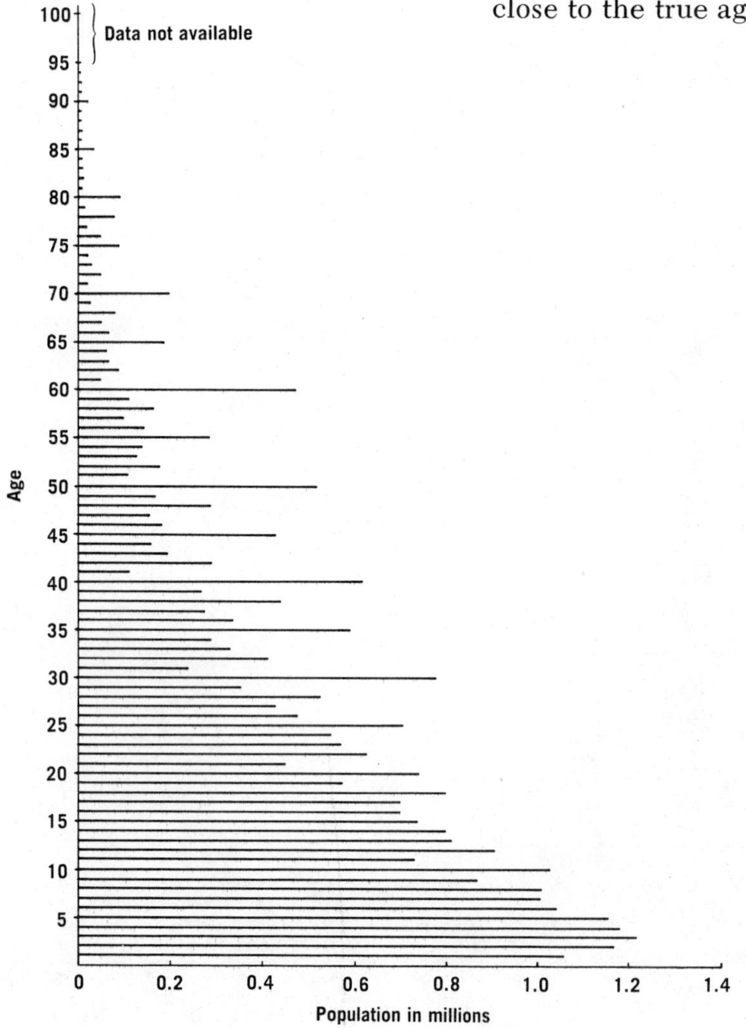

FIGURE 5.18 Population distribution of Mexico by single year of age in the 1960 census. (Data from *U.N. Demographic Yearbook,* 1962.)

not uncommon in underdeveloped countries. In such cases, use of data as reported leads to unreliable age-specific measures.

Other sources of error abound. For instance, the definitions of live birth, immigrants, and emigrants vary from country to country. Another problem in population enumeration arises because conflicting political claims over territory cause some people to be counted twice—once as subjects of one country, once as subjects of another. Other difficulties arise because of the habits of the group being counted. Nomadic bands, for example, are notoriously hard to count.

All the foregoing is to warn against hasty uncritical acceptance of a demographic "fact." It is not a plea to ignore data, for even if data are imprecise or sketchy, general trends in birth or death rates will almost always appear. In using demographic data, always consider what sources of error might be present, and that such error might change your interpretation of the subject under investigation.

5.8 DEMOGRAPHIC TRENDS IN THE UNITED STATES

(Problems 22, 28, 29)

American demographic history after World War II has been characterized by sharp changes in fertility patterns. In 1945, the total fertility rate was about 2.5. That is, on a period basis, the average American woman was giving birth to 2.5 children during her lifespan. By 1946, the TFR had risen to 3.2, reflecting post-war optimism and the reuniting of families. In 1957, at the height of the so-called "baby boom," the TFR had risen to 3.7. The 1960's and early 1970's have been years of rapid decline in TFR which, by 1973 was below 2.00 children, significantly below the "replacement level" of 2.1 (see Fig. 5.19). However, because of the large number of women in the child-bearing ages, American population continued to grow in 1972. What was remarkable was that the growth represented the smallest annual increment in population since 1945. The actual number of births declined nine per cent from 1971 to 1972. The total growth rate was only 0.78 per cent as compared to 1.83 per cent in 1956, and 0.71 per cent in 1945 (see Fig. 5.20).

Although this rapid decline in the TFR has been cited as evidence that the American population will cease growing, the actual growth depends not only on fertility patterns but also on the population distribu-

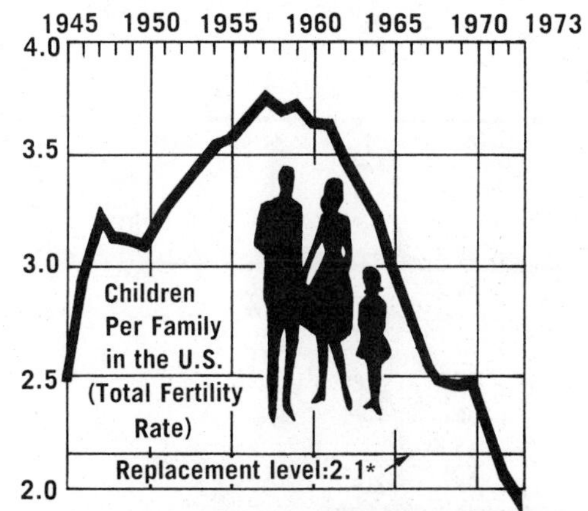

***Replacement level: The number of average births per woman over her lifetime necessary for the population eventually to reach zero population growth. This would take about 70 years.**

Sources: National Center for Health Statistics, Census Bureau

FIGURE 5.19 (© 1974 by The New York Times Company. Reprinted by permission.)

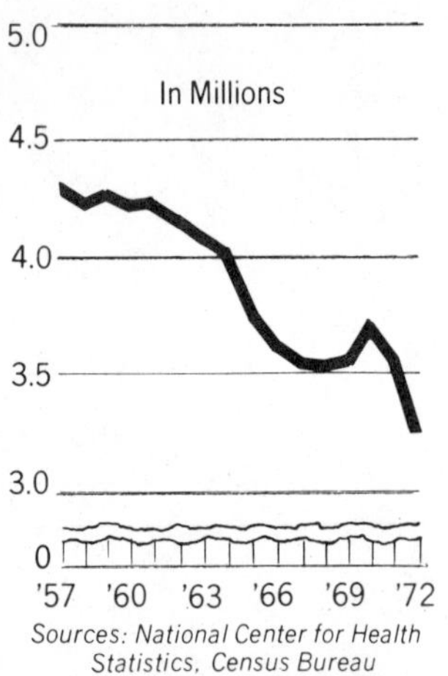

FIGURE 5.20 Births in the United States since 1957, the peak year. (© 1973 by The New York Times Company. Reprinted by permission.)

tion. It is apparent from the female population distribution of the United States that the future American growth will depend largely on the fertility patterns of the people born during the baby boom (see Fig. 5.21). The children born during the baby boom have not yet begun to produce babies. If they, and successive birth cohorts, bear children at replacement level, zero population growth will be realized in about 70 years, when U.S. population will level off at 320 million persons.

The premise that fertility rates will continue to be quite low is only an assumption. A rapid rise in population will result from even a small increase in the TFR for the baby-boom children. Accurate prediction of the behavior of fertility patterns depends on accurate assessment of future social patterns. Increasing age at marriage, increasing rates of divorce, the current rapid increase in the proportion of women at work, and the increasing popularity of the two-child family have all contributed to the decline in fertility in the U.S. If these trends continue, U.S. fertility rates will undoubtedly remain low. On the other hand, if economic and social pressures begin once more to encourage the values of homemaking, fertility rates may again rise steeply. The increased use of more effective contraceptive methods, the steady liberalization of abortion laws, and the increasing social acceptability of abortion are also important factors in limiting population growth. These factors, however, are probably responses to social trends discouraging births. In the U.S. as elsewhere, contraceptive devices are

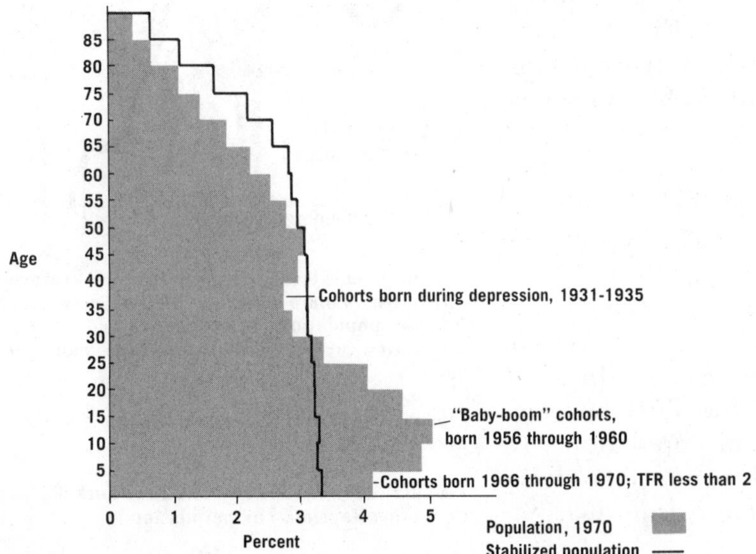

FIGURE 5.21 Per cent female age distribution, United States.

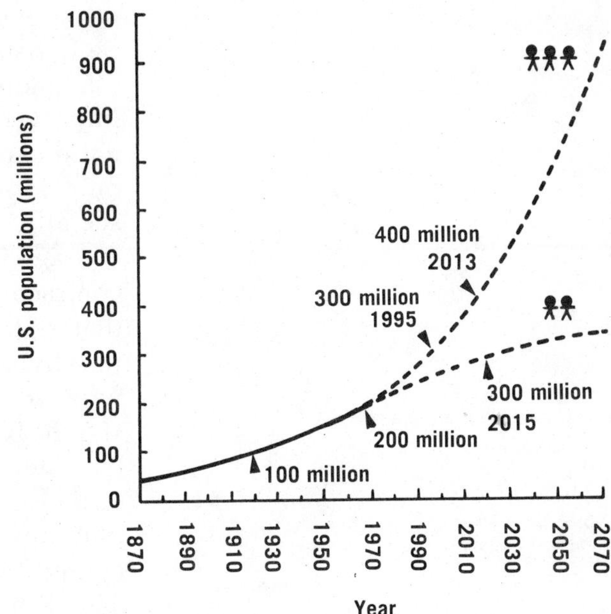

FIGURE 5.22 Projected population size of United States if average family size remains constant at two children or at three children.

most widely used and are most effective among married, motivated women. Indeed, recent studies show that about 53 per cent of unwed women do not use contraceptives when they have sexual relations.

If Americans continue to reproduce at replacement levels, the population will grow older. In 1970, 10 per cent of the population of the United States was aged 65 or older. By 2000, if the average family size continues to be approximately two per family, and if there is no migration, the figure will rise to 11 per cent. When population is stabilized, that is, when it stops growing, 16 per cent of Americans will be over 65. The implications of such a dramatic change in population will be profound. (Refer again to Figure 5.21.) Unless the age at retirement is raised to, say, 70, a large proportion of the population will depend on Social Security and other forms of old-age assistance. Priorities in medical care will have to change as more geriatric and less pediatric skills are demanded. Repercussions will be felt throughout the business world.

In spite of the economic and social adjustments a stable or decreased population will demand, they are generally believed to be small compared to the massive effects of a rapidly expanding population. The 1971 Report of the Commission on Population Growth and the American Future, commissioned by the President, concluded that the social and environmental consequences of increased population growth would be detrimental to the United States, and recommended, among other things, that population education be improved; that sex education be available to all; that child-care programs be improved; that children born out of wedlock be accorded "fair and equal status, socially, morally and legally"; that adoption be encouraged; that states eliminate legal restrictions to contraceptive information, procedures and supplies; that states adopt laws permitting contraceptive and prophylactic information to teenagers; that laws be written affirming that only the patient and the doctor should be involved in the decision to undergo voluntary sterilization; that abortion be permitted by the states; that employers of illegal aliens be punished; that immigration levels not be increased; and that comprehensive land-use planning be implemented. The President rejected the recommendations of his population commission, especially those which would legalize abortions and make contraceptive information available to minors. The importance of the President's position is that during the next few years the executive branch of the government will probably neither propose nor support measures encouraging population stabilization.

5.9 CONSEQUENCES OF POPULATION DENSITY

(Problems 19, 24)

As the population density in a given area increases, each person's share of the available supplies of land, water, fuels, wood, metals, and other resources must necessarily decrease. In the past, people in many parts of the world have raised their standard of living despite a rising population by using the available resources with increased efficiency. However, there are eventual limits to population growth. Since we cannot predict the future advances of technology accurately, we cannot predict the maximum possible human population. The total world food and energy supplies will be discussed in more detail in Chapters 6 and 7.

Before humans are eliminated by death through starvation and thirst, it is certain that the quality of life on earth will change. Many forests and wild places will disappear and be replaced by cities and indoor environments. Some individuals will welcome the change, some will be adversely affected, but what will happen to society as a whole as populations continue to increase?

Violence, disunity, political upheavals, and personal unhappiness are attributed by some to population density. In a series of experiments with strains of rodents, John Calhoun studied the effects of extreme crowding. He constructed cells supplied with enough food and water for many more rodents than the space would normally hold. A few animals were placed in each cell and allowed to breed. The population and the density grew quickly, and the animals began to act bizarrely. The females lost their ability to build proper nests or to care for their infants. Some of the males became sexually aggressive; most retreated from communication with others. In short, the normal processes of socialization were destroyed. Other literature presents evidence of decreased fertility and strange behavior in many animal species under conditions of overcrowding. However, we do not know in what way conclusions from animal experiments apply to human populations.

See John B. Calhoun in Scientific American, *206*:139, February, 1962, and "Environment and Society in Transition," *Annals of the New York Academy of Sciences*, June, 1971.

Available data do not support the hypothesis that high population density of a nation is necessarily associated with social upheavals, violence, or poverty. Research into the nature of the relationship between population density and social problems has yielded a morass of conflicting conclusions. No consistent patterns emerge when national population densities are defined as population size divided by total national area. Such densities are grossly misleading, for they do not measure population densities in populated areas. For example, the Netherlands, one of the most densely populated areas of the world, had 319 persons per square kilometer in 1970. By contrast, the population density in India was only 168/km² and in Algeria, only 6/km². Since, however, nearly all of the Netherlands is inhabitable, while much of India is jungle and most of Algeria is desert, the fact that the density of the Netherlands is so high and its society so stable does not by itself disprove a hypothesis that high population densities are socially detrimental. These examples are not isolated instances, for the 62 per cent of the earth's surface that is semi-arid, taiga, tropical jungle, arctic, tundra, or desert holds only one per cent of its population.

POPULATION DENSITY (1971)

REGION	*PERSONS/km²*
Africa	12
North America	11
Latin America	14
Asia	76
Europe	94
Oceania	2
U.S.S.R.	11
World average	27

The relation between population density on arable land and social problems is also confusing.

TABLE 5.7 MURDER RATES AND DENSITY IN THE TEN LARGEST AMERICAN CITIES (1970)*

CITY	*PERSONS PER SQ. MILE*	*MURDERS PER 100,000*
New York	26,343	14.2
San Francisco	15,764	15.5
Philadelphia	15,164	18.1
Chicago	15,126	24.1
Washington, D.C.	12,321	29.2
Baltimore	11,568	25.5
Detroit	10,953	32.7
Cleveland	9,893	36.1
Milwaukee	7,548	7.0
Los Angeles	6,073	14.0

*Data from *Statistical Abstract of the U.S.* (1972).

Japan, which supports 1700 persons per square kilometer of arable land, is an example of a very densely populated country which maintains a prosperous and relatively crime-free society.

Some studies of population density contend that the violence-producing effect of density is most apparent in urban areas, such as cities in the United States. However, the more densely populated cities are not necessarily those with higher crime rates. For example, New York City has a lower violent crime rate than Los Angeles, even though New York is five times as densely populated. Crime rates are probably more fundamentally related to poverty than to density. Table 5.7 compares murder rates and density in the 15 largest American cities.

The degree to which density is an important factor in breeding antisocial behavior is still unknown. One difficulty in studying the effects of population density on humans is that spatial requirements may in part be culturally determined. Much sociological research is needed to gain an understanding of the factors causing members of different societies to feel crowded.

We do know that density is associated with forms of government and with patterns of social life. In primitive societies, for example, population density is related to the kind of animals hunted. If the prey is large and strong, many men are needed for attack. This leads to concentrated bands of hunters. An example of such a society was the bison-hunting Plains Indians. On the other hand, if the prey is small enough to be caught by a single hunter, families tend to live isolated from each other. The Australian bushman is an example of such a pattern of isolation. Of course, the type of prey is only one of many factors which determine living patterns. In more modern cultures, very rural societies are often too dispersed for economical government services. Schooling, health care, and postal service, for example, are extremely expensive to deliver in sparsely populated areas. Some societal responses to injustices, manifested in labor unions or protest movements, are rare without a population density high enough so that large groups of people are regularly in contact with each other. On the other hand, in very densely populated areas, government cannot be immediately responsive to individual needs, and the civic role of most citizens is severely limited.

Change in density is not usually a useful measure in examining effects of population growth. A country with rich soil and advanced agricultural techniques can support many more persons per unit area than an agriculturally backward land; industrial nations which can afford importation of food can sustain a very dense population. The economic and social effects of population growth are far more important than the effects attributable to density itself.

5.10 THE DEMOGRAPHIC TRANSITION

(Problem 23)

Historically, much of the current population explosion is due to a rapid decrease in infant mortality without a decrease in birth rate. In Western Europe, United States, Canada, Russia, and Japan, both birth rates and death rates have decreased rather slowly over the past century. In many countries of Asia, Africa, and Latin America, introduction of modern medicine has led to changes in mortality patterns, while birth rates have remained relatively constant. Typically, medically primitive societies maintain a population balance by having both high birth and high death rates. When infant mortality drops, birth rates remain high for a while, leading to a period of time characterized by rapid population growth arising from continued high birth rates and newly achieved low death rates. This period is called the time of **demographic transition**. So-called developed societies are characterized by both low birth and low death rates. They have completed the transition by reducing birth rates. Sometimes, changes in the population age distribution cause an increase in the crude death rate after transition, even though the age-specific death rates may be continuing to decline.

The reader should understand that declines in birth rates have not usually occurred simply because of propaganda, statements of public policy, or even the availability of birth control. Rather, small families have been the result of complicated social and economic forces. The developed countries have been lowering their death rates for about a century. Thus, they have had many years to allow fertility patterns to change and have avoided explosive rates of growth. Furthermore, the technological and economic development of these countries has led to societal patterns which encourage, at least implicitly, a reduction in the number of births. One important example of such

TABLE 5.8 SOME TYPICAL VITAL RATES BEFORE, DURING, AND AFTER DEMOGRAPHIC TRANSITION*

		CRUDE BIRTH RATE/1000	CRUDE DEATH RATE/1000
Before transition	*Very high birth and death rates*		
	Afghanistan (1965–70)	50.5	26.5
	Angola (1965–70)	50.1	30.2
Transition period	*High birth and death rates*		
	Bolivia (1965–70)	44.0	19.1
	Indonesia (1965–70)	48.3	19.4
	High birth rates, moderate death rates		
	India (1965–70)	42.8	16.7
	Morocco (1965–70)	49.5	16.5
	Moderate birth rate, low death rate		
	Argentina (1967)	20.7	8.7
	Spain (1970)	19.6	8.5
	United States (1970)	18.2	9.4
After transition	*Low birth rate, low death rate*		
	Hungary (1970)	14.7	11.7
	Sweden (1970)	13.6	9.9

*Source: *U.N. Demographic Yearbook* (1970).

a social pattern is an economic structure encouraging women to work.

The underdeveloped countries remain today in a transition period. With the assistance of aid programs, missionaries, and others, these countries have been able to drop their death rates significantly in a few decades. Unfortunately, they have had only limited time and meager resources with which to develop the technological and social patterns which, in other countries, have preceded or coincided with a drop in birth rates. This lack of preparation is one reason why their current growth rates are generally viewed with alarm. Table 5.8 shows some typical birth and death rates for countries at various stages of demographic development.

5.11 WORLDWIDE DEMOGRAPHIC TRENDS

(Problem 30)

Because of the marked differences in mortality and fertility patterns between the developed and the underdeveloped nations, and because of the lack of reliable demographic data for much of the underdeveloped world, future projections of world population size are much more difficult to construct than are projections of the growth of an individual country. The developed nations are characterized by low mortality and fertility. Since these patterns are likely to persist for some years into the future, population growth is expected to be slow during the next century. The total population of the developed nations was 1.1 billion persons in 1970. If the TFR were to drop to 2.1 by 1975, the population would stabilize at 1.3 billion by 2050. Such a projection represents a minimum likely population. A maximum likely population is projected by assuming that the TFR will stabilize at 3.3 by 2000. In such a case, the total population of the developed nations would be 1.8 billion persons by 2050. Since the range between the likely maximum and likely minimum is small, if we assume no nuclear war or other catastrophe, the range of estimated population size, 1.3 to 1.8 billion persons, is a useful guide.

For the underdeveloped nations, the patterns are clouded. In the highly unlikely event that the TFR were to drop to replacement level by 1975, the population of the underdeveloped countries would rise sharply from 2.5 billion in 1970 to 4.0 billion in 2050 and continue to rise rapidly thereafter. Such striking population growth, even under conditions of replacement fertility, is a legacy of the second stage of the demographic transition, for the recent decades of low mortality and high fertility have left the underdeveloped countries with large numbers of young people. Barring catastrophe, we can regard 4.0 billion persons as the minimum number of inhabitants of the underdeveloped world in 2050. If however, the underdeveloped nations were to take as much time to complete the demographic transition as did Europe, the TFR will slowly drop until it reaches 3.3 in 2000. Such a condition would lead to a population of over 11 billion persons by 2050. The range between four billion and 11 billion is so large that accurate prediction is unrealistic. When we consider, moreover, that current data for the People's Republic of China and for continental Africa are lacking, the value of predictions of population size in the underdeveloped world is even more suspect.

We can conclude that, barring widespread famine or war, the proportion of people living in what are now called underdeveloped nations will rise rapidly

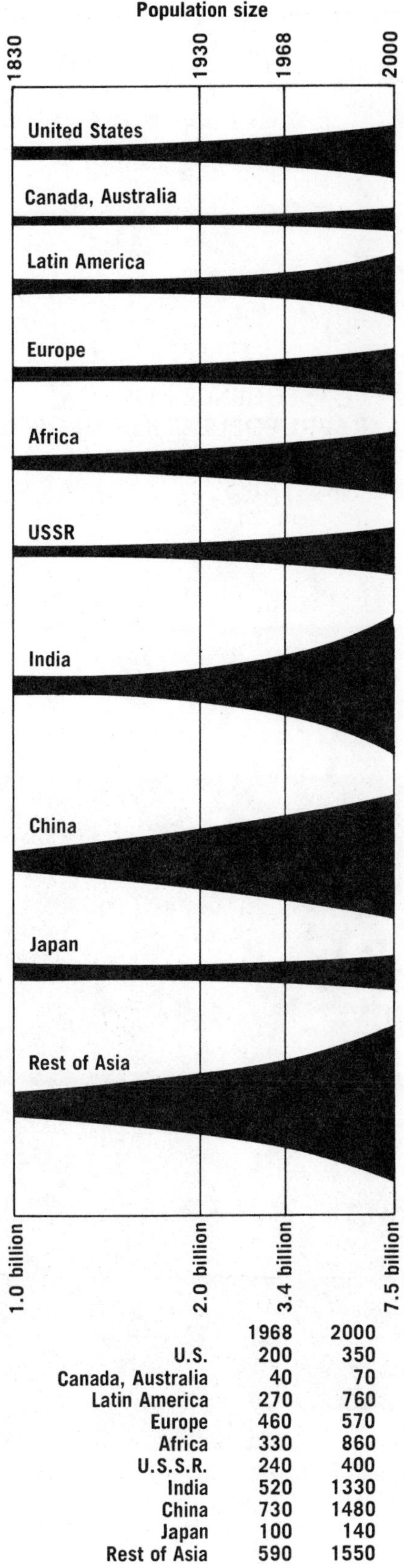

	1968	2000
U.S.	200	350
Canada, Australia	40	70
Latin America	270	760
Europe	460	570
Africa	330	860
U.S.S.R.	240	400
India	520	1330
China	730	1480
Japan	100	140
Rest of Asia	590	1550

FIGURE 5.23 Worldwide distribution of population: estimates and projections from 1830 to 2000.

during the next century, unless massive emigration occurs. Today, about 70 per cent of the world's population lives in underdeveloped nations. The figure will certainly rise to at least 80 per cent by 2050, and will continue to increase thereafter. This shifting balance of population will probably bring more abject poverty in many areas of the world, for the economic corollary to our demographic predictions is that the poor nations will become poorer as they become much more populous.

5.12 CONSEQUENCES OF RAPID POPULATION GROWTH IN UNDERDEVELOPED COUNTRIES

(Problem 25)

Rapid population growth, especially when it leads to increasing urbanization and rising unemployment, may be detrimental to the economic and social aspirations of an underdeveloped country. An underdeveloped country is often defined, arbitrarily, as one in which the average per capita annual income is under $600. Birth rates in such countries are well above 30 per thousand, and often more than 40 per thousand. If an underdeveloped country is growing rapidly, much of the economic effort which could have been expended in development to improve the per capita income is instead necessarily diverted to efforts for increasing agricultural production. Many schools are needed, because a very large proportion of the population is under age 15. The development of better schools, improved health care, greater industrialization, more modern housing, and pollution control can only be undertaken after an initial effort is made to assure minimal standards of nutrition. Although an adequate diet is often not attained, rapid population growth precludes raising the per capita caloric or protein content of the diet.

Houses in Brazil. (Courtesy of the United Nations.)

Problems caused by rapid urbanization are especially acute in the underdeveloped world. In the West, urbanization has often been associated with economic, industrial, artistic, and intellectual productivity. Migration in the West from rural areas to the cities has, in the past, been accompanied by fundamental changes in styles of life. Cities have been wealthier and more modern than the rural areas and have been characterized by a considerable occupational differentiation. By contrast, the patterns of urbanization in many underdeveloped countries are not associated with productivity. The modern Indian,

African, and South American city is often culturally rural. Migrants to the city may not be the upwardly mobile, but rather the farmers whose crops have failed. Cities in underdeveloped nations are often poor and agricultural; the literacy rate may be very low; housing is often shoddy; food is scarce. Dakarta, India, for example, is a coastal city which has five times more people now than it did before World War II. For food, it depends upon imported grain. Since India has not dispensed food efficiently in the past, the hungry flock to the seacoast cities to be near sources of food. To transform these poor cities to modern ones would require assured supplies of food, the introduction of profitable industry, and population control.

The economic success of some developing nations, even at great odds, is truly remarkable. An example of such success is Ghana. If there were no external investment, and if fertility were constant, the per capita income would probably fall eight per cent between 1960 and 1985. If, however, the fertility were to decline only one per cent annually, per capita income would rise nine per cent; a two per cent decline in fertility should lead to a 24 per cent rise in income. These figures mean that the per capita growth in productivity is very high. In most underdeveloped countries, the rapid rise in productivity rates has not been sufficient to prevent per capita incomes from diverging farther and farther from incomes in the developed nations. The race between population growth and increased agricultural productivity is discussed in Chapter 7.

As well as having large-scale societal and economic effects, rapid population growth has implications for individual family units. Since fertility patterns change slowly, the sudden recent decline in death rates has meant larger family sizes in many underdeveloped regions. In the Matlab region of Pakistan, for instance, the average woman of age 45 has already borne seven or eight children of whom two or three have died. If the current age-specific death rates had prevailed for the past 30 years, only one or two would have died. In other words, the average 45-year-old woman in Matlab has five living children. If the current generation of mothers were to bear babies at the same rate as their mothers did, an average woman of the current generation at 45 years would have six living children. The consequences within the family are multiple. Maternal and child health, nutrition, education, and even the intelligence of the child may all be adversely affected by large family sizes.

5.13 ARGUMENTS AGAINST STABILIZATION OF POPULATION

(Problem 31)

The governments of several developed nations have expressed alarm when population growth has fallen to or below the replacement level. The fears have centered on possible adverse economic effects and the problems associated with an aging population. In 1970, responding to several years of population growth below replacement level, the Japanese government instituted a program to encourage births.

Another example is Rumania. In 1957, when the birth rate was 24.2 births per thousand, the government officially approved induced abortions for fetuses under 12 weeks; the fee was the equivalent of less than $3.00. In 1958 there were 29 legal abortions for every 100 live births; by 1966 there were 400 abortions per 100 live births. The birth rate in 1966 was 14.3. At that time, abortion was the only important method of birth control. On October 1, 1966 the liberal abortion laws were repealed. Nine months later, in July 1968, the birth rate had reached 38.7 per thousand. (By 1970, it had declined to 21.1; presumably contraceptive methods were being used.)

This figure, though officially cited, has been attacked as unrealistically high. The figure for Hungary, which is demographically similar to Rumania, in that year was 136 abortions for every 100 live births.

Clearly, the poorer the nation or ethnic group, the more it can gain by population stabilization. A wealthy group can absorb large population increases without a large or significant decrease in standard of living. A poor nation, on the other hand, has great difficulty maintaining subsistence levels of support for all its citizens, and increasing size leads to further economic pressures. Yet many spokesmen for the underdeveloped nations, and several leaders of American blacks, view the popular call for population stabilization as a fundamentally racist or imperialist position. They claim that wealthy societies urge population limitation as a device for maintaining their own dominance. For example, if black reproductive rates were to continue, and white rates were to remain unchanged, or decrease, the proportion of blacks in the U.S. would increase from the current 12 per cent. Some American black spokesmen, feeling that increased per cent representation will have political

See, for example, the testimony of Rev. Jesse Jackson as quoted in *Population and the American Future*, pp. 72–73.

value, suspect that the current emphasis on population planning is a white ruse designed to prevent increased black influence. Similar arguments are expressed by some representatives of poor countries. The accusation is difficult to answer, for although population stabilization will certainly tend to maintain existing political inequities, allowing continued population growth will lead to all the problems mentioned throughout this chapter.

5.14 METHODS OF POPULATION CONTROL

(Problems 26, 27, 32)

Social trends leading to decreased fertility, such as the increasing role of the woman outside the home, do not in themselves prevent births. Instead, such patterns lead to the desire to control actively the number of births. Three important approaches to birth prevention are available—**contraception, sterilization,** and **induced abortion.** Sterilization procedures render a person incapable of fathering or conceiving a child. Examples of contraceptives, devices or techniques which prevent conception, are the condom, the intrauterine device (IUD), contraceptive pills, the diaphragm, and spermicidal agents. Induced abortion is the artificial termination of pregnancy. Sterilization and induced abortion are, by their very nature, effective methods of population control, while the effectiveness of various contraceptive methods is a function of both the theoretical efficacy of the method and the motivation of those who practice it. For example, the rapid decline in birth rate in nineteenth-century Europe has been credited in large part to the widespread use of *coitus interruptus,* intercourse with withdrawal before ejaculation. The historical success of this error-prone method has been attributed to the strong motivation of its users. By contrast, vaginal foam should theoretically be a highly effective method of birth control, yet in studies of contraceptives, many babies are born to women using foam. A likely explanation for its relative ineffectiveness is that highly motivated women chose other methods of conception control, and therefore, the less highly motivated women relying on foam use it improperly or often neglect to use it at all.

In addition to modern methods of contraception, many traditional practices tend to limit births. Perhaps the most important is breast-feeding, for fertility declines during lactation.

Most countries have laws concerning the teaching, advertising, dissemination, and use of various birth-control methods. The laws are often contradictory and widely disobeyed. In many countries which forbid birth-control methods, the well-to-do members of the society are able to circumvent the law by leaving their country or state for an abortion or sterilization, by purchasing smuggled contraceptives, or by going to a private physician who prescribes some contraceptive method as "therapy" or as a "disease preventive." The poor in any country have fewer opportunities to violate population-control laws. Until the recent Supreme Court decision allowing abortions throughout the United States, an unmarried pregnant American girl of moderate means could have gone to New York to have a safe, legal abortion, while many of her poor counterparts who wanted an abortion were forced to choose an illegal abortionist, where the operating conditions were often quite unsanitary. Out of every 100,000 American women who are pregnant and do not have an abortion, about 20 die of direct complications of pregnancy. Of every 100,000 who have a legal abortion in a hospital, only three die. By contrast, there are 100 deaths for every 100,000 women who undergo an illegal abortion; moreover, many of the survivors are rendered infertile.

Similar problems arise in other countries. In Japan, the laws are contradictory. The criminal code forbids abortion, but the Eugenic Protection Law permits it. In Eastern Europe, abortion is often prohibited except under special circumstances. In Bulgaria, for instance, women with at least three children and women over 45 are allowed abortions, others not necessarily. Again, there are many illegal abortions. Many countries allow abortion for socio-economic or humanitarian reasons, such as after rape.

Laws concerning contraceptives are varied. They are forbidden in Ireland and Spain but are required to be stocked by pharmacies in Sweden, and by pharmacies, department stores, and rural cooperatives in the People's Republic of China. Until July 1970, when the law was declared unconstitutional by the U.S. Court of Appeals of the First Circuit, Massachusetts forbade the sale of contraceptives and the dissemination of information about birth control to unmarried women. In Poland, information about contraception is mandatory before an abortion; in Denmark it is required after childbirth and abortions. In some North African countries, the manufacture and importation of contraceptive devices are illegal.

Sterilization laws are perhaps the most ambiguous. In the United States, 26 states permit compulsory sterilization to be performed on mentally infirm patients maintained at state institutions. Only five states specifically allow voluntary sterilization for therapeutic or socio-economic reasons. In the many states without any law concerning sterilization, the resulting legal ambiguity causes many doctors to be reluctant to perform sterilizations.

In 1968, the United Nations Conference on Human Rights unanimously proclaimed that family planning is a basic human right and declared that governments should abolish laws conflicting with the implementation of such rights and adopt new laws to further such rights. The legalization of contraception, abortion, and sterilization throughout the world would help reduce population growth and curtail maternal deaths from unsanitary, illegal abortions. It should be noted, however, that this proclamation of the United Nations is a recommendation which carries no legal weight anywhere in the world.

Among people of the developed countries, the choice of various methods of birth control is largely personal. Several methods may be tried and the most satisfactory chosen. In underdeveloped countries, the problem is quite different. Usually, a team of health workers makes a decision about the type of birth-control techniques to introduce. The wrong choice can lead to a rejection of family planning. Family planners in underdeveloped countries have learned that introduction of a highly reliable method of birth control to a relatively small group of couples is often more effective in reducing births than a moderately effective method introduced to many. Methods such as diaphragms, foams, and other techniques which require careful use are often not effective in the underdeveloped countries.

The acceptability of contraceptives varies by culture. The highly successful birth-control program in Puerto Rico is due to the widespread acceptability of sterilization there. The program in Taiwan has used the IUD, whereas Japan and Eastern Europe depend on abortion for limiting population size. However, no family planning program can be successful unless the families involved are motivated to practice birth control.

5.15 THE ULTIMATE HUMAN POPULATION

We cannot know what ultimately will limit the growth of human population. Pessimists predict mass starvation, total war, irremediable destruction of drinking water, a lethal upset in the oxygen-carbon dioxide balance. They cite evidence of cannibalism in overcrowded rats, suicide in lemmings. Optimists speak of rational self-limitation. However, it is reasonable to guess that limits, either cataclysmic or rational, will not occur on a worldwide basis but rather will depend on factors within smaller geographic units. Countries or areas with decreasing birth rates may achieve self-limitation with little disruption of life and environment. In countries with rapid rates of growth, where food shortages threaten, the mechanisms that will limit population will not be pleasant ones. Various setbacks are likely to precede significant decreases in birth rate and perhaps thereby avert mass death from actual starvation. For example, there may gradually occur a general state of undernutrition, leading both to increased susceptibility to disease and to lower fertility. Contamination of drinking waters may lead to widespread epidemics of contagious diseases. Technological development, if it causes increased air pollution, will lower the general health of a population. Moreover, these factors affect other forms of life as well. For example, if pure water becomes scarce, populations of many species decrease, lowering man's food supply.

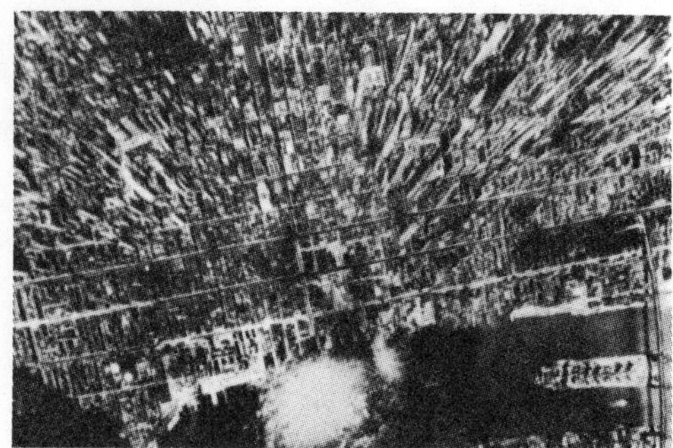

The exploding metropolis. (Courtesy of U.S. Department of Commerce. Coast and Geodetic Survey.)

The contrasts between the developed and the underdeveloped nations will undoubtedly become more marked. The most highly regarded projected growth rates predict that the less-developed regions of the world will nearly double in size by the end of the century, while the developed nations will grow to less than one and one-half times their present size. Economic and military disparities will magnify.

Traffic jam.

Predictions about man's potential growth must take into account the interrelationship of man and his environment. When we predict the amount of growth in some small period of time, it is reasonable to ignore the populations of other species. However, when we predict far into the future, we must account for other living things. We must not think of world population growing wildly until the time when there is not enough food, air, water, or room in which to stand. Rather, we would expect increasing population size to cause a gradual deterioration of levels of health and nutrition, and to lead to increases in the number of areas of the world beset by poverty and overcrowding.

PROBLEMS

1. **Population ecology.** Discuss some important factors relating to the population growth of species other than man. Contrast these factors with those important to man's population growth.
2. **Vocabulary.** Define the following terms: rate of growth; rate of change.
3. **Exponential growth.** Explain why world population size cannot grow forever.
4. **Population prediction.** Discuss some uses of predictions of population size.
5. **Growth curves.** Outline Malthus' theory of population growth. What is the distinction between arithmetic and geometric growth? Does the growth of money in a bank proceed by arithmetic or geometric progression? What about the growth in the age of a person?
6. **Growth curves.** What is a sigmoid curve? Discuss its relevance as a model for human population growth.
7. **Vocabulary.** Define demography; vital event; vital rate.
8. **Migration.** Immigration and emigration data are defined in terms of the stated intention of the migrant at the time he applies for the proper visa. Do you think such data are reliable measures of long-term migration? Define migration in terms that would be more demographically useful. Why is your definition not employed by data-collection agencies?
9. **Migration.** For which of the countries in Table 5.2 might we safely ignore migration in investigating population growth? In Japan, there were almost as many emigrants as immigrants. Can we therefore say that since net migration is low, we can ignore migration? Why or why not?
10. **Migration.** "Since there is no migration to and from earth, to predict population size we need only consider birth and death rates." Argue for or against this statement.
11. **Age-specific mortality rates.** Discuss possible explanations for the fact that the graphs of mortality rates in Figure 5.4 cross each other.
12. **Survivorship curves.** Compare the survivorship curves

for males in the U.S. and Sweden (Fig. 5.5). The probability that a person will survive *to any age* is higher for populations in Sweden than for populations in the United States. What does this information tell you about the relative average ages of men in the two countries? What about average life span? Explain.

13. **Infant mortality.** Explain the relationship between infant mortality and expectation of life.

14. **Population explosion.** Comment on the assertion that modern medicine, by prolonging the lives of elderly persons, has contributed greatly to the population explosion.

15. **Replacement value.** Under what conditions would the replacement value for the TFR be exactly 2? Can you think of a population with this value? Why is India's replacement value higher than Canada's?

16. **Family size.** What would happen to human population size if no family had more than two children? If the average family size were two children?

17. **Accuracy of data.** In Table 16 of the United Nations Demographic Yearbook (1970), the infant mortality rate in Angola for 1968 is cited as 15.9. What are the units of infant mortality rates? A footnote states, "Rates computed on number of baptisms recorded in Roman Catholic Church records." Comment on the accuracy of the quoted rate.

18. **Accuracy of data.** The U.N. cites that the crude birth rate in the Netherlands Antilles in 1968 was 23.0 and that the infant mortality rate was 20.2. These data exclude most live-born infants dying before registration of birth. How do these definitions affect international comparisons?

19. **Population density.** Population density is usually defined in terms of people per unit of land area. Density is often cited as an index of poverty. How useful is such a definition in a city with many luxury high-rise buildings, and large areas of decaying tenements? What alternative definitions would you suggest?

20. **Population distribution.** Explain why some declining animal populations have age distributions with large bases. (Consider populations where there is heavy predation of adults.)

21. **Population distribution.** How are age-sex distribution curves useful in predicting future growth? What are some of their limitations?

22. **Population distribution.** How would you expect each of the following to affect population growth? Consider which age groups are most likely to be affected by each event and the interrelationships between the event and the population change: (a) famine; (b) war; (c)

lowering of marital age; (d) development of an effective method of birth control; (e) outbreak of a cholera epidemic; (f) severe and chronic air pollution; (g) lowering of infant mortality; (h) institution of a social security system; (i) economic depression; (j) economic boom; (k) institution of child labor laws; (l) expansion of employment opportunities for women.

23. **Demographic transition.** Describe the demographic transition. In Bolivia (1964–66), the average food consumption included 1,760 Calories per person per day. In India (1968–69) the comparable rate was 1,940 Cal/day; in Spain (1970) 2,750 Cal/day; in Hungary (1969) 3,180 Cal/day. How do these caloric values relate to the stages of demographic transition? (See Table 5.8.) Consider social and economic factors as well as those relating to population structures. (For example, what segments of the population eat most? Least? How would you explain Argentina's average of 3,170 Cal/day? Sweden's 2,750 Cal/day?

24. **Urbanization.** New York, Chicago, and London have been called cities ecologically dependent upon the railway and the steamship; Los Angeles and the Boston-New York-Washington megalopolis are said to be ecological consequences of the automobile. Discuss.

25. **Rapid population growth.** In times of rapid population growth from excesses of births over deaths, what types of professions and services must become increasingly available very rapidly?

26. **Birth control.** The following measures have been proposed to measure contraceptive effectiveness: (a) number of failures (pregnancies) to 1,000 women on contraceptives; (b) number of failures per year to 1,000 women on contraceptives; (c) birth rate to population on specific contraceptives; (d) completed family size to women on specific contraceptives. Discuss their relative merits.

27. **Birth control.** What factors are most important in introducing birth-control methods to an underdeveloped nation?

28. **Trends in the United States.** What factors will tend to raise American birth rates? lower them?

29. **Social policies.** In the United States, families are allowed an income-tax deduction for each child. In most developed countries, each family is allotted an annual grant for each child. How do you think these tax laws affect family size?

30. **Social policy.** If you had the responsibility of discouraging population growth, would you consider curtailing income tax deductions if a family has more than four children? curtailing health benefits? Whom would such policies harm?

31. **Social policy.** What arguments and programs might allay the fears that population limitation programs are designed to preserve economic inequalities?

32. **Population control.** Joan P. Mencher, in an article entitled, "Socioeconomic constraints to development: The case of south India" (Transactions of the New York Academy of Sciences, 1973, pp. 155–167), points out that poor people are often treated shabbily in medical clinics. In south India, she says, "To have a baby does not require contact with hospital people, but to *avoid* having a baby requires contact with maternity assistants, doctors, etc., all of whom tend to treat the poor and low-caste people as 'animals.'" How do you think such treatment affects population control among the poor of south India?

The following problems require arithmetical reasoning and computations.

33. **Infant mortality.** Plot infant mortality rates (Table 5.3) against male or female expectation of life (Table 5.4) for Iceland, Sweden, the Netherlands, Japan, Denmark, France, New Zealand, East Germany, Canada, the United States, Czechoslovakia, West Germany, Israel, the Soviet Union, Poland, Yugoslavia, Mexico, and Guatemala. Describe and explain the relationships.

34. **Family size.** This problem investigates family size as it relates to population growth: Consider a population where 10 per cent of all women are childless. Roughly 10 per cent have only one child. Assume that all others have exactly two children and there there is no migration.

(a) How many children will a cohort of one thousand women produce?

(b) Suppose that 90 per cent of babies reach reproductive age. How many children reaching reproductive age will the cohort produce?

(c) How many children who reach reproductive age must a cohort of 1000 women produce in order to exactly replace itself? (Assume equal numbers of boy and girl babies.)

(d) How many children must be born to the cohort in order for it to replace itself, that is, to produce an eventual zero population growth?

(e) What must be the average number of babies produced per woman in order to achieve an eventual zero population growth?

(f) What must be the average number of babies produced by the 80 per cent of women who have more than one baby in order to achieve an eventual zero population growth?

(g) If women were to produce the average number of babies computed in part (e), why would a zero population growth not necessarily be achieved immediately?

35. **Population distribution.** In this problem we shall study vital rates and changes in population distribution. Let us consider a population on January 1, 1970 with the age distribution and vital rates shown at right. Assume there is no migration.

(a) What is the total population represented by the table?

(b) What is the overall (crude) birth rate per thousand per year?

(c) What is the crude death rate?

(d) What is the rate of natural increase, i.e., the difference in per cent, between crude birth and crude death rate?

(e) Suppose there are 3,000,000 14-year-olds, 1,000,000 44-year-olds, and 500,000 64-year-olds. Assume the death rates are the same for each age class. (For example, the mortality rate for 20-year-olds and for 44-year-olds is 10 deaths per thousand per year.) Finally, assume that all births occur on January 1 and all deaths occur on December 31. Show that the age distribution on January 1, 1971 is as shown at right.

(f) What is the total population on January 1, 1971?

(g) What has been the rate of growth during 1970? How is this answer related to your answer to part (d)? Explain.

(h) How has the population distribution curve changed shape during 1970? Discuss the implications of the decline in 1- to 14-year-olds. Which age group had the largest per cent of increase?

(i) In order to compute a population distribution for 1971, we made many simplifying assumptions in part (e). What were they? What would be the effects on your conclusions had more realistic assumptions been made?

(j) If there had been migration, how would the overall rate of population growth compare with the rate of natural increase, computed in part (d)?

Age	Population, both sexes, in millions	Death rate/1000	Birth rate/1000
<1	4	150	0
1–14	40	15	0
15–44	44	10	100
45–64	10	20	0
65+	2	100	0

Population Distribution: Jan. 1, 1971

Age	Population, both sexes in millions
<1	4.400
1–14	39.845
15–44	45.525
45–64	10.300
65+	2.290

36. **Stable population.** If a given combination of birth and death rates were to remain in effect for many years, and if there were no migration, a characteristic percentage composition would eventually develop. It would remain unchanged indefinitely so long as the vital rates remained unchanged. This population is the so-called **stable population.** The crude birth and death rates that would be computed if stability were achieved are called the **intrinsic birth rate** and **intrinsic death rate.** Their difference is the **intrinsic reproduction rate.** For Colombia, Hungary, and the United States, (a) compute all reproductive rates. (b) Which crude rates represent growing countries? (c) Which countries will continue to grow if the age-specific birth and death rates were to remain unchanged? (d) Discuss reasons for the difference between the crude and the intrinsic rates. (e) Why is the intrinsic death rate for Hungary so much higher than the crude death rate? (f) Discuss the

Vital Rates per Thousand (Females)

		Crude Rates		Intrinsic Rates	
Country	Year	Birth	Death	Birth	Death
Hungary	1965	12.3	10.0	10.5	17.6
United	1935	17.2	10.0	13.4	18.1
States	1965	18.8	8.0	21.3	8.6
Colombia	1964	36.9	9.4	38.4	10.1

demographic differences between the United States rates for 1935 and 1965.

37. **Survivorship curves.** Draw survivorship curves for each of the following situations: (a) When oysters first hatch, the mortality is extremely high, and predators consume almost 99 per cent of the young. Once an oyster has reached the adult stage it is relatively safe from predators and has a high probability of surviving to old age. (b) The American robin has a fairly constant death rate during its entire life span. (c) Newly hatched fruit flies placed in culture bottles without food survive for only about 70 hours. Very few individuals die at birth; almost all survive for nearly 70 hours, and then mortality is so high that the entire population dies within a few hours.

BIBLIOGRAPHY

This chapter has introduced demographic techniques for analyzing population growth. There are several valuable texts available for those interested in further study of demography.

Two excellent introductory texts requiring no calculus are:

Peter R. Cox: *Demography.* 4th Ed., Cambridge, England, Cambridge University Press, 1970. 469 pp.

Donald J. Bogue: *Principles of Demography.* New York, John Wiley & Sons, 1969. 899 pp.

More mathematical introductions are:

Mortimer Spiegelman: *Introduction to Demography,* Rev. Ed. Cambridge, Mass., Harvard University Press, 1968. 514 pp. (Spiegelman includes an extremely large bibliography covering a wide range of topics related to population size, control, measurement, and so forth.)

Nathan Keyfitz: *Introduction to the Mathematics of Population.* Reading, Mass., Addison-Wesley Publishing Co., 1968. 450 pp. (This highly technical and mathematical text is especially careful in its presentation of interrelationships among various measures of population composition and vital rates.)

Sociological factors, as we have noted, are of crucial importance to the study of population growth. A useful introductory text which combines sociology and demography is:

William Petersen: *Population.* 2nd Ed. New York, Macmillan Co., 1969. 735 pp. (Petersen includes a fine annotated bibliography at the end of each chapter.)

A more advanced sociological discussion is presented by:

James M. Beshers: *Population Processes in Social Systems.* New York, Macmillan Co., The Free Press, 1967. 207 pp. (This book is useful for learning the interrelationships between social systems and patterns of demographic transition, fertility, migration, and mortality.)

Several volumes of collected papers afford most interesting reading in many areas of importance to the student of population. A fascinating collection of essays is found in:

Stuart Mudd, ed.: *The Population Crisis and the Use of World Resources.* The Hague, Dr. W. Junk, Publishers, 1964. 562 pp.

Another recommended reader is:

Garrett Hardin, ed.: *Population, Evolution, and Birth Control.* 2nd Ed. San Francisco, W. H. Freeman & Co., 1969. 386 pp.

Several books sound a tocsin for our crowded planet. One of the most popular of these is:

Paul R. Ehrlich. *The Population Bomb.* New York, Ballantine Books, 1968. 201 pp. (Ehrlich includes a bibliography of similar discussions.)

On the other hand, there is an important argument for encouraging moderate population growth expressed in a very provocative work:

Alfred Sauvy: *General Theory of Population.* (Translated by Christophe Compos.) New York, Basic Books, 1969. 550 pp.

Arguments pointing to an implicit elitist attitude among people advocating birth control are cogently presented in:

Richard Neuhaus: *In Defense of People.* New York, The Macmillan Co., 1971. 315 pp.

For the student interested in data sources, two works are highly recommended. The most useful and complete source of world population data is the *United Nations Demographic Yearbook,* published annually since 1948. For many nations and areas of the world the *Yearbook* includes the most recent available information on population sizes, vital rates, and many more specialized demographic statistics.

Another interesting source of data is:

Nathan Keyfitz and Wilhelm Flieger: *World Population: An Analysis of Vital Data.* Chicago, University of Chicago Press, 1968. 672 pp. (The authors summarize data for several countries, extrapolate various demographic measures into the future, and perform many calculations useful to the student of population.)

For up-to-date trends in population studies, refer to *Studies in Family Planning,* a monthly bulletin published by *The Population Council,* New York City.

Finally, two recent reports of commissions are especially valuable as summaries of current trends in population. The first is *Population and the American Future* (1972), the Report of the Presidential Commission on Population Growth and the American Future, and is available through the U.S. Superintendent of Documents. The second is a two-volume work entitled *Rapid Population Growth,* published for the National Academy of Sciences by Johns Hopkins Press, Baltimore, Maryland, 1971. (Vol. 1, 105 pp.; Vol. 2, 690 pp.)

6

ENERGY: RESOURCES, CONSUMPTION, AND POLLUTION

6.1 ENERGY CONSUMPTION

(Problems 40, 41)

In a natural ecosystem, radiant energy that is received from the sun and trapped in the form of potential energy in plant tissues flows through the food web which consists of various animals and decay organisms, until it is degraded completely into heat and radiated back into space. We showed in Chapter 1 how the system depends upon a continuous inflow of energy from the sun. Moreover, the total amount of sunlight puts a limit on the total metabolism of a system, for the biotic community cannot utilize more energy than it receives. The ecosystems that have existed in the past evolved to survive within these constraints and were very long-lasting.

It is clear that technological man no longer lives within this ancient energy flow pattern. Before the advent of fire, our ancestors needed only 2000 kcals of energy per day per person. The energy used was in the form of food. Later, people domesticated animals, engaged in agriculture, and used fuel for cooking and heating. Per capita energy requirements rose by a fac-

Remember, 1 kcal = 1 kilocalorie = 1000 calories = 1 Calorie (Capital C). The energy values in food tables are given in Calories.

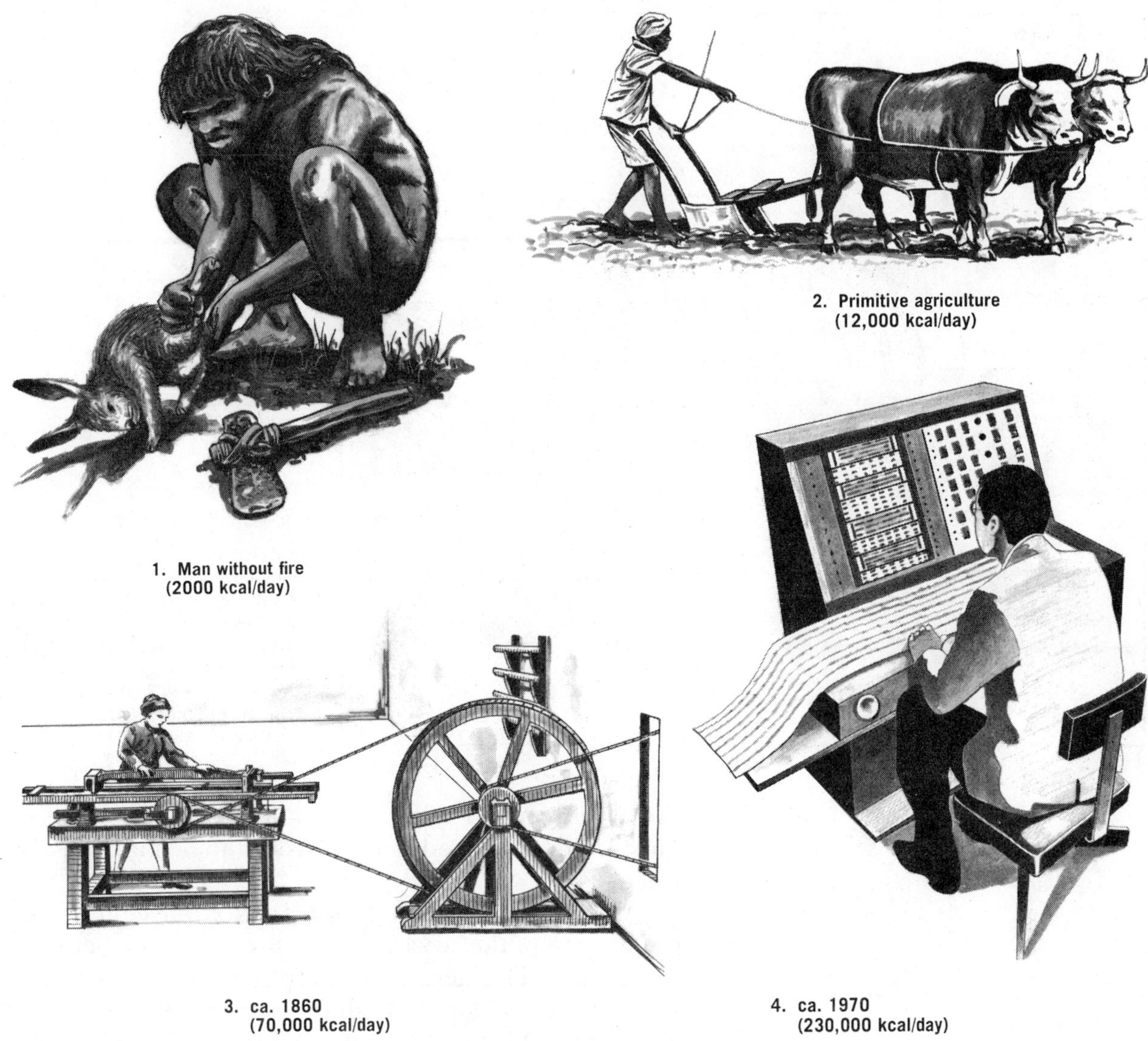

1. Man without fire (2000 kcal/day)

2. Primitive agriculture (12,000 kcal/day)

3. ca. 1860 (70,000 kcal/day)

4. ca. 1970 (230,000 kcal/day)

Energy consumption by man.

tor of roughly six, to 12,000 kcals per day. By 1860, small amounts of coal were being mined, heat engines had been invented, and a resident of London used about 70,000 kcals per day. In Western Europe at this time, the total population and the per capita energy requirements were so high that man needed more energy than could be simultaneously replenished by the sun. He began to use reserves of energy stored as fossil fuels. Such a situation is inherently unstable, for consumption cannot be forever higher than production. In the United States in 1970, man's per capita utilization was 230,000 kcals per day, a rate which greatly accelerates the exhaustion of fossil fuels. This prodigious rate is unique in the history of

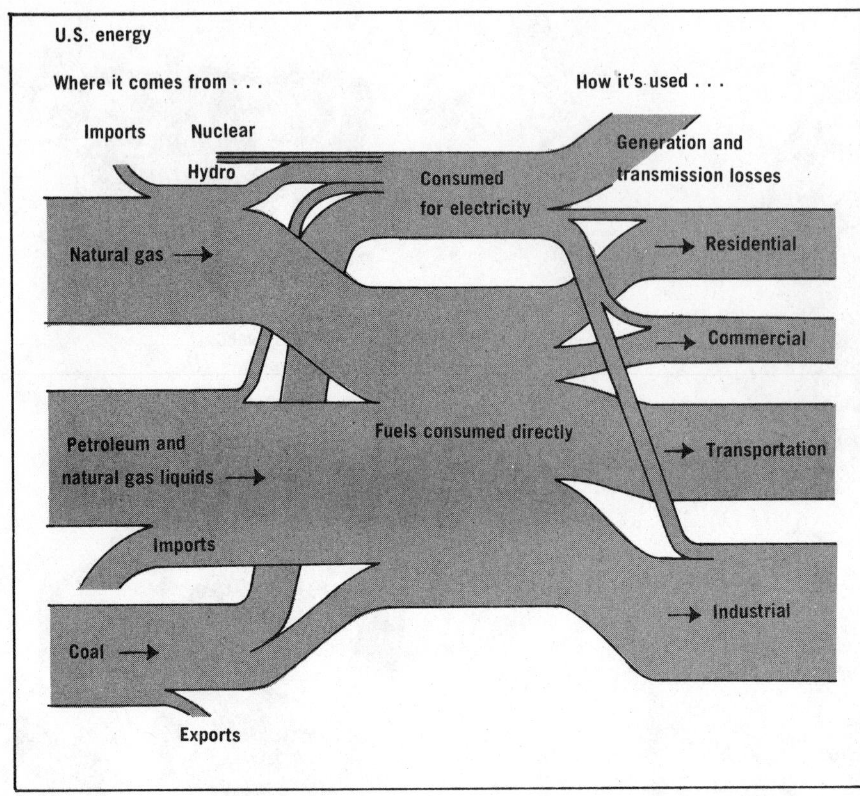

NOTE: Based on graphical scheme by Earl Cook, Texas A&M University. Relative scale is based on 1972 study by Office of Science and Technology of energy consumption in 1968 in Btu's. (From *Chemical & Engineering News,* November 13, 1972, p. 21.)

the world. At no other time, and in no other place, have people utilized energy faster than Americans do today.

The United States, home for six per cent of the population of the world, is responsible for 35 per cent of the world's energy consumption. (See Table 6.1.) What does it mean to say that the average U.S. citizen uses 230,000 kcals of energy daily? Recall from Chap-

TABLE 6.1 ENERGY CONSUMPTION FOR SELECTED NATIONS*

	PER CAPITA DAILY ENERGY CONSUMPTION IN THOUSANDS OF KCALS	*PROPORTION OF WORLD'S ENERGY CONSUMPTION*	*PROPORTION OF WORLD'S POPULATION*
United States	230	35%	6%
Canada	165	3%	0.6%
United Kingdom	145	6%	1.5%
Germany	110	5%	1.5%
U.S.S.R.	85	16%	7%
Japan	40	3%	3%
Mexico	30	1%	1.3%
Brazil	15	1%	3%
India	6	2%	15%

*Data from *Scientific American,* September 1971, p. 142 and *U.N. Demographic Yearbook,* 1970.

ter 1 that energy isn't a material, like steel, that you can hold in your hand, but rather it is the ability to do work or to heat a body above the temperature of its surroundings. (See Table 6.2 for units of energy.) The calorie was originally defined as a unit of heat. Translated into American units, 230,000 kcals is the energy needed to heat 10,000 gallons of water approximately 11 Fahrenheit degrees. But technology is primarily concerned with doing work, not heating water. In these terms 230,000 kcals is also equal to the work of lifting 710,000,000 pounds one foot. Since people are more versatile in their daily activities, the 230,000 kcals of energy are therefore spread out in a variety of ways: a man heats his home, drives his car, bears responsibility for the fuel used by the trains, trucks, and tractors that serve him; he also burns lights and runs electric appliances; indirectly he uses the energy consumed by the industries that supply him with goods. This utilization therefore seems less personal than the 2000 daily kilocalories of food for primitive people, or even the 12,000 kcal rate for early agriculturalists. In fact, industry consumes about 41 per cent of the total energy used in the United States. Furthermore, 230,000 kcals is the *average* utilization per person. The poor in America consume only a fraction of this amount daily; the wealthy consume much more.

6.2 ENERGY AND POWER

(Problems 1, 2, 3, 4)

We must understand the relationship between energy and power. Briefly, energy is heat or work, while **power** is the amount of heat or work delivered

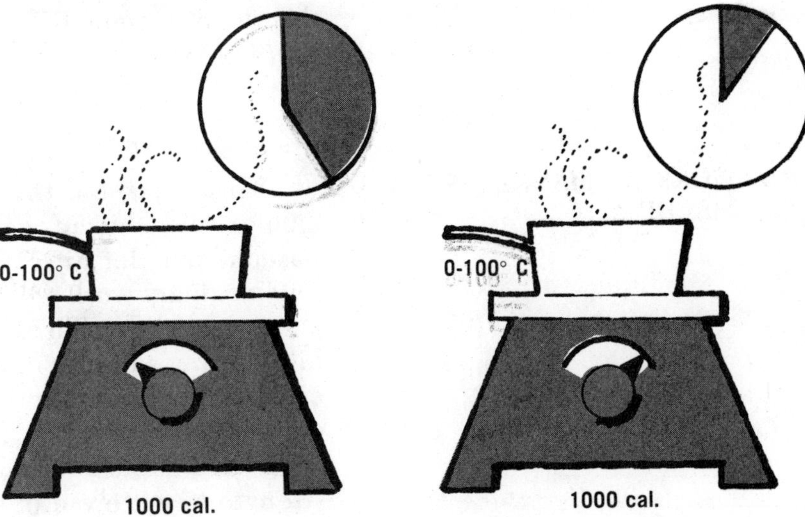

Equal energies at low power (left) and high power (right).

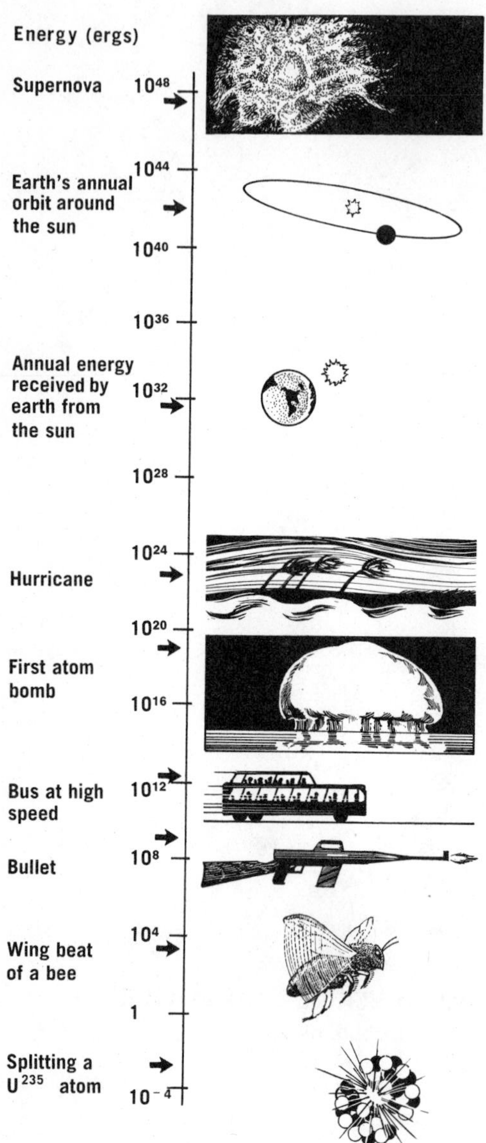

in a given time interval. For example, 1000 calories of energy are needed to heat 10 grams of water from 0°C (freezing) to 100°C (boiling). The correspondence between energy and the temperature rise of a quantity of water is an inherent property of water and is independent of how we do the operation. We could put the water on a hot plate, or rub two sticks together under the water, or spill the water down a long pipe and allow friction to heat it, or do any other conceivable operation involving heat or work. The energy required to heat 10 grams of water from freezing to boiling would always be 1000 calories. If we used a hot plate, we could choose to turn the switch to "low" and wait a while, or turn it to "high," and accelerate the process. When turned to "low," the hot plate uses small quantities of energy per unit time, but more time is required to deliver the 1000 calories. In the "high" position, large quantities of energy per unit time are delivered for a short while. The total energy delivered to the water is the same, but the power delivered in the two operations is different, simply because power is defined as the energy per unit time. (See Table 6.2.)

Where does energy go after it has been used? A lump of coal turns to ash, but what becomes of its energy? This non-substance, called "the ability to do work," is elusive, hard to define, and hard to keep track of. A piece of steel may wear out with use and get bent up a bit, but at least it remains recognizable. Could we not find used energy and reuse it? Or better yet, if energy isn't really matter, could we find some for nothing? These two questions plagued scientists for a long time, and the search for the answers led to the development of the science of heat-motion, or **thermodynamics.**

6.3 WORK BY MAN, BEAST, AND MACHINE

To understand the first two Laws of Thermodynamics and the answers to the hopeful questions posed above, let us search for solutions to the following imaginary work problem:

A landslide causes a large boulder to roll downhill; it comes to rest in a position where it blocks the entrance to a cave, and a man wishes to remove the obstruction. What can he do? Primitive man might have tried to push it, roll it, or drag it. If he lacked the strength to get it out of the way, he would have been forced to search for a new cave. However, if he could

TABLE 6.2 COMMON UNITS OF ENERGY AND POWER

UNIT	MEANING
	Energy
calorie	The heat required to raise the temperature of one gram of water one degree Celsius.
kilocalorie	1000 calories.
Btu (British thermal unit)	The heat required to raise the temperature of one pound of water one degree Fahrenheit.
foot-pound	The work required to raise one pound one foot in the air.
watt-hour	The energy released when one watt of power is delivered for one hour.
erg	A fundamental metric unit, the work required to accelerate one gram of mass at a rate of one centimeter per second per second through a distance of one centimeter
	Power
horsepower	The power required to lift 550 pounds one foot in the air in one second.
watt	The power produced when ten million ergs are supplied in one second (equal to 0.0013 horsepower).
Btu/hr.	The rate of power at which one Btu will be delivered in one hour.

cooperate with his fellows, perhaps a group of men, working together, could have been able to push or roll or drag it out of the way. The situation was still further improved after people had learned to domesticate large animals and use them as beasts of burden. But a much more far-reaching development for doing work was the invention of machines. The lever, the wheel, the roller, the screw, the inclined plane, and, later, the gear and the block and tackle were all devices that could increase the force that a man or his animals could exert.

These three techniques – cooperation among people, the use of animal power, and the application of machines – enabled man to build (and sometimes to destroy) such wonders as the pyramids of Egypt, the Great Wall of China, and the cities of Athens, Rome, and Carthage.

We will now discuss other ways to get work done. But do not forget the cave man and his boulder; we shall return to him.

6.4 THE FIRST LAW OF THERMODYNAMICS (OR "YOU CAN'T WIN")

As machines became more and more effective in extending the force of living muscle, inventors saw no reason to doubt that, by providing machines that

(Problems 5, 7. We recommend especially that you try Problem 7 before you laugh at what you think was the naiveté of the attempts at "perpetual motion.")

would produce work indefinitely just because their mechanisms were so clever, continued improvement in design would eventually free man and his animals entirely from their labor. If our cave man had such a device (which we call a "perpetual motion machine of the first kind"), he could move all the boulders he wished at his leisure. It may be difficult for the modern reader to appreciate the fact that this objective seemed entirely reasonable, apparently requiring only continued progress along the lines that had already been so successful. However, all attempts failed. The failures have been so consistent that we are now convinced that the effort is hopeless. This conviction has been stated as a law of nature and is called the **First Law of Thermodynamics.**

This law can also be expressed in terms of conservation of energy by the statement "energy cannot be created or destroyed."

Don't ask for a proof of the First Law. There is none. The First Law is simply a concise statement about man's experience with energy. If it is impossible to create energy, then it is hopeless to try to invent a perpetual motion machine, and we may as well turn to some other method of doing work.

6.5 HEAT ENGINES AND THE SECOND LAW OF THERMODYNAMICS (OR "YOU CAN'T BREAK EVEN")

(Problems 6, 8)

Modern technology demands far more energy than can be supplied by men and beasts. Instead, it makes use of heat engines, which consume fuel to produce heat, and convert the heat into work.* The idea that heat could be converted into work was far from obvious. In fact, heat engines have been used successfully only during the past 200 years or so. (James Watt developed his steam engine in 1769.) The first experimental proof that energy can be converted, without gain or loss, from one form to another was supplied by James Joule in 1849. Heat and work are two forms of energy, and it is therefore possible to convert heat into work or work into heat. Thus one can create heat by rubbing two sticks together, and the heat produced is exactly equivalent in energy to the work required to rub the sticks. It was also discovered

*Of course, a person (or an ox) is also a heat engine. People consume food, which is their fuel, and convert its energy into the work of muscle contraction. Mechanical heat engines, however, can be much larger than animals, and can consume fuels such as gas, oil and coal that animals do not eat. The result is that the total amount of work done and heat produced is increased to an extent that has new effects on the environment.

that fuels contain potential heat. Thus, a pound of coal contains stored energy which can be released by combustion. But heat is also stored in substances that are not fuels – in such ordinary substances as water. Water loses energy when it turns to ice; therefore,water must *have* energy. Why not, then, use this energy to drive a machine to do work? Such a machine, although seemingly not quite so miraculous as the perpetual motion machine of the first kind, would extract energy from its surroundings (for example from the air or from the ocean) and convert it into useful work. The air or water would then be cooled by the extraction of energy from it, and could be returned to the environment. Automobiles could then run on air, and the exhaust would be cool air. A power plant located on a river would cool the river while it lighted the city. Such a machine would *not* violate the First Law, because energy would be conserved. The work would come from the energy extracted from the air or water, not from an impossibly profitable creation of energy. Such a device is called a "perpetual motion machine of the second kind"; alas, it too has never been made and never will be.

Let us return now from the impossible to heat engines that use fuel, where the situation continues to be discouraging. It was learned through experiment that the potential energy inherent in a fuel could never be completely converted into work; some was always lost to the surroundings. We say "lost" only in the sense that this energy was no longer available to do work; what it did, instead, was warm the environment. Ingenious men did try to invent heat engines that would convert *all* the energy of a fuel into work, but they always failed. It was found, instead, that a heat engine could be made to work *only* by the following two sets of processes: (a) Heat must be absorbed by the working parts from some hot source. The hot source is generally provided when some substance such as water or air (called the "working substance") is heated by the energy obtained from a fuel, such as wood, coal, oil, or uranium. (b) Waste heat must be rejected to an external reservoir at a lower temperature.

A heat engine cannot work any other way. The original form of this negative statement, as made by Lord Kelvin (1824–1907) is, "It is impossible by means of inanimate material agency to derive mechan-

William Thomson (later Lord Kelvin) was a British physicist who proposed the absolute scale of temperature (the "Kelvin" scale) in 1848. His major contributions to science were in the field of thermodynamics.

ical effect from any portion of matter by cooling it below the temperature of the coldest of the surrounding objects." This is an expression of the **Second Law of Thermodynamics.**

To help us gain further insight into this very fundamental concept, we return to our cave man. Imagine that he discovers he can move his boulder by wedging a bar of copper between it and the cliff and then heating the bar with a flame. Because heated metal expands, and the cliff is stationary, the boulder will move. He has constructed a basic heat engine. Now let us assume the following circumstances:

(a) The cave man has several copper bars, each 10 meters long.

(b) It is very awkward to build a fire under the copper bar between the boulder and the cliff, but it is convenient to have a fire nearby to heat the copper bars to a temperature of 100°C.

(c) The outside temperature is 20° C.

(d) The bar expands 0.17 millimeter for every degree of temperature rise.

The cave man, thinking he has discovered that hot copper bars can do work, heats one of his bars on the fire, wedges the hot bar between the cliff and the rock, then waits and watches. The bar cools and contracts, and no work is done. This failure teaches him his first lesson in the design of heat engines: Work is done only by heating materials and allowing them to expand. Therefore, temperature differences rather than absolute temperatures are important. Having learned this, our cave man now decides to wedge a *cold* bar between the rock and the cliff and to bring hot bars to the cold one. Let us say that he heats three bars to 100°C (we will ignore heat losses to the air). He places the first hot bar on the cold one, as shown in Figure 6.1. The cold bar gets warmer and expands, while the hot one cools and contracts. When both reach the same temperature, all heat transfer (and hence all work) stops. The temperature of the working bar, which was originally 20°C, rises 40 degrees while the temperature of the heating bar drops 40 degrees. The final temperature of both bars is 60°C. In order to move the rock further, he places the second hot bar on top of the working bar and observes that the working bar warms from 60° to 80°, and the new hot bar cools from 100° to 80°. Repeating the process a third time, the working bar heats from 80° to 90°. Recall that for

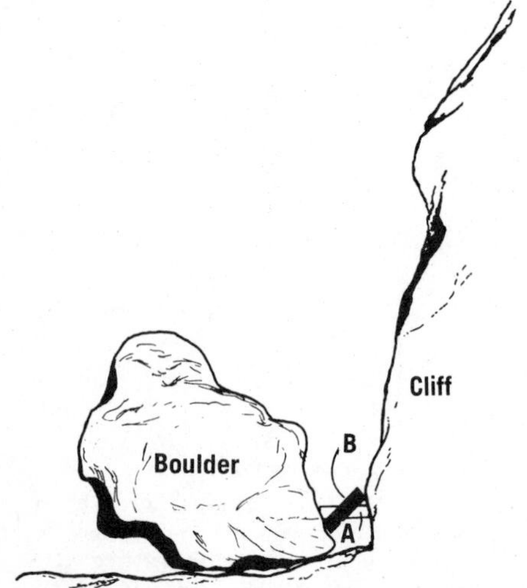

FIGURE 6.1 Bar A, originally at 20°C, warms up to 60°C. Bar B, originally at 100°C, cools down to 60°C.

every degree rise, the working bar expands 0.17 mm. Thus, the first time he heats the bar the rock moves 6.8 mm, the second time it moves 3.4 mm, and the last time it moves only 1.7 mm. Therefore, the amount of work that can be extracted from a given quantity of heat depends on the temperature difference between the hot and the cold bars. This limitation is *not* imposed by the First Law; if the hot bar were to cool to 20°C and the cold one warm to 100°C, their total energies would be conserved. But such an event is contrary to all experience. In fact, the universal observation that two bodies in thermal contact eventually reach a common temperature can be taken as another expression of the Second Law.

Of course, if our cave man were clever, he would not need to tolerate a diminished output of work after the use of each successive hot bar. Instead, after each heating he would cool his working bar back to 20° in some nearby stream, replace it on the cliff (with an additional stone wedge to make it tight), and move the boulder a full 6.8 mm for every hot bar he used. The excess energy would be dissipated as heat and the stream would become warmer. But because the stream would be heated only a small amount, no useful work could be obtained from it, and, as we shall see in Section 6.18, it would become thermally polluted. Thus, only some of the energy from the firewood could be used to produce work; the rest would be lost to technological processes. This observation leads to a restatement of the Second Law: "Energy cannot be completely recycled."

6.6 OUR FOSSIL FUEL SUPPLY

(Problems 9, 10, 11)

In 1970, 95 per cent of the total energy used in the United States was derived from fossil fuels—coal, oil, and natural gas. Only four per cent was from a renewable supply—mostly power from falling water, plus some wood and solar energy. The remaining one per cent was from nuclear fission. The Second Law assures us that someday our reserves will be depleted. How much time do we have? For a realistic estimate of the number of years before man has used all the earth's energy reserves, we must forecast man's population growth rate (see Chapter 5), estimate the quantity of the remaining reserves, and predict accurately man's future rate of consumption. All such forecasts are subject to large errors.

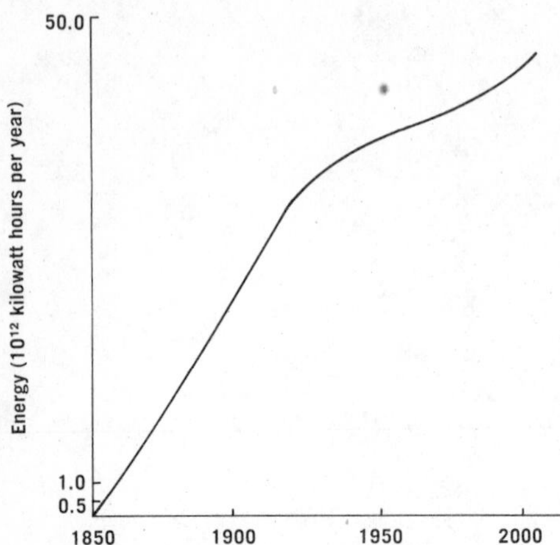

FIGURE 6.2 Growth of U.S. energy production. (Adapted from The Energy Resources of the Earth, by M. King Hubbert.)

Geologists have used several methods to estimate the remaining reserves of fossil fuels. Fuels that have been positively located and identified are called *proved* reserves. Moreover, from seismic recordings and other data, the location and size of *probable* sources have been recorded. Educated hunches and preliminary data give us a third category, *possible* deposits. By adding to the proved sources a reasonable fraction of the probable ones, and assuming that new, as yet unthought-of sources will balance disappointments among the possible deposits, we can get a rough guess at the world's supply of fuel.

To estimate the energy requirements from the 1970's into the early twenty-first century, we start by graphing the past energy consumption, and then try to guess how the curve will continue in the future. Figure 6.2 shows the production of energy in the United States from 1850 to the present. From 1850 to about 1915, the average growth in energy production was 6.9 per cent per year, compared to a population growth of only four per cent. During this period, then, the quantity of energy used doubled every ten years. Between 1915 and 1955 the consumption rate increased only 1.85 per cent per year, corresponding to a doubling in use roughly every forty years. In the period of time between 1961 and 1970, consumption has been increasing about 4.5 per cent per year, corresponding to a 15- to 16-year doubling period. During the same period the population was rising at a rate of 1.1 per cent. Total world use of energy is increasing at a rate of roughly seven per cent per year. If that rate remains constant, world energy requirements will double every 10 years.

It is obvious that the world cannot double its consumption of a depletable resource every ten years for very long, for each doubling corresponds to an increasingly large increase in the magnitude of energy. Think of it this way: If energy consumption is doubling every 10 years, then in the past 10 years we have used as much energy as in our whole previous history. Or another way: If energy consumption is doubling every 10 years, then by the time we have used up half our total reserves, we will have only enough left for another 10 years. Even if there were a very large reserve of fuel, the environmental side effects of energy production would eventually limit expansion. In that case, a reasonable prediction for growth with time is

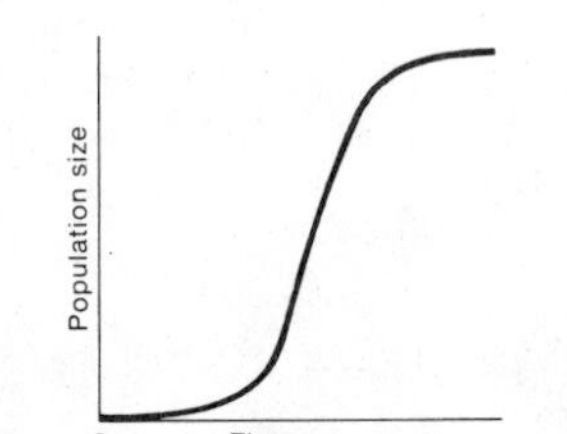

FIGURE 6.3 Sigmoid growth curve.

depicted in Figure 6.3. This resulting curve, called a sigmoid function, shows an initial growth rate, followed by a rapid rise, and finally a leveling off. This level portion would correspond to a condition in which the environmental cost of increased use would be higher than the expected gain.

See Chapter 5 for a discussion of the relationship between growth rates and doubling times and for a further discussion of curves and projection.

In the real world it is reasonable to suspect that the combination of environmental side effects, and depletion of resources will lead to the leveling off and eventual decline of consumption of a depletable fuel. We can expect that for some time consumption of coal, gas, and oil will increase rapidly as technology improves the methods of exploration, mining, and transmission. As poorer reserves must be tapped and pollution control becomes more expensive, the cost of power will rise and people will consume less.

It is generally believed that natural gas is our least abundant fossil fuel. Moreover, government price regulations in the United States have had the effect of discouraging exploration. The peak consumption rate will probably be reached about 1985, or before many of the readers of this book have reached middle age. Already, contractors in New Jersey and other areas are unable to supply gas lines for heating new houses because the industry claims it will not have enough fuel for new customers. The scarcity of natural gas is environmentally unfortunate since it burns cleaner than any other widely used fuel.

Figures 6.4 and 6.5 combine estimates of world supply with the predicted rates of use to display past and predicted coal and oil production. We have barely begun to utilize our coal reserves. We do not mean to imply that there will be sufficient energy available

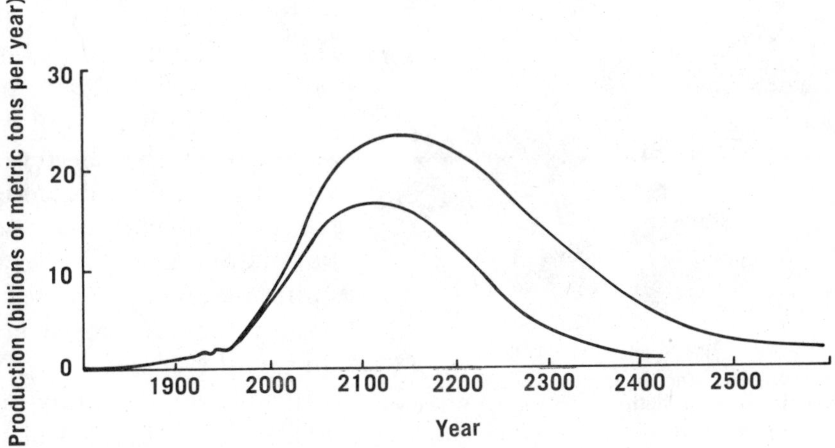

FIGURE 6.4 **Past and predicted world coal production based on two different estimates of initial supply. (From *Scientific American*, September, 1971, p. 69. Copyright © by Scientific American, Inc. All rights reserved. Adapted from The Energy Resources of the Earth, by M. King Hubbert.)**

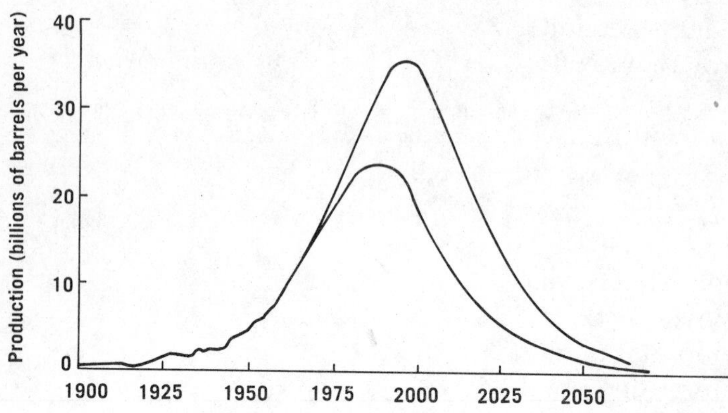

FIGURE 6.5 Past and predicted world oil production based on two different estimates of initial supply. (From *Scientific American*. September, 1971, p. 69. Copyright © by Scientific American, Inc. All rights reserved: Adapted from The Energy Resources of the Earth, by M. King Hubbert.)

from coal in the year 2300 to supply all of our needs. Figure 6.4 tells us simply that world production of coal will be around 10 billion metric tons per year at that time. Probably the world will require even more fuel and people will be exploiting other sources. The fossil fuels cannot be the primary source of energy for a rapidly expanding population for more than a few generations. When 90 per cent of the reserves have been removed, the energy needed to mine and extract the fuel approaches the energy released when it is burned, so that further exploitation is uneconomical. According to reliable estimates, natural gas in the United States will be 90 per cent depleted around 2015; the world supply of oil will be 90 per cent depleted between 2020 and 2030, and world coal between 2300 and 2400.

Articles in the popular press often present estimates quite different from these, for they often use two other, much less accurate ways of projecting the world's supply of fuel. One is to predict the length of time the reservoirs would last at the current rate of consumption. However, since it is unreasonable to assume that both population and per capita consumption will not increase above present levels, this viewpoint is unrealistically optimistic. A survey in 1966 showed that at present rates of consumption coal reserves would last 1440 years. On the other hand, some authors present alarming predictions of depletion based on the current rate of acceleration in use. Just as population cannot be expected to grow at a rapid pace until one fine day we wake up to find that each person has one square foot of land area allocated to him, so the world society will not continue to accelerate utilization of a resource right up to the day that it is all gone. Using the alarmist prediction, coal reserves

"Be practical—people always need coal."

(From: The Energy Crisis. Science & Public Affairs Book Bulletin of the Atomic Scientists, p. 63.)

would reach 90 per cent depletion by the year 2088. The more realistic arguments already presented predict depletion of gas reserves in 35 years, oil in 50 years, and coal in 350 to 400 years. Will we be driving teams of horses to town when the oil runs out? Will the technological age die completely with the depletion of our coal? The answer is almost surely no, for several stopgap measures are available, and a few more long range solutions may be found.

Until now, most of our petroleum has been extracted from underground wells or "lakes" made up of a viscous, but pumpable liquid. Underground deposits of shale in Colorado, Wyoming, and Utah have been found to contain large quantities of petroleum impregnated in the pores of the rock, and sand fields, laden with oil, have been discovered in Alberta, Canada. The quantity of oil that can be obtained from oil shales and tar sands is unknown. The richest deposits, containing about 100 gallons of oil per ton of ore, are believed to add another 25 per cent to the world's initial supply of liquid petroleum. At present, importation of oil is usually more economical than mining the shales, although a Canadian firm is currently extracting oil from tar sands, and a large pilot plant for handling oil shale is in operation in western Colorado. As imported oil becomes more costly and as the technology needed for extraction becomes more highly developed, shale deposits will be used more. Many shales have smaller quantities of fuel per ton of rock. If all the shales containing from five to 100 gallons per ton are considered to be recoverable for fuel, the total reserves are equal to *100 times* the initial estimated supply of coal. We do not know how much of this can in fact be used. Present technology would require more than five gallons of fuel to mine, transport, extract, and refine a ton of shale. Thus, it is clearly uneconomical to mine some shales. But what about the shale with 50 gallons of oil per ton? Twenty-five? As the cost of fuel from other sources increases and as the technology of retrieving oil from shale advances, oil shale exploitation will probably increase, for lower-grade ores will become relatively more economical. The break-even point, and hence the available reserves, cannot now be predicted.

Another method of increasing the supply of oil and gas is to convert coal to liquid fuel. Today, chemists can convert organic chemicals from one form to

A Fable

Said the scientist to the King, "Our only lake, from which we get all our water and our fish, is growing a terrible slime on its surface. Only a few square feet are covered now, but the area is doubling every day."

"Don't bother me yet," replied the King. "Come back when half the lake is slimed over and then we will start to look into it."

another to synthesize new chemical products. The state of this art is so highly advanced that many medicines and plastics are presently synthesized from coal. Attempts have been made to convert coal, which is relatively plentiful, into methane, which is scarce. Methane is a very desirable fuel because it burns cleanly and is simple to transport, while coal is dirty and imposes handling problems that are particularly difficult for small users. The theory of the conversion of coal to methane is well understood, and the practice has been worked out in the laboratory. The United States Bureau of Mines has been searching for a contractor to construct a pilot plant to convert coal to gas. The only major problem now is cost. Synthetic methane can be expected to cost two to three times as much to produce as the natural fuels now cost to extract. But that does not mean that the cost of the gas to the consumer would be increased proportionately. Initial cost is only a small fraction of the price of natural gas. If transportation costs, distribution costs, and profits were held constant, the total proportional price rise would be much less. Moreover, the present price base for natural gas is rising.

$$\text{coal} \xrightarrow[\text{high pressure}]{\text{hydrogen}} \text{liquid fuels}$$

During World War II, the Germans converted coal to gasoline and, surprisingly, the cost was not unreasonable. Coal can also be converted to kerosene, motor oil, and other petroleum products. Unfortunately, research in the United States in this area has not been pursued vigorously enough in previous years. However, the cost of refining crude oil is growing while new technical improvements are reducing the cost of converting coal to oil. By the end of the century we shall probably see the conversion of solid to liquid fuels with little change in the prevailing market prices of the liquid products.

6.7 ENVIRONMENTAL PROBLEMS CAUSED BY COAL MINING AND OIL DRILLING

(Problem 12)

Coal can be mined either in the traditional underground tunnels or in open pits. The former method is more expensive and less efficient, for coal seams lying between tunnels are often left untouched. Compared to open pits, underground coal mines are more prone to fire, and thus more dangerous to the miners. However, the surface contours of the land are left undisturbed and the topsoil is not removed, although occasionally buildings situated above old coal mines have collapsed as the land above the tunnels settles.

To improve the efficiency of mining, coal companies are switching an increasingly large portion of their operations to open pits, otherwise called **strip mines.** In the operation of a strip mine, the surface layers of topsoil, subsoil, and rock are first stripped off by huge power shovels or draglines, exposing the underlying coal seam. The soil is piled alongside the cut or in an adjacent cut which has already been thoroughly mined. The coal seam is then mined and a new cut is started. By 1965, 1.3 million acres of land in the United States had been mined in this unsightly way. Less than half of this land has been reclaimed to its former agricultural productivity. The credit for this rejuvenation must be shared about equally between natural processes of succession and remedial actions by man. The least harm is done to flat, fertile land, which can be mined, reclaimed, and farmed again within five years. On the other hand, hillsides in West Virginia have been ravaged so badly that they cannot be expected to be useful for farming or recreation within our lifetimes. Some of the land that has been reclaimed by green plants has not been returned to its original contours and its former productivity, but instead low-quality trees or grasses have replaced hardwoods or corn.

Strip mining operation. (Photograph by Arthur Sirdofsky.)

Unreclaimed strip mines are a multifaceted environmental insult. Neither facts nor figures are needed to convince a discerning person that open, rootless, dirt piles are uglier, less useful, and more liable to water erosion than the natural forest or prairie. The uglification is an esthetic loss to all of us. Erosion silts streams and reservoirs, and kills fish. Furthermore, sulfur deposits are often associated with coal seams. This sulfur, generally present as iron sulfide, FeS_2, reacts with water in the presence of air to produce sulfuric acid, which runs off, together with the silt, into the stream below.

The United States has roughly 40 million acres of coal deposits which can be strip mined in the near future. Such vast areas magnify the problems outlined above. A much greater effort at land reclamation should be made. If the true environmental cost, the ruined streams and land, are levied against the price of coal, an incentive would be provided to rework and replant the land. Naturally, the cost will be reflected in higher fuel bills, but it is a question of direct cost in dollars and cents or indirect cost in terms of our recre-

Strip mined land.

Lye is any strongly alkaline (basic) solution, especially that which is formed by extraction of ashes. A caustic substance is one that is destructive of living tissue; chemically the term refers to a strong base, particularly NaOH (caustic soda, or sodium hydroxide) and KOH (caustic potash, or potassium hydroxide).

ation and food reserves. The price must be paid in some form of currency.

There are various possible reclamation procedures available. The simplest technique is to regrade the old mine to produce pleasing contours, and then cover it with topsoil and replant. Land reclaimed in this manner will necessarily be lower than the original surface, and this change may upset the water table of the area.

Strip-mined areas can be refilled with various solid wastes, such as ashes. A one-million-watt electric generator fueled by coal produces 850 to 1000 tons of ashes every day, and thus provides an abundant source of filler. Unfortunately, rainwater reacts with coal ash to produce lye, which is a strong caustic and therefore a water pollutant. However, lye seepage can be prevented by sealing the pit with an impervious material.

Another procedure is practiced by the Aloe Coal Company of Pennsylvania, which fills its pits with compacted garbage from the city of Pittsburgh, thus reducing two environmental problems in one economical operation.

Alternatively, we could return to tunnel mining, but, aside from the waste and increased cost, more men would die to provide a cleaner environment for the rest of us.

In contrast to the disruption caused by coal mining, oil drilling in an area of flat land, such as the plains of Texas, is relatively innocuous. As our easily accessible oil fields are being pumped to depletion, however, man has invaded the delicate ecosystems of the continental shelves and the Arctic for more petroleum.

Offshore drilling platform, Gulf of Mexico, 1974. (Photo courtesy of Tenneco, Inc., Houston, Texas.)

Despite careful precautions, accidents seem to occur periodically in all industrial operations, and drilling for oil is no exception. Broken drill pipes, excess pressure, or difficulties in capping a new well have repeatedly led to blowouts, spills, and oil fires. When these have occurred on land, the problems have been locally contained and the environmental disruptions have been minimized. However, when they occur on offshore rigs such as those currently in operation in the coastal waters of Louisiana and Southern California, the result is disastrous. A notable incident of this type occurred off the coast of Santa Barbara, California, in 1969; it is estimated that between 20,000 and 50,000 gallons of oil per day were released for 11 days.

6.8 TRANSPORTATION OF FUELS: A CASE STUDY OF THE ALASKA PIPELINE

(Problems 13, 14)

After fuels are extracted from the earth, they must be transported to the refinery or the power plant. Thus far, the shipment of coal or uranium by train, and oil or gas by pipeline, has produced a minimum of environmental disruption. As the more accessible sources become exhausted, however, fuels are increasingly sought in more inhospitable places, where transportation may cause environmental problems. Our example concerns the large quantities of oil and gas that have been discovered in the north slope of Alaska. Because fossil fuels are needed by our civilization, these deposits will eventually be exploited. However, the Arctic tundra is such a delicate ecosystem that transporting the oil may be very disrupting to the environment. The Alaskan tundra and forests are very much larger than any wilderness in the contiguous United States, and any pipeline across this region will necessarily affect its wilderness quality.

The proposed pipe would run from Prudhoe Bay on the north slope to the ice-free port of Valdez in southern Alaska (see map), a distance of 789 miles. Much of the route travels through ecosystems of the northern tundra, while the southern section travels through forests. Each day the pipeline would carry about two million barrels of oil, which would be heated to 145°F (63°C) for practically all of the jour-

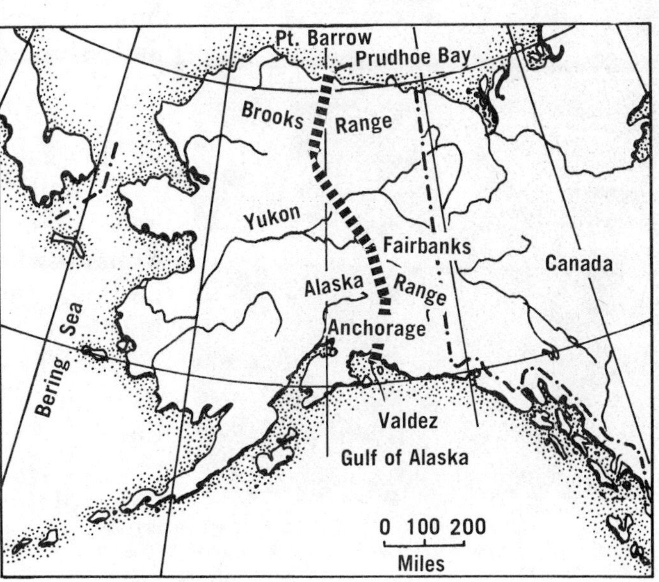

Proposed pipeline route.

ney. Heating the line would be necessary because at the prevailing frigid temperatures of the surroundings the cold oil would be too thick to flow easily.

The following are some of the important environmental side effects of the pipeline:

1. During pipeline construction, new roads would be constructed in areas previously inaccessible to motor vehicles. These roads would provide easy access for hunters and tourists, necessarily leading to deterioration of the wilderness.

2. Construction workers riding in jeeps, skimobiles, or airplanes have run caribou to death. Even though these practices are illegal, importation of thousands of additional workers for pipeline construction would virtually assure continuation of such violations of the wilderness.

3. The roadbuilding and pipeline projects would require roughly 80 million cubic yards of sand, gravel, and rock. The strip mining of hillsides and stream beds near the construction route to obtain this fill would almost certainly cause silting of previously pristine wild rivers.

4. The land of the tundra consists of a thin layer of topsoil over a layer of permanent frozen subsoil, or **permafrost.** During the summer, the hot line would melt the ground ice and kill vegetation close to the line. The resulting wastelands, though very small in area, would be in the shape of a continuous band. In turn, removal of the plant cover would alter the insulation properties of the soil adjacent to nearby permafrost, causing uneven freezing and thawing. When mud is in contact with ice, soil slippage is probable, and slipping soil can cause a break in the pipeline. Even with emergency shutdown procedures, a break could easily cause spillage of about two and a half million gallons of oil.

5. Although some southern areas of the tundra have been successfully replanted after being disrupted by heavy equipment, experimenters have thus far been unable to replant tundra vegetation in many northern areas.

6. The caribou is a migratory animal that spends its summers near the northern seas and its winters in the southern forests. Herds numbering about 150,000 individuals still group in the tundra every spring at their calving grounds, and many of the animals live a life that is, as yet, largely undisturbed by humans. The effect of a narrow band of soft, unfrozen, barren earth

Picture of caribou on tundra. (From Bernard Stonehouse: *Animals of the Artic.* **Holt, Rinehart & Winston, 1971, p. 19.)**

on the migration of caribou is unknown. They might be able to cross it, or they might falter when their feet sink into unknown turf, and not be able to complete their journey.

7. To reduce the probability of a break caused by ground slippage, it is proposed to elevate about half of the pipe either on wooden stilts or on a bed of gravel. Such structures might impose another serious barrier to the migrating caribou.

8. The proposed route travels across an area prone to earthquakes, and the probability is high that a large quake would occur sometime during the operation of the pipeline.

9. If, during the operation of the pipeline, a break does occur, the consequences may be severe. At the very least, hundreds of acres of delicate tundra would be destroyed and rehabilitation would be quite slow, but widespread economic or environmental perturbations would be minimal. At worst, however, a break in the pipeline could spill millions of gallons of oil onto the Gulkana River, a commercially valuable salmon spawning ground. The Gulkana flows into the Copper River, which presently supports heavy summer populations of wild ducks and geese.

10. Because of routine losses and possible accidents large quantities of oil would probably spill into the ocean at Valdez. The pipeline corporation has announced that it will build special ballast treatment tanks to reduce the amount of oil discharged. But we do not now know how much oil would be spilled, how much reduction would be realized, or how much oil could be tolerated by the aquatic life without severe damage to the northern oceans.

The Alaskan Pipeline presents complicated environmental problems. Conservationists have proposed alternate routes which they feel would reduce the destruction to the tundra. One suggestion is to run the oil pipeline adjacent to a natural gas pipe which is proposed to run across the Canadian Yukon and along the Mackenzie River. Since the same roads would serve both lines, this plan would reduce migratory disruptions and other environmental stresses. No tankers would be needed at Valdez, since the pipe would go directly to the central United States. However, no solution will leave the environment unscathed.

After a lengthy battle between conservationists and the oil companies in the United States Congress

Cranes used in the unloading of barges which bring bulk supplies to Prudhoe Bay during the summer from Houston, Seattle, and other points. The ice moves off the Arctic coast for less than 2 months each summer, and the barges have to slip through the moving ice and hurriedly deliver their supplies to avoid being frozen in for the winter. (Courtesy of BP Alaska, Inc., Anchorage, Alaska.)

(Courtesy of BP Alaska, Inc., Anchorage, Alaska.)

and in the courts, legislation was passed which virtually assures the pipeline construction. Americans today want both a clean environment and the benefits of a sophisticated but cheap technology. To achieve both, compromises must be made. To take an absolute stand that the Alaskan pipeline is "good" or "bad," or that the final route chosen is "right" or "wrong," involves, at least in part, a judgment about what is valuable. Regardless of what we cherish most, however, we must understand that the probing of delicate ecosystems for oil, gas, or minerals involves the real danger of their irreversible disruption.

Recent legislation passed by the U.S. Congress in the summer of 1973 has greatly reduced the chance for a delay in the construction of the pipeline.

There are no people alive today who can remember seeing large buffalo herds travel across the North American prairie. Very few have paddled a canoe along the Mackenzie, the Gulkana, or the Perth Rivers and seen moose in the northern forests, or geese nesting in the muskeg country, or caribou darkening the tundra. The construction of the pipeline is not likely, by itself, to bring about the extinction of any species, but it will certainly reduce the herd and flock sizes; it will reduce the remoteness and desolation of the North American wilderness.

6.9 NUCLEAR FUELS

(Problem 16)

Oil shales, tar sands, and chemical conversions will undoubtedly extend the life of the Fossil Fuel Age, but such stopgaps cannot be the basis for a long-range continuation of our technological existence. Instead, we must turn either to renewable resources or to nuclear energy.

In 1970, only one per cent of the energy used in the United States was provided by nuclear fuels, but this fraction is expected to increase rapidly in the near future. By the year 2000, 25 per cent of the total energy of this country will probably be supplied by nuclear plants, and, of course, the percentage will continue to rise. We must therefore consider the availability of ^{235}U, the isotope now used as a fuel in nuclear reactors. Once again, we are concerned with both abundance and concentration in natural ores, since both factors determine cost. Presently, the uranium from which the fissionable uranium-235 is extracted costs about eight dollars per pound, but the high-grade ores which make such a low price possible are scarce. Table 6.3 shows the relative amounts of ore available at various prices. Authorities predict that by the year 2000 the low-cost ores will be substantially consumed, and hence we will be paying higher electric bills. We

Atoms of a given element may differ in atomic weight; these different species are called *isotopes*, and the nearest whole number to the atomic weight is the *mass number*. For example, one isotope of uranium has a mass number of 235; it is designated uranium-235 or ^{235}U. Another is uranium-238 or ^{238}U. These values are related to important differences in nuclear properties.

TABLE 6.3 AVAILABLE URANIUM RESERVES

COST IN DOLLARS PER POUND	*TONS OF URANIUM AT THIS OR LOWER PRICES*	*INCREASE IN COST OF ENERGY, IN MILS/KWH, RESULTING FROM INCREASE IN FUEL COST*
8	594,000	0
10	940,000	0.1
15	1,450,000	0.4
30	2,240,000	1.3
50	10,000,000	2.5
100	25,000,000	5.5

will complain and perhaps curtail our consumption, but most of us will not change our lives much. Moreover, since expensive low-grade ores are relatively plentiful, this high-priced energy will be available for a thousand years or more.

Within this century breeder reactors, which use plentiful non-fissionable ^{238}U or ^{232}Th to synthesize fissionable materials, are likely to be developed. These fertile fuels are present in the currently mined uranium ores, and additional reserves exist in the granites of the Sierra Nevada mountains, in the White Mountains of New Hampshire, and in many of our states just west of the Appalachian Mountains. Granite in the Sierra Nevada contains 50 p.p.m. of non-fissionable uranium. Converted into fuel, one pound of granite could be expected to produce a heat equivalent of 70 pounds (32 kg) of coal. If we used all the mountain granites and other sparsely concentrated reservoirs, we might obtain 100 to 1000 times the energy supplied by the original fossil fuel reserves.

P.p.m. means "parts per million." For every million pounds of Sierra Nevada granite there are 50 pounds of uranium.

Beyond the breeder lies the possibility that man will someday harness the fusion reaction as a power base. Some authorities expect fusion never to be functional, while others predict the first feasible demonstration within 10 years. Most members of the scientific community feel that development of such a power plant is highly unlikely before the year 2000. If fusion is found to be practical, the power available to us will be enormous. Probably the first to be developed will be the deuterium-tritium reaction which requires lithium for the synthesis of tritium. We have only enough lithium to provide energy equal to our original supply of fossil fuel, but if the deuterium-deuterium fusion reaction can be used, fuel can be obtained from ocean waters, and power will be available to man for a million years or so.

6.10 RENEWABLE ENERGY SOURCES

(Problem 17)

Simple device for using solar energy.

At present, our efforts to explore various renewable energy sources are meager indeed. We have seen that the renewable sources on which we used to depend will no longer suffice. In the early nineteenth century, the power base for most of the world was wood, a renewable energy source. Forests cannot grow fast enough to provide fuel for people's varied needs today.

In this section,we shall discuss harnessing energy from other renewable sources: the sun, the rivers and streams, the tides, the wind, and the burning of garbage.

Solar Energy

Every 15 minutes, the solar energy incident on our planet is equivalent to man's power needs for a year at his 1970 consumption level. Thus, if we could

Top: Solar collectors for heating water. Bottom: Parabolic solar heaters oriented to receive maximum sunlight. (Courtesy of Prof. Harry Lustig, Department of Physics, the City College of the City University of New York.)

trap, concentrate, and store solar energy, man's power needs would be fulfilled. Solar energy can be used either directly as heat or more indirectly for the production of electricity. Small scale solar heaters are relatively cheap to build and are in common use in many parts of the world for water and space heating (see page 216); however, they are unpopular in the United States. Even during the past few years, when fossil fuels have become increasingly expensive, the sales of solar heaters have actually decreased, and companies manufacturing these units have gone out of business.

Energy in falling water.

At present, the cost involved in converting solar energy to electricity is huge. For example, in 1970 the investment in an electrical generating power plant amounted to about $500 per kw (kilowatt) for a fossil fuel station, $750 per kw for a nuclear facility, and $2,000,000 per kw for an electric solar cell collector. Experiments with new materials suggest that the cost may be reduced to $2500 per kw. This cost and the amount of land needed are prohibitively large. If research can provide new materials, perhaps the cost of energy from the sun can be greatly reduced and a virtually limitless supply of power will be available. The successful utilization of solar energy would solve all our energy needs during the life of the planet, for a renewed supply arrives daily.

Hydroelectric Power

Today, only four per cent of the power used in the United States is derived from hydroelectric sources. The sun evaporates water at sea level and the winds bring it to the mountains, where it collects into streams with potential to fall and release energy. Thus, hydroelectric energy is an indirect form of solar energy. As more sites are exploited, more power will be tapped by harnessing falling water. As the total power need of the country continues to rise, however, the relative importance of hydroelectric power will probably not increase much. There simply are not enough good locations for building large dams, and those that are available are often in recreational areas. Furthermore, damming the Grand Canyon or the Snake River Gorge would be a great aesthetic loss.

Geothermal Energy

Energy derived from the heat of the earth's crust, called *geothermal energy*, has gained increased attention in the past few years. In various places on the

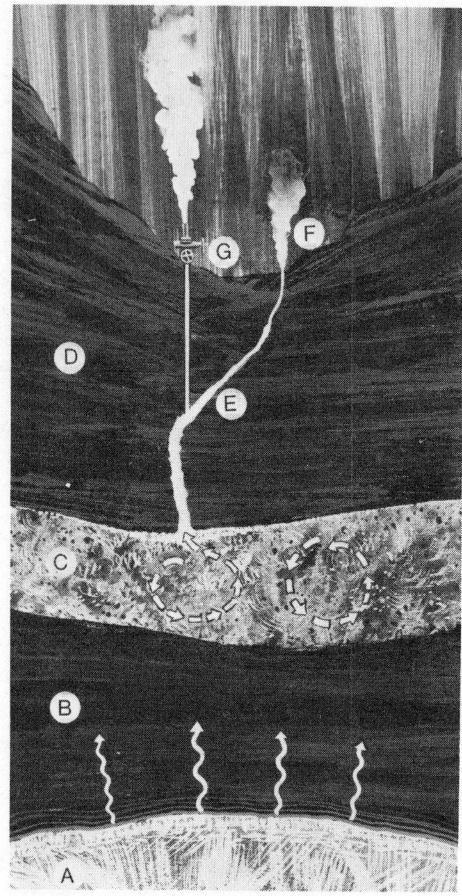

Geothermal field: *A*. Magma (molten mass, still in the process of cooling). *B*. Solid rock; conducts heat upward. *C*. Porous rock; contains water that is boiled by heat from below. *D*. Solid rock; prevents steam from escaping. *E*. Fissure; allows steam to escape. *F*, Geyser, fumarole, hot spring. *G*, Well; taps steam in fissure. (Photos courtesy of Pacific Gas and Electric Company.)

globe, such as in the hot springs and geysers of Yellowstone National Park in Wyoming, hot water is produced near the surface. Although no one is suggesting the harnessing of Old Faithful for generating electricity, there are several places where hot (500°F) underground steam is available for the price of a deep well. The Pacific Gas and Electric Company has connected a generator to well holes in central California and is presently producing electricity from them on a small commercial scale. This project is new in the United States, but an Italian facility at Larderello has been in operation since 1913 and is presently working at a capacity of 400,000 kw. Even optimistic supporters, however, do not expect a large portion of our power to be supplied in this manner, for there are not enough hot springs at or near the surface. Additionally, continuous exploitation for more than a century or two is expected to exhaust the water or heat content of these wet wells.

A more ambitious proposal to drill one to four miles through the earth's crust into its molten core has much greater potential. New types of drills have been devised which melt their way through bedrock. If the core could be reached, water could be poured down the hole and the resulting steam withdrawn. If successful, geothermal tappings of this type could provide man's requirements for energy far into the future. Great care must be exercised in such drillings, for the deep wells and large quantities of water flowing through the holes might cause slippage of rock layers, and hence earthquakes.

Energy from the Tides

Section of the geothermal steam field where Pacific Gas and Electric Company generates 396,000 kilowatts of electricity from underground steam.

The force of the tides as a source of power has long intrigued man. Imagine a turbine which is immersed in the sea near shore. Electricity could be generated as the tides turn the turbine blades, but the output for a single generator would be small. Moreover, lining the coast with millions of small generators would be uneconomical.

An alternative practical solution is to build a tidal dam across a bay or estuary where there is a large rise and fall of the tides. By damming and funneling the water, significant concentrations and economical power output can be realized. A tidal generating plant in France on the coast of Brittany generates 240,000 kw, about the output of a small fossil-fueled plant. If we should exploit this resource further, however, we

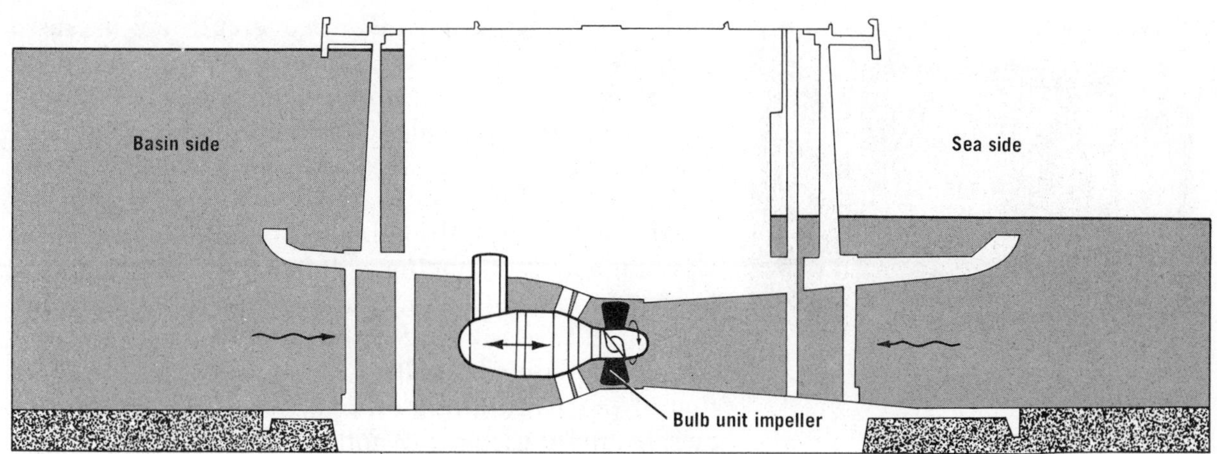

Tidal dam and turbine.

shall have to reckon with the consequences on aquatic life that would inevitably result from the construction and maintenance of dams across the estuaries. Also, there are not enough properly oriented bays and estuaries to provide more than a small percentage of our energy requirements.

Energy from the Wind

The power of wind has been used since antiquity to pump water and grind grain. Construction of large-scale electric generating stations of this type would pose serious engineering problems because of their huge size, but small windmills as an auxiliary source of home electricity would be economical and would conserve other sources of power. Of course, the wind does not always blow, and therefore some form of energy storage would be needed to provide power during calm periods.

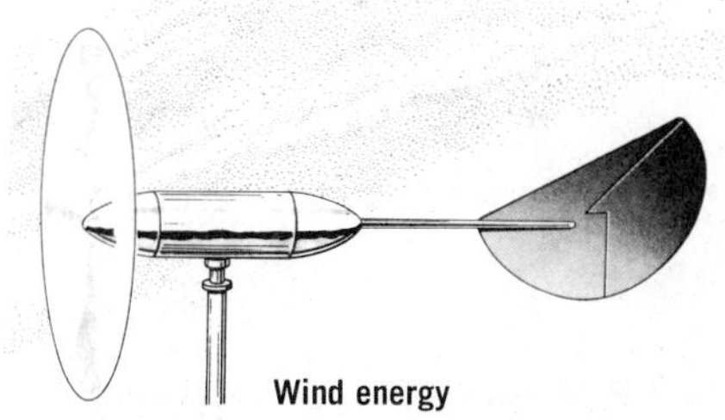

Energy from the Burning of Garbage

Another practical proposal is to burn garbage as a renewable energy source. In recent years half of the solid waste in the United States has consisted of paper. Burning it would both provide energy and significantly reduce the problem of solid waste disposal. In fact, a Parisian company has been burning solid wastes to generate electricity for 50 years. At the present level of operation, a total of 1.5 million tons of garbage is consumed per year as an auxiliary fuel source. In the United States in 1971, there was an estimated 880 million tons of dry solid wastes, of which only 135

"I mean, like, don't you think it kind of weird to get electricity from our windmill-powered generator and then use it to watch reruns of 'I Love Lucy'?"
(Drawing by W. Miller; © 1974 The New Yorker Magazine, Inc.)

$$\left(\begin{array}{cc} \mathrm{H} & \mathrm{H} \\ -\mathrm{C}- & \mathrm{C}- \\ \mathrm{H} & \mathrm{Cl} \end{array}\right)_n + \text{oxygen}$$

$$\rightarrow CO_2 + H_2O + HCl$$

million tons were actually recoverable for use as a fuel, the remainder being difficult and uneconomical to collect and transport. But this 135 million tons could provide energy equivalent to 170 million barrels of oil (15 per cent of the 1971 consumption in the U.S.), or 1.35 trillion cubic feet of natural gas (38 per cent of the total). Saving 170 million barrels of oil in a year while simultaneously reducing solid waste pollution is a healthy approach to environmental problems. Why, then, aren't we burning garbage now?

One of the difficulties of using trash as fuel is that of persuading large numbers of householders to classify their garbage. Another problem results from the use of large quantities of plastics in packaging. Plastics burn with more smoke than wood or paper; furthermore, chlorinated plastics, like polyvinyl chloride, generate hydrochloric acid and thus produce additional air pollution. Engineers could, at a price, remove noxious smokestack emissions, and housekeepers could, if they tried, (and perhaps if they were given economic incentives) keep plastics out of the wastebasket.

Clearly, the burning of garbage is not a panacea. But if garbage, solar, geothermal, tidal, and wind energies were all exploited along with the implementation of various social remedies outlined later, the total contribution could be very substantial. In an age during which our fossil fuel reserves are rapidly being depleted and the transition to a fusion-based power grid may be very slow, we would be wise to buy a few more years' time by exploiting alternate energy sources.

6.11 THE ENERGY CRISIS

If we have enough oil to last 60 years, enough coal to last 350 to 400 years, enough uranium to last well into the twenty-first century, and if the potential exists for harnessing large amounts of power from solar, geothermal, and wind sources, why are there energy shortages throughout the world? The complexity of the problem is staggering. It involves all the following factors: the chemistry of both fossil fuel utilization and pollution control; the physics of nuclear reactions; the engineering problems in building mammoth pipelines, tankers, refineries, and nuclear reactors; the geology of oil, coal, gas, and uranium reserves; the ecology of the world's ecosystems; the sociology of people's needs and wants; the psychology of ancient hatreds; the domestic politics of every nation of the world; the international politics of the twentieth century; and the economics of our civilized world. Of course, a complete analysis of the energy crisis is beyond the scope of human ability, let alone that of this book. Nevertheless, we can probe a few features of the problem.

First, the fuel reserves are not distributed uniformly throughout the Earth, but are concentrated in small areas. For example, the United States has enormous reserves of coal, and the Middle Eastern nations are noted for their supplies of crude oil. Thus, a situation has developed in which a small number of people control that which a large number of people need—or want. In many cases conflicts of interest arise. For example, European nations import almost all their petroleum, most of it from the Middle East. If unlimited supplies of inexpensive petroleum are made available to Europe, the Middle Eastern reserves are estimated to last only about 23 years. Consider the positions of a European industrialist, on one hand, and of an Arabian oil sheik on the other. The European businessman is in competition with other manufacturers for sources of supply and for available markets and sales. He needs oil; if his supply runs out, he must modify his operation in some way: He may reorganize his factory to use another fuel, or switch to another process that uses much less energy, or go out of business. The oil sheik, on the other hand, has a guaranteed market, and he might feel that he would be better off by either raising prices or curtailing production, so that his oil can survive for generations. Both men are acting in their own interests, and this conflict is one root of the energy crisis.

Another problem arises as a direct result of the population explosion. Naturally, if the per capita fuel consumption remains constant and the population increases, fuel utilization will increase proportionally. In many instances, however, dense populations require more fuel per person than sparse ones. Imagine that there are 1000 acres of land available to support 100 people. Everyone can eat well even if agriculture is inefficient. If fertilizers are unavailable, or too expensive, and the land is producing at only 75 per cent of capacity, there is still enough food for all. However, if 1000 acres must support 2000 people, imported fer-

I don't believe it! That Mickey Mouse power company of ours is having another power crisis!
(Reprinted by permission of the Chicago Tribune. Copyright 1973. All rights reserved.)

tilizers would be essential for survival. Since energy is needed to manufacture fertilizers, the people in the densely populated community would depend on outside sources of energy, while the people in the sparsely settled community would need no outside source at all. This problem will be discussed further in Chapter 7.

The magnitude of our needs poses other problems as well. The United States alone consumes 17,248,000 barrels (450,000,000 gallons) of petroleum per day. If a war erupts in the Middle East or if an oil embargo is suddenly imposed, it is difficult to organize other large sources of supply quickly. The Commonwealth Edison Company of Chicago alone consumed 20 million tons of coal in 1970. That rate is approximately 55,000 tons per day. It takes such a tremendous industrial capacity to process all that coal and ship it to the power plants that it would require considerable time to provide enough *extra* capacity to take care of emergencies. Therefore, any breakdown, strike, or similar disruption can put Chicago out of power. Storage capabilities cannot maintain very many days' reserve and if, for example, the railroad goes on strike, no alternate source of transportation for such quantities can be mustered. Or suppose that a heat wave suddenly causes a 10 per cent increase in power requirements for air conditioning. Such a surge would call for 16,000 tons of additional fuel per day (assuming that power is delivered at 35 per cent efficiency; see page 214). If the fuel supply cannot be stepped up rapidly enough, a power shortage results.

The energy requirements of the world are so great, and the mining, transportation, generation, and transmission of power are so complex that the power industries cannot move quickly in response to changes or emergencies. A new nuclear plant takes 6 years to complete and will probably become obsolete 30 years later. Fossil-fueled generators cannot be built much faster. A coal mine requires four years to open, and an oil field may take up to 10 years to reach production. Therefore, if power consumption rises faster than had been predicted a decade earlier, the supplier's capability in an area will not be able to supply the demand. The unpredicted, rapid rise of home air conditioners in New York City has been partially responsible for the power shortages experienced there in recent summers. There is enough coal in the ground, and enough turbines being built in the factories to supply the required power, but they are not in use now.

Similarly, people's habits are slow to change. For instance, if costs are figured over a 10 year period, solar water heaters are unquestionably cheaper than electric water heaters in most regions, and are just as easy to operate, yet there is little demand for these units in many nations. Or as another example, many men wear jackets and ties during the summer, and then turn the air conditioners on "high." As the energy crisis worsens, perhaps people's behavior will change, but these changes are occurring slowly, and during the time lag a great deal of energy is being used needlessly.

In the preceding section we have shown how alternate methods of power production might, in the future, provide energy for our civilization. We should not underestimate the lag that will accompany any large-scale transition to new sources of energy. It is also important to reemphasize that although coal will be commercially available in the future, other fuels will soon be required in quantity. Quite possibly, the new sources of energy won't be available in time to replace fossil fuels, and the next 50 years may see a very serious fuel shortage. Because our civilization needs a continuous flow of energy to survive, we must encourage research, conservation, and innovative engineering now to avert future disaster.

Let us now look at the energy situation from a different viewpoint. Why are these huge quantities of fuel needed in the first place? In other words, what is the nature of our energy consumption, and can we use less energy instead of increasing the supply? Of course, our answer will depend on which area of the world we choose to examine. In many of the less developed nations a reduction in energy use could lead to increased starvation. In technologically advanced or energy-rich countries, on the other hand, conservation would affect people's lives less drastically.

In the following sections we will examine energy utilization in the United States and discuss feasible conservation practices. Naturally, the principles outlined apply to other developed nations as well. In 1971, 25 per cent of the total energy used in the United States was applied toward transportation (about two-thirds of which was consumed by automobiles), 41 per cent was used industrially, and 34 per cent was used in household and commercial applications. We shall consider each in turn.

6.12 ENERGY CONSUMPTION FOR TRANSPORTATION

(Problems 19, 20)

The two *least* efficient modes of transportation, the automobile and the airplane, are the two fastest-growing industries in the transportation field. Consider seven means of moving people: walking, bicycles, buses, trains, automobiles, "jumbo" jet passenger planes, and supersonic transport (SST) planes. If the energy required to drive each of these is converted into passenger miles (pm) per gallon of petroleum fuel, we obtain the results shown in the margin of page 211.

The automobile has a peculiar hold over people. Many would rather drive through rush hour traffic at 10 miles per hour in their own personal cars than speed along at 40 miles per hour on public transportation. During the oil embargo of the winter of 1973–74, people waited for hours in gas lines rather than riding bicycles, subways, trains, or buses, or joining together in car pools. One reason is clearly economic; the car is sometimes the fastest, most economical means of getting from one particular place to another. However, the relationship between man and his machines goes beyond the practical. An automobile is no longer simply a means of transportation. Remember the mythical magic lamps and their genies? One had only to rub the lamp and speak one's wish, and it was done. The modern "fully equipped" automobile scarcely seems to ask for more. A touch of the finger, hardly even a push, brings warming, or cooling, or music; it opens or closes windows, locks or unlocks the car doors, or even the garage doors. A touch of the toe brings a surge of acceleration or a swift stop. The average efficiency of cars driven in the United States decreased from 14.28 miles per gallon in 1960 to 13.49 miles per gallon in 1972, and much of this decrease is directly attributable to accessories which do not help the car go from one place to another. For example, in 1960 only 6.9 per cent of the new automobiles were equipped with factory installed air conditioners; the percentage in 1972 was 68.6. Much of the increase occurred in northern latitudes (some even in Alaska) where the number of uncomfortably warm days are fewer than in areas where air conditioning first became popular.

The automobile, especially, is remarkably addictive. I have described it as a suit of armor with 200 horses inside, big enough to make love in. It is not surprising that it is popular. It turns its driver into a knight with the mobility of the aristocrat and perhaps some of his other vices. The pedestrian and the person who rides public transportation are, by comparison, peasants looking up with almost inevitable envy at the knights riding by in their mechanical steeds. Once having tasted the delights of a society in which almost everyone can be a knight, it is hard to go back to being peasants. I suspect, therefore, that there will be very strong technological pressures to preserve the automobile in some form, even if we have to go to nuclear fusion for the ultimate source of power and to liquid hydrogen for the gasoline substitute. The alternative would seem to be a society of contented peasants, each cultivating his own little garden and riding to work on the bus, or even on an electric streetcar. Somehow this outcome seems less plausible than a desperate attempt to find new sources of energy to sustain our knightly mobility.

From K. E. Boulding, "The Social System and the Energy Crisis." Science, 184 (Apr. 19, 1974), pp. 255–257. Copyright 1974 by the American Association for the Advancement of Science.

Although some efforts are being made in the United States to conserve energy used in transportation, the most efficient modes of transportation are still used sparingly. Most of the automobile trips in this country cover less than five miles, yet bicycles are still unpopular, and adequate bicycle paths are absent in most cities. Currently most mass transportation systems in the United States are grossly inadequate. No passenger trains operate between Los Angeles and Denver, or between Denver and St. Louis. A bus ride from a city on the northern boundary of the Los Angeles megalopolis to a city on the southern edge of the basin can require eight hours, even though the distance actually traversed is only about 75 miles. Amtrak, the nation's national passenger railroad system does not serve any cities in South Dakota, Arkansas or Maine. Improvements are being made on existent lines, but these advances are slow. The Department of Transportation recently cut back funding of research on many aspects of a rapid ground transportation system, and many energy-saving devices which have already been developed are not being incorporated into the design of new railroad cars. Existing tracks are being abandoned, and passenger and freight routes are being taken out of service.

Bicycle	1000 pm/gal
Walking	660 pm/gal
Intercity bus	85 pm/gal
Train	50 to 200 pm/gal, ranging from poorly patronized intercity routes to commuter trains.
Automobile	$10n$ to $40n$ pm/gal, depending on model and condition of car, where n = number of passengers.
Jumbo jet	20 to 25 pm/gal, for estimated average passenger loads.
SST	13.5 pm/gal, for estimated passenger loads.

Pipeline	93 mile/gal
Railroad	63 mile/gal
Waterway	62 mile/gal
Truck	11 mile/gal
Airplane	1 mile/gal

Consider also various means of moving freight. We will express these values in relative units that are dimensionally similar to those cited above, namely the number of miles that a given amount of freight can be transported by a gallon of fuel, using as a comparison standard an air-freight rate of one mile per gallon.* The data are shown in the margin.

The implications of these data are clear: reliance on inefficient methods of transportation contributes to the rapidly accelerating depletion of our fossil fuels. Even if the technology of the twenty-first century should provide us with vast supplies of energy, such wastage of fuel is harmful for several important reasons:

1. Every gallon of fuel consumed represents a multiple source of environmental disruption. Land is preempted and defaced in mining and in the establishment of transportation routes. The refining of petroleum and the consumption of fuel in internal combustion engines produce serious air pollution problems.

2. While new sources of power, such as nuclear

*The actual value for air freight is 42,000 Btu per ton mile.

reactors, are becoming increasingly available, they are not mobile sources. Fossil fuels are particularly valuable because they can be used in small engines. There are, of course, various possible ways to compensate for the dearth of fossil fuels by converting some of the abundant energy from stationary sources to more mobile forms of energy. The most obvious of these is the use of rechargeable batteries. Another appealing alternative is hydrogen, available in quantity from the dissociation of water: $2H_2O + \text{energy} \rightarrow 2H_2 + O_2$. Hydrogen is a convenient, mobile, and clean fuel that burns in air: $2H_2 + O_2 \rightarrow 2H_2O + \text{energy}$. Note that this equation is just the reverse of the preceding one. This chemical reciprocity means that hydrogen can serve as a medium for the transfer of energy from large stationary sources to small mobile engines. The problem is that we do not expect to be able to construct enough electric generating stations to produce hydrogen in quantity in this century, and that cheap petroleum reserves will be virtually depleted by 2020 or 2030. For a world that runs on oil, that's a slim margin.

3. As natural fibers such as wood and cotton and metals such as iron and copper become scarcer, the economic pressure to shift to plastics continues to increase. Commerical plastics, by and large, are organic substances whose energy content is roughly at the level of fossil fuels. (Coal burns; so does polyethylene). The conversion of fossil fuels to plastics and to other valuable organic chemicals, therefore, does not demand unreasonable consumption of energy (although it does require ingenuity). If fossil fuels were exhausted, chemists could synthesize organic materials from energy-poor carbon sources such as atmospheric carbon dioxide and carbonate minerals, but the manufacture would require a replacement of the energy deficit at a correspondingly higher cost. Any shift to "abundant" metals will necessarily incur a similar deficit. The best example is aluminum, which is abundant in nature in the form of its oxide, Al_2O_3, the major constituent of bauxite ore. Aluminum oxide is an extremely stable (energy-poor) substance, and its conversion to metallic aluminum is comparable in energy demand to the conversion of carbonate rocks to plastics.

On November 7, 1973, President Nixon proposed several approaches to conserve fuel used for transportation. He recommended a national speed limit of 50 mph, voluntary car pooling, and possible rationing of fuel. Such methods do indeed reduce energy consumption, but some of them are difficult to enforce. No references were made to mass transportation, or to the curtailing of road building.

Although the petroleum crisis will become acute during our lifetimes, little attempt is being made to remedy the situation. Some of the fuel used for transportation could be conserved if one or more of the following measures were adopted:

1. Construct large-scale, efficient, mass transport systems, such as high-speed trains, in and between major cities.
2. Curtail road building. When public transportation becomes more convenient than driving private cars over crowded highways, public habits will change.
3. Place an environmental tax on fuel consumption, with the highest rates directed at the least efficient modes of transportation.
4. Accelerate the development of methods for obtaining oil by extracting it from shale or manufacturing it from coal.

6.13 ENERGY CONSUMPTION FOR INDUSTRY

(Problem 21)

Industrial power consumption involves a wide variety of products and processes. In Table 6.4 industrial consumers are listed in decreasing order of their energy demands. Manufacturing processes are using increasingly large amounts of energy because both the demand for goods and the energy needed per item are increasing. One reason why the energy needed per item is so high today is the trend toward increased automation to offset labor costs. Another factor, however, is more foreboding. As a natural resource like iron ore becomes depleted, lower-grade ores become commercially more attractive. The mining, purification, and metallurgy of low-grade ores requires more energy than the same operations performed on high-quality deposits.

One way to reduce industrial energy consumption is by increased recycling of material. In general, production of manufactured goods from recycled wastes consumes less energy than the production from raw materials. However, total manufacturing costs reflect the price of labor, transportation, and capital investment as well as fuel bills, and unfortunately, under our present economic structures, recycling operations are often more expensive than primary production. Unless industries are offered economic incentives or penalties, we can expect little immediate increase in industrial recycling.

TABLE 6.4 INDUSTRIAL ENERGY CONSUMERS IN THE UNITED STATES

TYPE OF INDUSTRY	TOTAL ENERGY (TRILLIONS OF BTU PER YEAR)
Primary metals	5298
Chemicals and chemical products	4937
Petroleum refining	2826
Food and kindred products	1328
Paper and allied products	1299
Stone, clay, glass and concrete	1222
All other industries	8050

Many aspects of the problem of industrial energy

consumption are closely linked with the problems of solid waste disposal, for most manufactured goods are ultimately discarded. Thus reevaluation of planned obsolescence, reduction of paper products in packaging, and general reduced consumption would greatly help alleviate environmental stresses caused by manufacturing and waste disposal.

6.14 ENERGY CONSUMPTION FOR HOUSEHOLD AND COMMERCIAL USE

(Problems 22, 23, 24)

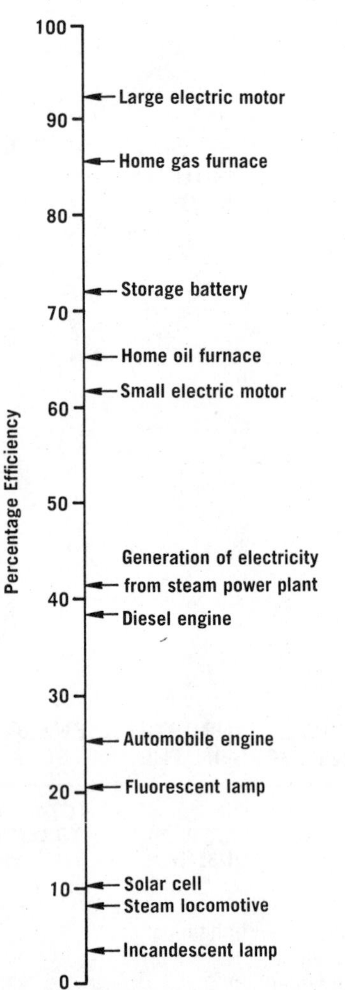

FIGURE 6.6 Efficiencies of some common machines, devices, and processes. The values for electrical devices do not include the energy losses at the generating station.

Most of the residential and commercial uses of power involve the burning of fossil fuels to heat air and water, and the consumption of electricity for other purposes.

Fire for heating was the most ancient use of fuel. Without space heating, people could not live in some areas they now inhabit. Today, space heating accounts for only 12 per cent of our total energy consumption (compared to 16 per cent for the automobile, for example) and remains one of the most efficient uses of fuel. Home furnaces can be built to operate at 65 to 85 per cent efficiency, a figure well above that for electric generating stations or automobiles. (See Figure 6.6.) However, these efficiencies are seldom realized in the home because most furnaces are poorly adjusted. Moreover, most homes are poorly insulated. Building in most areas in the United States is regulated by strict legal codes. Yet the insulation requirements of the Federal Housing Authority are below the recommendations of the insulation manufacturers. Compliance with the manufacturers' suggestions would reduce fuel consumption and hence fuel bills, by about 33 per cent. The increased investment in insulation would be recovered in the first or second year.

Despite the high efficiency of home furnaces, gas and oil units do pollute the air, and the control of pollution on a house-by-house basis would be very costly. Therefore, there is an increasing trend toward the use of "clean, efficient, electric heat," as expressed by electric utility advertising slogans. If "electric heat" is taken to mean heat generated by electrical resistance, as in a giant electric oven, then the process in the home is, in fact, clean and 100 per cent efficient. Of course, the pollution is simply released at the site of the electric generating station instead of in the home. As we will see in Section 6.16, commercial electric generating stations are only 41 per cent effi-

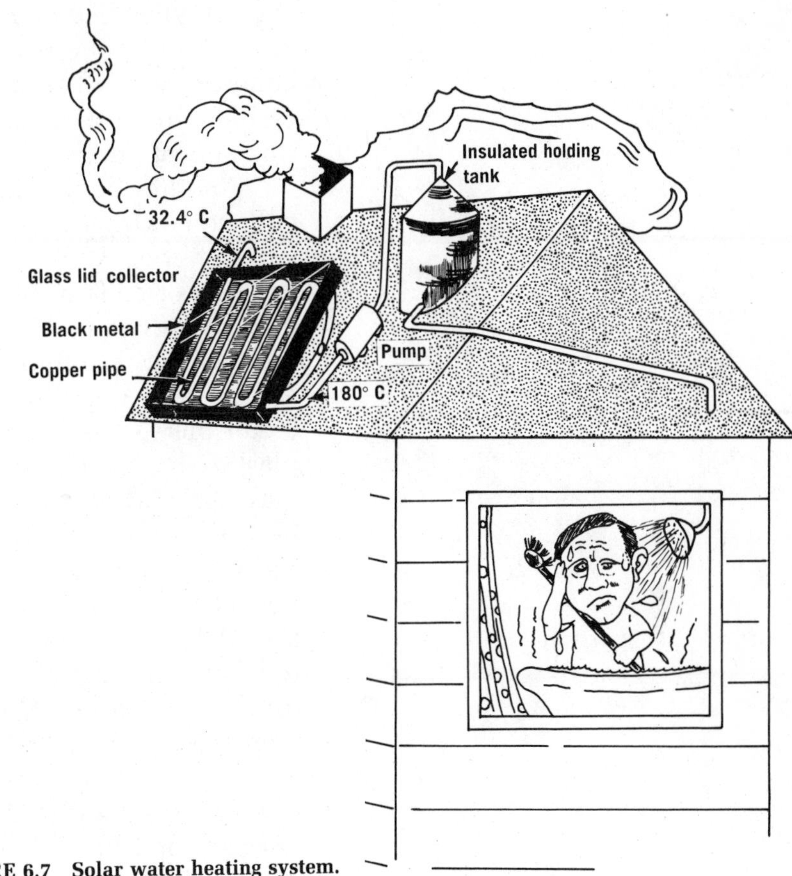

FIGURE 6.7 Solar water heating system.

cient at best, and, if we include losses in transmission, we cannot expect more than about 33 per cent efficiency as delivered to the home.

The technology is available to make improvements toward the realization of clean, efficient electric heat, if we are willing to pay the price. Consider the household refrigerator, which we normally think of as being only a cooling machine, but which actually is a device that uses fuel to pump heat from the cold interior to the warmer exterior. Therefore, the refrigerator is just as much a warming machine as a cooling machine. The fuel is needed because, according to the Second Law of Thermodynamics, heat would normally flow the other way, from the hot outside room into the cold refrigerator, as in fact it does when the electricity fails. If one stands behind the refrigerator while is is operating, one can feel the warm air being pumped out. Heat pumps that can heat a room by cooling the outside are, in fact, commercially available, and can be designed to achieve efficiencies three times as great as electric resistance heaters. Thus, the overall efficiency of an electric heat pump, including calculation of the inefficiencies of the power plant, is competitive with home furnaces. However, high installation costs have discouraged the wide application of heat pumps.

Another device for heating the home is the solar collector, which must collect, store, and transmit the sun's energy. As shown in Figure 6.7, such a collector consists of coils of pipe welded to a flat surface. The pipe carries some working agent, usually water. Pipe and plate are painted black to absorb as much sunlight as possible. A few inches above the surface of the collector plate, one or more layers of glass or transparent plastic sheet are installed. Since the outer layer of glass or plastic is transparent to most of the sun's rays, they are absorbed by the blackened pipes, and the water is thereby heated. Normally, much of the heat absorbed by a surface is lost to the air by reradiation. However, the blackened surface radiates infrared, invisible light. Glass and plastics are opaque to infrared radiation. The result is that, as with a common greenhouse, the heat cannot escape and remains in the system. Flat plate collectors of this sort can heat water to boiling even in temperate zones. Under normal conditions, heating to 180°F can be attained routinely. Water at this temperature is ideal for use in home heating applications. Of course, a lot of hot water must be stored for use when there is no sunlight—at night or during cloudy weather. The required storage capacity to provide for prolonged spells of sunless weather would be prohibitively large. But, if our demands are more modest, and we construct solar units to store heat only through the night and one cloudy day at a time, and we use conventional furnaces during prolonged spells of bad weather, then solar heat becomes quite practical, despite the high initial cost. Table 6.5 compares the cost of solar with other forms of heating for several American cities, assuming that the initial outlay is depreciated over a period of 20 years and all operating costs, including interest, are taken into consideration. Of the cities studied, only in Santa Maria, California is solar heat the cheapest available form, but in all other cities studied it is cheaper than electric heat.

Solar units can certainly provide "clean, efficient heat," yet there is no present market in the United States. The reason is unclear; perhaps the

TABLE 6.5 A COST COMPARISON OF SOLAR, ELECTRIC, AND GAS-OIL HEATING*

	COST OF HEAT IN DOLLARS PER MILLION BTU		
Area	*Solar*	*Resistive Electric*	*Gas-oil Average*
Santa Maria, Calif.	1.10–1.59	4.36	1.52
Albuquerque, N.M.	1.60–2.32	4.62	1.48
Phoenix, Ariz.	2.05–3.09	4.25	1.20
Omaha, Neb.	2.45–2.98	3.24	1.18
Boston, Mass.	2.50–3.02	5.25	1.75
Charleston, S.C.	2.55–3.56	4.22	1.26
Seattle, Wash.	2.60–3.32	2.31	1.92
Miami, Fla.	4.05–4.64	4.90	2.27

*Reprinted from *New Energy Technology* by Hottel and Howard by permission of The M.I.T. Press, Cambridge, Massachusetts. Copyright © 1971 by The M.I.T. Press.

advertising has been ineffective, perhaps the large, flat collectors on a roof would be considered ugly, or perhaps the high initial capital investment outweighs the prospects of low fuel bills. Worldwide, solar space heating has not been popular, but solar water heating is finding increasing popularity in Japan, Australia, and Israel. In the United States a solar water heater would cost between \$100 and \$500, depending on the location, compared to \$50 to \$125 for an electric or gas-fired unit.

Personal comfort depends not only on temperature, but also on the humidity and air flow in a room. Additionally, certain forms of radiation can produce comfort even if the ambient temperature is low. Thus, a person sitting in bright sun on a winter day can feel warm even at temperatures at which the moisture from his breath condenses. Likewise, a man standing in front of a cast-iron stove might feel as comfortable at 68°F as he feels at 75°F in a room heated by other means. If the best available human engineering were used in home heating, fuel bills could be lowered.

The relationship between temperature and comfort is also dependent on habits and customs. According to research carried out by The American Society of Ventilating Engineers in 1932, the preferred room temperature during the winter for a majority of subjects was 66°F. Similar research at later dates showed that the comfort range had risen to 67.5°F in 1941 and up to 68°F by 1945. In 1971 most homes, schools, and offices were heated to 70–72°F. Part of the change is due to the fact that 35 years ago people wore sweaters and long underwear indoors. If one lives through several winters of 66°F, then 70°F seems uncomfortably warm.

6.15 ELECTRICITY

(Problem 26)

Few issues in the field of environmental science have received so much attention as the problem of generating electric power. The production of electricity leads to water pollution, air pollution, solid wastes, radioactive wastes, land uglification, and thermal pollution. Yet our present society could not operate without electric power, because lighting, entertainment, home conveniences, many industrial processes, and thousands of cheap, relatively silent and trouble-free motors are dependent on it.

Electric consumption has risen faster than any

other form of energy consumption in the past few years and is expected to continue to accelerate rapidly (see Fig. 6.8). One cannot reap the benefits of electricity without some environmental side effects, but the use of innovative technology can certainly reduce the environmental insult.

Presently, most of our electricity is produced in steam turbines. The operating principle here is uncomplicated. Some power source, such as coal, gas, oil, or nuclear fuel, is used to heat water in a boiler and produce hot, high-pressure steam. This steam expands against the blades of a turbine and the spinning turbine operates a generator which produces electricity. The steam, its useful energy now spent, flows into a condenser where it is cooled and liquified, and the water is returned to the boiler to be reused. The cooling action of the condenser is vital to the whole generating process.

Let us return now for the last time to the caveman with his boulder, and the Second Law of Thermodynamics. The caveman learned that maximum efficiency was realized only when the temperature difference between the cold bar and the hot bar was greatest. Modern electric generators are much more complicated than the hot-bar engine, but the thermodynamic principles are the same. The steam, which is the working substance, expands against the turbine blades to do work. The steam cools as it expands, but it cannot cool below the temperature of its surroundings. Because the work output is proportional to the

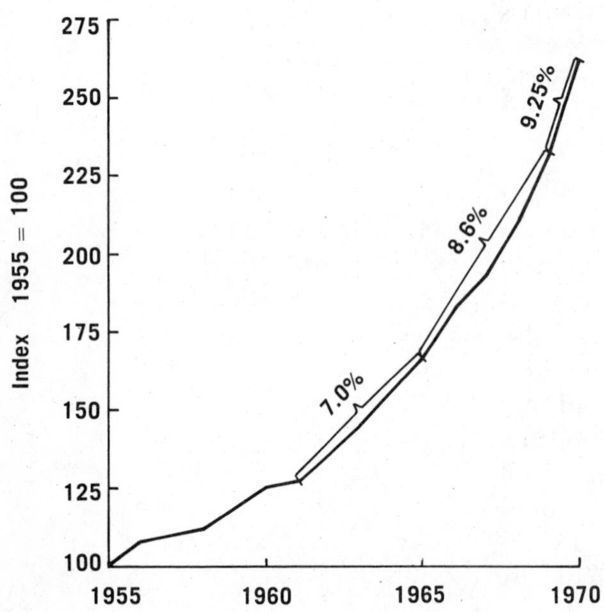

FIGURE 6.8 Consumption of electricity in the U.S. The percentages refer to growth rates over the periods indicated. (From *Scientific American*, September, 1971, p. 140. Copyright © by Scientific American, Inc. All rights reserved. From The Flow of Energy in an Industrial Society, by Earl Cook.)

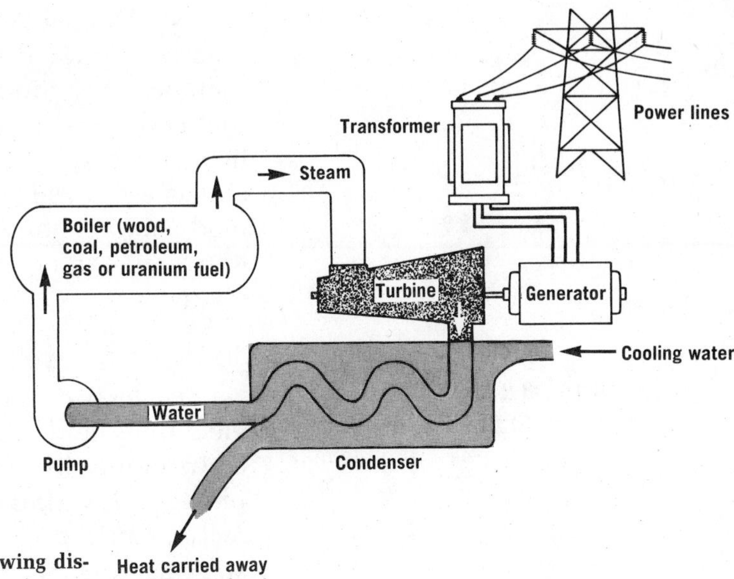

FIGURE 6.9 Power generator (schematic), showing discharge of waste heat.

degree of expansion of the steam, and the expansion is proportional to the temperature difference between the hot and cold ends of the turbine, some provision *must* be made to maintain a cool environment at one end of the engine. These relationships are shown schematically in Figure 6.9. In practice, cooling is accomplished by circulating water around the condenser. It is obvious that maximum efficiency is reached with very hot steam and a very cool condenser. In practice, the nature of the metals in the turbine limits the temperature to a maximum of about 1000°F (538°C). The low temperature is limited by the cheapest coolant, generally river, lake, or ocean water. Within the constraints of these two limits, an efficiency of 60 per cent is theoretically possible, but uncontrollable variations in steam temperature, and miscellaneous heat losses, reduce the efficiency to about 40 per cent, even in the best installations. This level means that for every 100 units of potential energy in the form of fuel, 40 units of electrical energy are available as useful work and 60 units of energy are dissipated to the surroundings as heat. Moreover, because refrigeration of the condenser is too expensive, further increases in efficiency depend on the use of hotter steam. Since the maximum upper temperature is limited by the ability of metals to withstand the heat stress, efficiency cannot be expected to increase appreciably unless some new breakthrough in metallurgy takes place.

Nuclear-fueled power plants are less efficient than fossil-fueled units operating at the same cooling temperature. Since the metals which contain the fuel cells must survive radioactive stresses as well as heat stresses, lower operating temperatures are used, and as we have learned from the Second Law of Thermodynamics, the smaller the temperature difference between engine and coolant, the less efficient the engine.

6.16 TRANSMISSION OF ELECTRICITY

(Problem 27)

Electricity is carried by both low-voltage service lines and high-voltage trunk lines. Service or distribution lines are carried along familiar "telephone poles" and can be run underground for a minimal extra cost. High-voltage transmission lines, strung across large metal towers, carry current through a potential difference of 135,000 to 500,000 volts and require a right-of-way at least 100 feet wide. The transmission network of the United States is quite complex, for not only must power be brought from distant generating stations to local centers for distribution, but the major generating facilities must be interconnected into a unified grid to prevent major blackouts in the event a single plant breaks down or requires repairs. One can guess that the first high-voltage transmission lines to enter a rural county were a welcome harbinger of the comforts of electricity. But today, the crisscross of large electrical towers and cables is viewed as an eyesore, and the rights-of-way are seen as a waste of land. In fact, an area of land equal to that of the state of Connecticut is now removed from productive use because it is tied up in rights-of-way. Power companies find it expensive to buy 12 acres of land per mile of right-of-way in urban areas. The obvious solution, putting the lines underground, is fraught with technical problems.

Because the earth is a more efficient power sink than the air, if high-voltage transmission wires of the type now used overhead were buried underground, power losses would be prohibitive. Specially insulated wires must be used, but these cost roughly $840,000 per mile or 16 times as much for materials and installation as an overhead wire. However, underground lines require no land, for they can be located along roads or under farmlands.

Research is presently being conducted in the

field of refrigerated underground lines. The principle here is twofold; at very cold temperatures power losses due to heat will be substantially reduced, and the conductivity of metals rises sharply at very cold temperatures. Such extremely cooled wires are believed to be able to operate at 200 to 300 million volts, and therefore to provide an extremely large capacity for transmission of power.

6.17 THERMAL POLLUTION

(Problems 28, 29, 30, 31, 32, 33, 36)

In the previous section we mentioned that an electrical generating plant uses a body of water as a coolant. A one-million-watt facility, running at 40 per cent efficiency, would heat 30 million gallons of water by 8.5 Celsius degrees (15 Fahrenheit degrees) every *hour.* A one-million-watt nuclear plant, operating at about 32 per cent efficiency, would require 50 million gallons of coolant water every hour to extract the same quantity of heat. It is not surprising that such large quantities of heat, added to aquatic systems, cause ecological disruptions. The term **thermal pollution** has been used to describe these heat effects.

The careful reader will note an apparent discrepancy here. If an operating efficiency of 40 per cent from a fossil-fuel plant yields 30 million gallons of warm water, then a nuclear plant at 32 per cent efficiency should warm 42.5 million gallons, not 50 million. The difference is accounted for by the fact that some of the heat from the fossil-fuel plant is not absorbed by water, but instead warms up a lot of *air,* by means of the hot gases discharged from the stacks. This is ecologically less disrupting than the warming of streams and lakes. Therefore, the difference in thermal pollution between the two types of plants is greater than that implied by thermodynamic considerations alone.

The processes of life involve chemical reactions, and the rates of chemical reactions are very sensitive to changes in temperature. As a rough approximation, the rate of a chemical reaction doubles for every rise in temperature of 10 Celsius degrees (18 Fahrenheit degrees). We know that if our own body temperature rises by as much as 5°C (or 9°F, which would make a body temperature of $98.6 + 9 = 107.6$°F) the fever may be fatal. What then happens to our system when the outside air temperature rises or falls by about 10 Celsius degrees? We adjust by internal regulatory mechanisms that maintain a constant body temperature. This ability is characteristic of warm-blooded animals, such as mammals and birds. Thus, the body temperature of a man or dog in a room cannot be determined by reading the wall thermometer. In contrast, non-mammalian aquatic organisms such as fish are cold-blooded: that is, they are unable to regulate their body temperatures as efficiently as warm-blooded animals.

How, then, does a fish respond to temperature increase? All its body processes (its metabolism) speed up, and its need for oxygen and its rate of respiration therefore rise. The increased need for oxygen is especially serious since hot water has a smaller

For many years, the word pollute has meant "to impair the purity of," either morally* or physically. The terms air pollution and water pollution refer to the impairment of the normal compositions of air and water by the addition of foreign matter, such as sulfuric acid. Within the past few years two new expressions, thermal pollution and noise pollution, have become common. Neither of these refers to the impairment of purity by the addition of foreign matter. Thermal pollution is the impairment of the quality of environmental air or water by raising its temperature. The relative intensity of thermal pollution cannot be assessed with a thermometer, because what is pleasantly warm water for a man can be death to a trout. Thermal pollution must therefore be appraised by observing the effect on an ecosystem of a rise in temperature. Similarly, noise pollution has nothing to do with purity: foul air can be quiet, and pure air can be noisy. Noise pollution is the impairment of the environmental quality of air by noise.**

*(1857.) Buckle, *Civilization*, I., viii, p. 526: "The clergy . . . urging him to exterminate the heretics, whose presence they thought polluted France."

**(1585.) T. Washington, trans. *Nicholay's Voyage*, IV; ii; p. 115: "No drop of the bloud should fall into the water, least the same shuld thereby be polluted."

capacity for holding dissolved oxygen than cold water. Above some maximum tolerable temperature, death occurs from failure of the nervous system, the respiratory system, or essential cell processes. According to the Federal Water Pollution Control Administration, almost no species of fish common to the United States can survive in waters warmer than 93°F. The brook trout, for example, swims more rapidly and becomes generally more active as the temperature rises from 40° to 48°F. In the range from 49° to 60°F, however, activity and swimming speed decrease, with a consequent decline in the trout's ability to catch the minnows on which it feeds. This inactivity is more critical because the trout *needs* more food to maintain its higher metabolic rate in the warmer water. Outright death occurs at about 77°F. In addition, spawning and other reproductive mechanisms of fish are triggered by such temperature changes as the warming of waters in the spring. Abnormal changes, to which the fish is not adapted, can upset the reproductive cycle.

In general, not only the fish, but entire aquatic ecosystems are rather sensitively affected by temperature changes. Any disruption of the food chain, for example, may upset the entire system. If a change of temperature shifts the seasonal variations in the types and abundances of lower organisms, then the fish may lack the right food at the right time. For example, immature fish (the "fry" stage) can eat only small organisms, such as immature copepods. If the development of these organisms has been advanced or retarded by a temperature change, they may be absent just at the time that the fry are totally dependent on them. It may be very difficult to predict such effects by studying stream temperatures, because a very large portion of the total flow of a given stream may be by-passed through a power plant to carry away its waste heat. While fish are easily excluded by screens from such undesirable detours, it is not easy to keep out microscopic organisms that do make up an important part of the food chain. These organisms are subjected to temperatures that will exceed the maximum temperature of the stream. We do not yet know how serious the consequences may be—the indications of various studies range from a zero effect to a 95 per cent kill of the plankton.

Higher temperatures often prove to be more hospitable for pathogenic organisms, and thermal pollution may therefore convert a low incidence of fish disease to a massive fish kill as the pathogens become more virulent and the fish less resistant. Such situations have long been known in the confined environments of farm and hatchery ponds, which can warm up easily because the total amount of water involved is small. As thermal pollution in larger bodies of water increases, so will the potential for increased loss of fish by disease.

Aquatic ecosystems near power facilities are subject not only to the effects of an elevated average temperature, but also to the thermal shocks of unnaturally rapid temperature changes. Power generation and heat discharge vary considerably from a peak in the afternoon to a low in the hours between midnight and daybreak. Additionally, a complete shutdown of a day or longer occurs occasionally. Thus, the development of cold-water species is hindered by hot water, and the development of hot-water species is upset by the unpredictable flow of heat. In addition, both types of organisms are adversely affected by rapid temperature changes.

Additional disruptions can occur because hot water has a reduced oxygen content. Power plants are usually located near population centers, and many cities dump sewage into rivers. Since sewage decomposition is dependent on oxygen, hot rivers are less able to cleanse themselves than cold ones. The combination of thermal pollution with increased nutrients from undecomposed sewage can lead to rapid and excessive algal growth and eutrophication. Therefore, thermal pollution imposes the unhappy choice of dirtier rivers or more expensive sewage treatment plants.

Manmade poisons, too, become more dangerous to fish as the water temperature rises. First of all, toxic effects are accelerated at higher temperatures. Second, as we just mentioned, warm water favors increased growth of plant varieties such as algae. The algae tend to collect in the power-plant condensers and reduce water-flow efficiency. The electric company responds by periodically introducing chemical poisons into the cooling system to clean the pipes. These poisons are then mixed with the downstream effluent. Additionally, domestic and industrial water consumers are more apt to discharge treatment chemi-

"It's not the humidity—it's the thermal pollution." (Amer. Scient. Sept–Oct. 1971.)

cals (such as copper sulfate) into water with high algae concentrations than into clean water. Thus, in warmer water, not only are fish less likely to resist poisons, but they are also likely to be exposed to them more.

The ecological balance in lakes can be particularly delicate, and the true biological effects are easily masked by misleading temperature data. Since cold water is denser than warm water, it sinks, and the bottom of a lake is almost always cooler than the surface. It is therefore advantageous for power companies to draw cold water from these bottom layers and discharge heated water onto the surface. The natural temperature difference between the two layers, which is about 10 Celsius degrees, is roughly equal to the temperature rise imposed on the water as it circulates through the power plant. Therefore, the power plant operation can be effected without changing the *surface* temperature of the lake. This does not mean, however, that thermal pollution has somehow been miraculously avoided, for the relative volumes of the hot and cold layers, and therefore the *average* temperature of the lake, do change. The hot discharge increases the yearly plankton production by prolonging the growing season. However, benthic and deep-water species are adversely affected. The temperature of the hypolimnion is unchanged by the power plants, and therefore the *concentration* of oxygen will remain constant, but as the volume of the hypolimnion is decreased, the *total quantity* of dissolved oxygen is reduced. The lake becomes increasingly discolored and eutrophic as the increased surface production and the decreased underwater consumption cause organic matter to accumulate.

Studies of the actual effects of thermal pollution are contradictory. A study of the 10-mile stretch of the Connecticut River above and below the Connecticut Yankee atomic power plant has shown little environmental effect of thermal pollution. An average of 22 million gallons of river water enters this plant every hour and is warmed by about 11 Celsius degrees (20 Fahrenheit degrees) before it returns to the river. A comprehensive environmental study of the situation yielded the following conclusions:

1. Except for concentrations of algae near the hot discharge, there was little alteration of plankton populations in the river.

2. Adjacent to the intake pipes, the diversity and quantity of benthic life was markedly decreased, probably because of silt removal by the pumping operation.

3. Near the heated discharge, the quantity and diversity of benthic species increased relative to the undisturbed system, and there was a shift in the dominant species. Clams yielded to insect larvae and sea worms.

4. The population of the resident fishes—catfish, white and yellow perch, sunfish, shiners, bass, suckers, and killifish—did not seem to be seriously affected by the discharge.

5. Catfish caught near the hot discharge plume appeared to be in poorer physical condition than other river catfish.

6. A commercially important spawning fish, the shad, escapes harm by swimming under the hot water.

This report, showing only minor disruptions of river life at the Yankee atomic power plant, is certainly reassuring. However, large fish kills have been reported at or near other generating facilities. For example, 150,000 fish were killed in January, 1970 at a Consolidated Edison unit on the Hudson River; another 120,000 perch at the same location died three months later; "two truckloads" of fish were killed by thermal discharge near Port Jefferson, New York. Some of these fish were killed, not by heat, but by mechanical trauma at the intake pipes.

It is impossible from these data to predict with assurance the effects of hot water on any other river or the cumulative effects of two power plants on a river from data about one plant. For example, yellow perch, sunfish, and suckers can survive warmer water than trout, grayling, and salmon. The discharge of waste heat into natural waters carries the threat of ecological damage; therefore, the potential environmental effects should always be appraised before new thermal loads are imposed on any aquatic system.

As we have already explained, the electric power industry, and particularly nuclear power plants, necessarily produce waste heat, and such heat is most conveniently discharged into flowing waters. It is estimated that by the year 2000 the rate of cooling water needed by the power plants will be equivalent to one-third of the total rate of freshwater run-off in the United States. Since this heat could not easily be distributed uniformly among all the large and small bodies of water in the country, but would be concentrated either at shorelines or along rivers of sufficient capacity to accommodate large plants, the result will be that serious ecological damage will become more widespread.

6.18 WASTE HEAT AND CLIMATE

(Problem 37)

The burning of fossil fuels and the fission or fusion of atomic nuclei add heat to the environment above and beyond the energy received from the sun.

Will direct heat discharged into the atmosphere or waterways affect global or local climate? Table 6.6 shows that the sum of all human activities does not even begin to compete with the energy received from the sun, and even in the years to come a significant direct-heating effect is unlikely on a global scale. But, in many metropolitan areas, the heat output of man's activities already exceeds the heat received from the sun. The climate of some metropolitan areas is also influenced by the fact that the buildings and roadways do not absorb or reflect the sun's rays to the same degree as did the forests and prairies on which they were built. It is not surprising then that the climates of many large cities are measurably different from the climates of the surrounding countrysides. (See Table 6.7.)

The magnitude of these climatic changes will increase in the future, and the effect will undoubtedly be measurable over larger areas as cities merge into megalopolises.

TABLE 6.6 SOME HEAT SOURCES AND THEIR MAGNITUDES

DESCRIPTION OF HEAT SOURCE	*MAGNITUDE (watts/meter^2)*
Net solar radiation at earth's surface	approx. 100
Energy production in 1970 distributed evenly over all continents	0.054
Energy production in 1970 distributed evenly over whole globe	0.016
Heat production of man's activities averaged over all urban areas	12
Heat production of man's activities in:	
Cincinnati	26
Los Angeles	21
Moscow*	127
Manhattan	630

*Net solar radiation over Moscow is only 42 watts/meter^2.

6.19 SOLUTIONS TO THE PROBLEM OF THERMAL POLLUTION

(Problems 34, 35, 38, 42)

Recall that we cannot invent a process to avoid thermal pollution. The generation of power produces excess heat, and we must deal with this problem by non-miraculous means. Some suggestions are described in the following paragraphs.

How to Decrease the Thermal Load

As mentioned previously, mechanical losses account for a 15 to 20 per cent loss of efficiency in conventional power plants. Thus, thermal pollution could be reduced if power plants operated at higher efficiency. A promising new design that increases efficiency is based on the principle of magnetohydrodynamics (MHD). In this system, air is heated directly and is seeded with metals like potassium or sodium

$$\underset{\text{potassium atom}}{K} + \text{heat} \longrightarrow \underset{\text{potassium ion}}{K^+} + \underset{\text{negative electron}}{e^-}$$

which lose electrons at high temperatures. This hot, electrified air stream is allowed to travel through a large pipe that is ringed with magnets. It is the movement of these charged particles that constitutes the generated electric current. At the end of the passage of the hot air through the magnets, the expensive seeding materials must be recovered. Furthermore, the exhausted hot air can operate a conventional turbine with the production of additional electricity. The advantages of this system are twofold. First of all, the overall efficiency of the system is expected to reach 60 per cent, and secondly, most of the waste heat is dissipated directly into the air rather than into aquatic ecosystems. Lack of funds has led to the stagnation of the MHD program in the United States, but the feasibility of the technique has been shown by the Russians, who now operate a 250,000-watt MHD generator near Moscow.

Since solar energy does not impose any additional heat on the environment, increased reliance on sunlight will decrease thermal pollution. Sunlight can generate electricity directly from some materials by releasing electrons from their surfaces. Such materials are said to be photovoltaic, and devices that make use of this phenomenon are called **photovoltaic cells.** Such direct conversion of light to usable power is an ideal environmental goal—there are no moving mechanical parts, and there is no thermal pollution. Some photovoltaic materials are known and have been used successfully in spacecraft, where cost is not the controlling factor. However, they are as yet too expensive to be used in competition with the more familiar but "dirtier" energy sources. Considerable effort is needed to develop new, cheap materials that can make photovoltaic power generation competitive. However, little money is available for research. Even if large sums of money are allocated, developmental success of solar cells cannot be assured, but the benefits of success would be high. The decision to fund such research would be a measure of our willingness to pay for a clean environment.

TABLE 6.7 AVERAGE CHANGES IN CLIMATIC ELEMENTS CAUSED BY URBANIZATION*

ELEMENT	***COMPARISON WITH RURAL ENVIRONMENT***
Cloudiness:	
cover	5 to 10% more
fog-winter	100% more
fog-summer	30% more
Precipitation, total	5 to 10% more
Relative humidity:	
winter	2% less
summer	8% less
Radiation:	
global	15 to 20% less
duration of sunshine	5 to 15% less
Temperature:	
annual mean	0.9°F to 1.8°F more
winter minimum (average)	1.8°F to 3.6°F more
Wind speed:	
annual mean	20 to 30% less
extreme gusts	10 to 20% less
calms	5 to 20% more

*Reprinted from *Inadvertent Climate Modification* by SMIC by permission of The M.I.T. Press, Cambridge, Massachusetts. Copyright © 1971 by The M.I.T. Press.

Let us imagine that competitive solar cells were available. A one-million-watt plant would require about 16 square miles of collector area, depending of course on the latitude and cloudiness of the region. Put in other terms, 35,000 square miles of land would supply all the projected power needed for the continental United States in 1990. It has been suggested that "waste land" be used productively by building solar generating stations there. A marsh such as the great swamp in northern New Jersey is one possible site. (See page 74 in Chapter 2.) As we have already learned, it is impossible to generate electric power without some adverse environmental effects. In this case, destruction of that swamp, or other "wastelands" may lead to vast ecological disruptions.

Studies are under way to determine the feasibility of building a large photovoltaic collector in space and beaming the collected power to earth by microwave transmission.

How to Put Waste Heat to Good Use

Large quantities of hot water can be used for diverse purposes. Hot water can improve growth conditions in large greenhouses. The costs of the greenhouse and water handling systems are high, and the economic feasibility might ultimately depend on the accommodation that can be made between the farmer and the power companies. Unfortunately, greenhouse owners don't need hot water in the summer at a time when power demands for air conditioning are high and the deleterious effects of thermal pollution on aquatic life are apt to be most severe.

Success in utilizing hot water for irrigating open field crops has been reported in the State of Washington. Alternatively, hot water circulating in closed pipes can be used to heat the soil without irrigating it; the resulting benefit is a higher agricultural yield, as shown by the data in Table 6.8.

TABLE 6.8 EFFECTS OF SOIL HEATING WITHOUT IRRIGATION ON VEGETABLE PRODUCTION IN MUSCLE SHOALS, ALABAMA

CROP	*YIELD IN TONS PER ACRE*	
	Heat	*No Heat*
String beans	6.9	2.7
Sweet corn	6.2	3.2
Summer squash	20.6	17.6

The practice of adding chemical poisons to kill algae in the condensers is incompatible with the use of that water for irrigation, although not incompatible with closed-system soil heating. If a power plant were to shut down during a cold wave, the flow of hot water would suddenly halt and a whole crop could be lost. In order to mitigate the high average cost of raising food under these conditions, a substitute source of heat would have to be available on a standby basis.

The growth rate of certain fish can be enhanced markedly by raising them under carefully controlled temperatures. Proponents of aquaculture claim that high yields of food can be realized if cheap hot water is available. (Japanese aquaculturists support up to three million pounds of carp per acre!) Thus, the heated discharge that can be detrimental to a whole ecosystem can be advantageous to a single species. However, fish farmers have the same sorts of problems as greenhouse operators. In addition, they must somehow dispose of considerable quantities of warm water heavily polluted by fish wastes.

Pumping hot water into home radiators is attractive but, again, economically impractical in many instances. First of all, city residents prefer not to live close to power plants, and the costs and heat losses involved in piping hot water any significant distance are prohibitive. Second, the installation of an underground steam system would be an extremely complex task in a large established city. Such a proposal would be practical only for newly built areas. Third, the maximum power demands do not coincide with maximum heating demands. Peak electric utilization is in the afternoon, while late-night use is low. Some storage facility or auxiliary steam generator would be needed to supply heat in the evening and at night. Finally, home heating would not remove waste heat during warm seasons or in warm climates. Despite these difficulties the prospect of "free" heat is appealing, and the proposals are being studied in many cities.

Other proposals are to use the heat to speed the decomposition of sewage (without contact between the hot water and the sludge), or to desalinate seawater. Both these proposals require careful cost analysis.

An interesting example of the use of waste heat is provided by the relationship between the Bayway (New Jersey) refinery of the Humble Oil and Refining Company and the Linden (New Jersey) Generating Station. The Linden power plant is capable of producing electricity at 39 per cent efficiency. For the past 15 years, this efficiency has been lowered by a less than optimum cooling of the condenser, and some of the waste heat has been sold as steam to Humble. If we consider the two-plant operation as a single energy unit, the overall efficiency of power production has been raised to a level of 54 per cent.

The process is beneficial to many: the companies save money, our fuel reserves are conserved, and thermal pollution of waterways is reduced 15 per cent. Similar operations are being planned or are in effect in other countries.

Though the technical problems of waste-heat utilization are not insurmountable in theory, the solutions are often economically discouraging. Typically, the waste water is hot enough to damage an aquatic ecosystem but not hot enough to be attractive for commercial use. Perhaps if the total environmental cost of thermal pollution were considered, waste-heat utilization would be more attractive.

How to Use the Atmosphere as a Sink

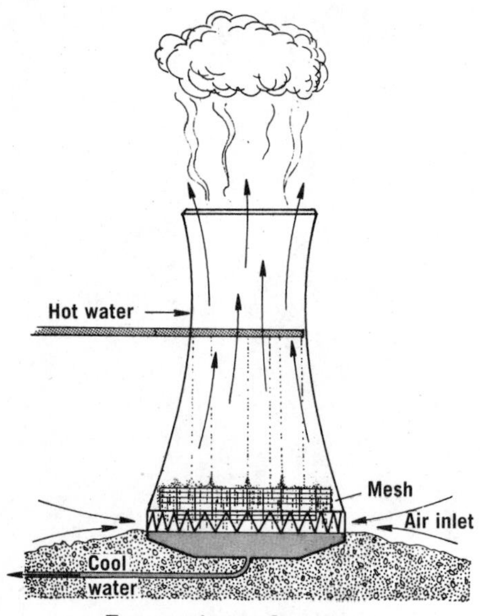

Evaporative cooling tower.

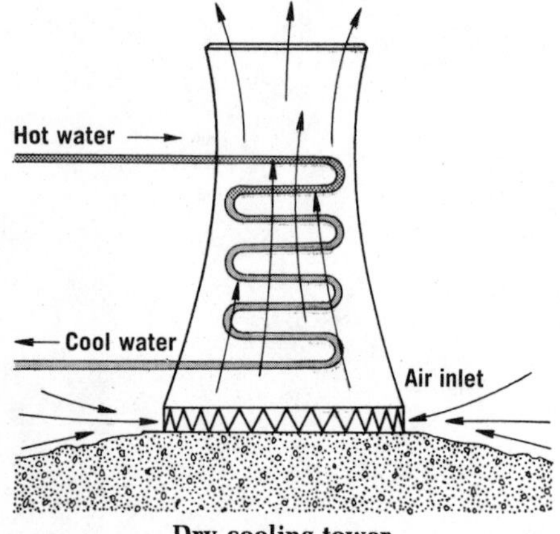

Dry cooling tower.

Another approach to the problem of thermal insult to our waterways is to dispose of the heat into the air. Air has much less capacity per unit volume for absorbing heat than water does, so the direct action of air as the cooling medium in the condenser is not economically feasible. For this reason power plants must still be located near a source of water, the only other available coolant. However, the water can be made to lose some of its heat to the atmosphere and then can be recycled into the condenser. There are various devices available that can effect such a transfer.

The two cheapest techniques are based on the fact that evaporation of water is a cooling process. Many power plants simply maintain their own shallow lakes, called cooling ponds. Hot water is pumped into the pond where evaporation as well as direct contact with the air cools it, and the cool water is drawn into the condenser from some point distant from the discharge pipe. Water from outside sources must be added periodically to replenish evaporative losses. Cooling ponds are practical where land is cheap, but a one-million-watt plant needs one to two thousand acres of surface, and the land costs can be prohibitive.

A cooling tower, which can serve as a substitute for a cooling pond, is a large structure, about 600 feet in diameter at the base and 500 feet high. Hot water is pumped into the tower near the top and sprayed onto a wooden mesh. Air is pulled into the tower either by large fans or convection currents, and flows through the water mist. Evaporative cooling occurs and the cool water is collected at the bottom. No hot water is introduced into aquatic ecosystems, but a large cooling tower loses over one million gallons of water per day to evaporation. Thus, fogs and mists are common in the vicinity of these units, reducing the sunshine in nearby areas. Reaction of the water vapor with sulfur dioxide emissions from coal-fired power plants can cause the resultant air to carry sulfuric acid aerosols.

Environmental problems can be reduced if dry cooling towers are used instead of evaporative wet ones. A dry tower is nothing more than a huge version of an automobile radiator installed into a tower to promote a speedy flow of air past the cooling pipes. Dry towers are uneconomical because of the cost of the prodigious amount of piping required. The relative costs of the three cooling facilities are shown in Table 6.9.

TABLE 6.9 COST OF THERMAL POLLUTION CONTROL FOR A FOSSIL-FUELED PLANT

TYPE OF CONTROL	*ESTIMATED AVERAGE COST mils per kwh**
Cooling Pond	0.08
Wet Tower, mechanical draft	0.10
natural draft	0.18
Dry Tower, mechanical draft	0.81
natural draft	0.99

*The cost of electricity is about one cent per kwh for consumers who use 1000 kwh per month, assuming that non-recycled water is used for cooling. The figures in this column therefore represent the *added* cost for control of thermal pollution. One mil = one-tenth of a cent.

6.20 ENERGY—AN OVERVIEW

(*Problem 39*)

From fuel mining to fuel transportation to fuel consumption to energy transmission, the environment always loses when electric power is generated. The problems are magnified by population density and increased demand on electricity. An electrical generating facility in Great Falls, Montana, in 1920 would have found all the problems inherent in a large modern facility, but space, air, and clean water were so plentiful, and the population was so sparse, that it would have been difficult to detect any environmental effects. No one would have believed that a serious problem could exist. A cowboy standing on a butte two days out of town could scan a circle a hundred miles in diameter and see almost no sign of man. This concept of inexhaustibility gave rise to what economist Kenneth Boulding calls "cowboy economics." Early Americans thought that since you can neither damage nor deplete the environment, you may as well take the cheapest route to your goal. Today we know that we can damage and deplete our planet. Old attitudes must therefore be re-evaluated.

The possibility of a serious crisis looms ahead. Three factors have combined to enhance the potential severity of this crisis: (1) Power companies have not always employed the best available technology to reduce pollution. Traditionally, they have been slow to employ environmentally satisfactory mining practices, to use the most effective air-cleaning de-

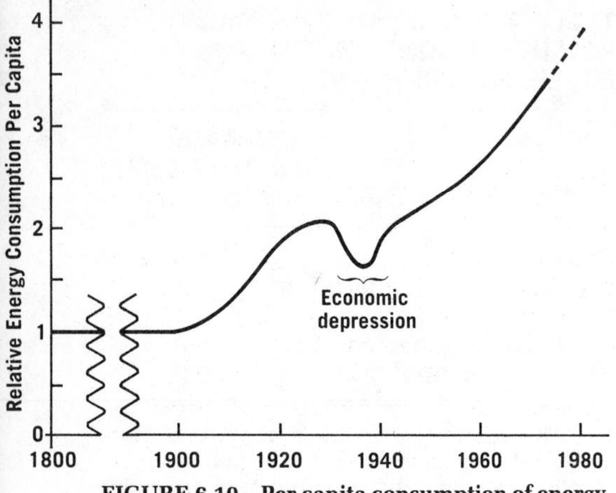

FIGURE 6.10 Per capita consumption of energy in the United States. (Vertical scale is based on: 1 unit = 10^8 Btu's per capita per year.)

vices, to construct cooling ponds or towers, and to engage in innovative research on environmental problems. (2) Population has risen. (3) Per capita consumption of electricity has also been steadily rising, as is shown graphically in Figure 6.10. Some of the components of this per capita increase are listed in Table 6.10. For example, in 1971, 51 per cent of the homes in the United States used electric blankets, rather than ordinary insulating ones; 44.5 per cent of the homes used room air conditioners, compared to 1.3 per cent in 1952; and 29 per cent of the homes had electric dishwashers, compared to three per cent in 1952. The severity of the energy crisis could be reduced by reversing any or all of these trends, but voluntary solutions to such problems are not notably successful. An alternative is some indirect or direct form of compulsion, such as a rise in the cost of energy, or

TABLE 6.10 HOMES WITH SELECTED ELECTRICAL APPLIANCES: 1952 to 1971*

ITEM	*1952* NUMBER	*1952* PER CENT	*1960* NUMBER	*1960* PER CENT	*1965* NUMBER	*1965* PER CENT	*1970* NUMBER	*1970* PER CENT	*1971* NUMBER	*1971* PER CENT
Total number of wired homes	42.3	100.0	51.7	100.0	57.6	100.0	64.0	100.0	65.6	100.0
Air-conditioners, room	0.6	1.3	7.8	15.1	13.9	24.2	26.0	40.6	29.2	44.5
Bed coverings	3.6	8.6	12.2	23.6	20.0	34.7	31.7	49.5	33.5	51.1
Blenders	1.5	3.5	4.2	8.0	7.5	13.0	23.4	36.5	26.2	40.0
Can openers	(NA)†	(NA)	2.5	4.8	14.2	24.7	29.1	45.5	31.5	48.1
Coffeemakers	21.6	51.0	30.2	58.3	41.3	71.7	56.7	88.6	59.7	91.0
Dishwashers	1.3	3.0	3.7	7.1	7.8	13.5	17.0	26.5	19.4	29.6
Disposers, food waste	1.4	3.3	5.4	10.5	7.9	13.6	16.3	25.5	18.6	28.4
Dryers, clothes (includes gas)	1.5	3.6	10.1	19.6	15.2	26.4	28.6	44.6	31.2	47.6
Freezers, home	4.9	11.5	12.1	23.4	15.7	27.2	20.0	31.2	21.4	32.7
Frypans	(NA)	(NA)	22.5	43.4	28.3	49.2	36.0	56.2	38.0	58.0
Hotplates and buffet ranges	9.0	21.2	12.5	24.2	13.1	22.7	15.7	24.5	16.3	24.8
Irons, total	37.9	89.6	45.7	88.4	57.1	99.1	63.8	99.7	65.4	99.8
Steam and steam/spray	8.3	19.5	30.6	59.2	44.6	77.5	56.5	88.2	59.1	90.1
Mixers	12.6	29.7	29.0	56.0	41.9	72.8	52.8	82.4	55.3	84.4
Radios	43.7	96.2	50.3	94.3	58.2	99.3	63.9	99.8	65.4	99.8
Ranges: Free-standing	10.2	24.1	16.0	30.9	18.5	32.1	25.9	40.5	27.9	42.6
Built-in			3.3	6.4	5.9	10.3	9.6	15.0	10.3	15.7
Refrigerators	37.8	89.2	50.8	98.2	57.3	99.5	63.9	99.8	65.4	99.8
Television: Black and white	19.8	46.7	46.2	89.4	55.9	97.1	63.2	98.7	65.4	99.8
Color	(x)‡	(x)	(NA)	(NA)	5.5	9.5	27.2	42.5	33.5	51.1
Toasters	30.0	70.9	37.2	72.0	48.1	83.6	59.3	92.6	61.7	94.2
Vacuum cleaners	25.1	59.4	38.4	74.3	48.1	83.5	58.9	92.0	61.9	94.4
Washers, clothes	32.2	76.2	44.1	85.4	50.3	87.4	59.0	92.1	61.8	94.3
Water heaters	5.8	13.8	9.8	18.9	13.5	23.4	20.2	31.6	22.2	33.8

*Wired homes in millions, as of December 31, 1971. Percentages based on total number of homes wired for electricity. Prior to 1970, radio data based on total homes as follows: 53,300,000 in 1960, and 58,566,000 in 1965. (From *Merchandising Week,* annual statistical issues. New York, Billboard Publications, Inc., 1972.)

†Not available.

‡Not applicable.

legislation that regulates the excessive use of power. Among the many possible specific approaches, the following ones, though as yet unpopular, may soon be considered palatable:

1. More funds could be made available for research in various fields related to energy. This includes funding of "long shots" such as large-scale solar energy installations as well as more conservative programs such as the breeder reactors. Ultimately, it makes little difference in total cost whether the power companies support the research and the cost is reflected in higher electric bills or whether government funds are used and taxes are raised. Depending on the tax structure, however, the two methods of funding ultimately shift the burden of research cost onto different groups of people.

2. Low-cost loans could be made available for construction of solar heaters or installation of any device or improvement which conserves fuels.

3. The recent very rapid increase in the use of room air conditioners has accounted for much consumption of power. There are various ways to reduce this consumption without sacrificing comfort. Buildings could be designed to promote convectional flow of air through them. This means more than just add-

"The egg timer is pinging. The toaster is popping. The coffeepot is perking. Is this it, Alice? Is this the great American dream?" (Drawing by H. Martin; © 1973 The New Yorker Magazine, Inc.)

ing windows. The use of novel roof designs, the proper orientation of the house itself, and the installation of larger attic louvers can all increase natural cooling and reduce the need for power. Small air-conditioning units are half as efficient as central systems in apartment houses. All new apartments could be required to install central air conditioning, and those wishing to plug into it could pay to do so. There could be a ban on cooling below 75°F, and heating above 70°F. Finally, perhaps fashions could somehow be changed so men would no longer feel constrained to wear jackets, ties, and long pants in business offices during the summer!

4. Resistive electric heating could be made illegal or could be very heavily taxed in order to encourage more efficient heat pumps or home furnaces.

5. Fluorescent lights are three times as efficient as incandescent lighting, but most of our lighting fixtures accommodate only the conventional screw-in type incandescent bulbs. The substitution of screw-in fluorescent bulbs would save a lot of energy without the necessity of converting most of the existing fixtures.

6. Excessive commercial power consumption for advertising and displays could be limited.

7. According to current pricing practice, electricity becomes cheaper per kilowatt hour as demand is increased. In a typical example, the first 50 kwh cost six cents per kilowatt hour, while any consumption above 900 kwh costs less than one cent per kilowatt hour. This rate structure naturally encourages additional use and should be re-evaluated. If the system were reversed, and prices per kilowatt hour rose with increased demand, people would be more reluctant to use electricity to perform tasks that were done manually 10 years ago. As a result, such a system would undoubtedly slow the rapid rise of per capita consumption. Naturally, different rate structures would be necessary for industrial consumers, or many manufactured goods would become too expensive to produce.

8. Of course, if population growth were stabilized, the energy requirements of the world would rise less sharply.

Even if these proposals to conserve power were adopted, our complex civilization would still consume large quantities of fuels, and many of the energy-related problems discussed in this chapter would remain with us. The most serious problems of power use—depletion, air pollution, radioactive waste disposal, thermal pollution, and destruction of wilderness—must be faced by innovative thinking in the physical sciences, economics, and social philosophy.

PROBLEMS

1. **Energy and power.** What is energy? Compare energy with heat, work, power.
2. **Energy and power.** Three farmers are faced with the problem of hauling a ton of hay up a hill. The first makes twenty trips, carrying the hay himself. The second loads a wagon and has his horse pull the hay up in four trips. The third farmer drives a truck up in one load. Which process—manpower, animal power, or machine power—has performed more external work? Which device is capable of exerting more power?
3. **Energy and power.** Your electric bill reflects the number of kilowatt hours that you have consumed. Are you being charged for power or energy? When we speak of the capability of a power plant we speak of its wattage. Are we speaking of power or energy? Explain.
4. **Energy and power.** Define foot-pounds per minute. Is this a unit of energy or power? What about pounds of ice melted? Pounds of ice melted per hour?
5. **First Law.** Write four statements of the First Law of Thermodynamics.
6. **Second Law.** Write four statements of the Second Law of Thermodynamics.
7. **First Law.** The drawing at right shows a perpetual motion machine based on osmosis. The two identical membranes are permeable to water but not to sugar. Water from the reservoir passes up through membrane 2 to a height that is determined by the osmotic pressure of the solution. Water also permeates membrane 1, falling back to the reservoir and doing work on the way down. Do you think this machine will work? If not, why not?

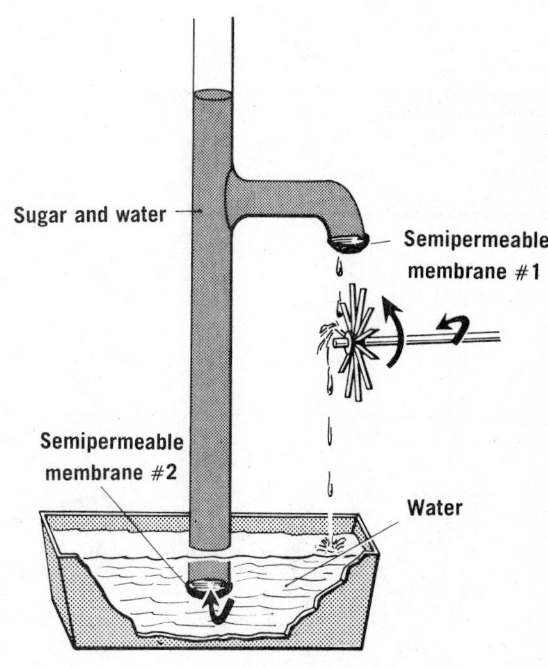

8. **Second Law.** Problem 8 in Chapter 1 reads, "We speak about nutrient cycles and energy flow. Explain why the words cycle and flow are used in their present context." Armed with the message of the Second Law, can you better explain the meaning of energy flow?
9. **Resources.** Predict the general shape of the curve showing world production of iron through the years. Explain your prediction.
10. **Oil shales.** Predict some environmental side effects of mining oil shales.
11. **Natural gas.** If the wholesale price of natural gas in

Wyoming increases by a factor of three, how will that affect the retail price in New Jersey? Explain.

12. **Coal mining.** Discuss the relative advantages and disadvantages of strip and tunnel mining.

13. **Alaska pipeline.** Analyze critically the statements below.

"[The Department of] Interior estimates that if performance of the oil tankers on the Valdez run were no better than the worldwide average, we could anticipate spills averaging 384 barrels a day." (From "The Living Wilderness," Summer 1970, page 8.)

"The Copper River Basin has the highest density [of waterfowl]; however these high densities occur a mile or more from the route." (From "The Environment," published by the Alyska Pipeline Service Co.)

14. **Alaska pipeline.** Explain why a pipeline across the tundra is more difficult to construct and more disruptive to the local ecology of the area than a pipeline across Texas.

15. **Energy supply.** Outline the predicted problems of energy production and use in the year 2100. Compare with the factors outlined in the answer to Problem 14.

16. **Uranium mining.** Predict some environmental side effects of mining Sierra Nevada granites for their uranium and thorium content. Compare your answer with the answer to Problem 10.

17. **Alternate energy sources.** Discuss briefly the prospects for solar, geothermal, tidal, wind, and garbage energies. What problems do these methods entail?

18. **Energy shortages.** In 1970, Consolidated Edison Company of New York was forced to cut the power delivered to consumers by 5 per cent on six different occasions. What do you think were some of the factors that caused these cutbacks?

19. **Transportation.** Presently, gasoline taxes are used for the construction of new roads. This practice has been considered fair because the roads are paid for by those who use them most. Increasingly, economists and social philosophers feel that many of our traditional concepts of fairness must be reevaluated in the light of environmental problems. Do you feel that there should be a reevaluation of road tax use? If so, how would you allocate funds? If not, explain.

20. **Hydrogen economy.** Explain how a hydrogen economy would function differently from a petroleum economy.

21. **Recycling.** Discuss some factors which might make recycling economically attractive.

22. **Space heating.** Why can a furnace be built to be more efficient than a steam engine?

23. **Heat pumps.** It has been suggested that small air conditioners be placed in the center of a room rather

than in a window for more efficient central cooling. Do you think this is a good idea?

24. **Heat pumps.** What is the common name for a heat pump that heats the outdoors at the expense of the indoors?

25. **Solar energy.** Describe the construction and function of a flat plate collector.

26. **Electric power generation.** Explain the function of a condenser in a steam turbine. Is a condenser needed at a hydroelectric facility?

27. **Electrical transmission.** Explain the difference between transmission and distribution lines.

28. **Thermal pollution.** Define thermal pollution. How does it differ in principle from air or water pollution?

29. **Nuclear vs. fossil fuels.** Explain why nuclear-fueled power plants require more cooling water than fossil-fueled plants.

30. **Thermal pollution.** Dogs live on all parts of the Earth from the tropics to the Arctic. Trout, on the other hand, are confined to waters no warmer than about 60°F. Why are the permissible temperatures for trout so much more limited?

31. **Thermal pollution.** Since marine life is abundant in warm tropical waters, why should the warming of waters in temperate zones pose any threat to the environment?

32. **Thermal pollution.** Warm water carries less oxygen than cold water. This fact is responsible for a series of disturbances harmful to aquatic organisms. Discuss.

33. **Thermal pollution.** What would happen if cooling water were drawn from the surface and discharged into the hypolimnion? Would this change be more or less useful to the power companies than the current practice? How would it affect aquatic life?

34. **Thermal pollution.** On page 222 of this book we stated that the heat discharged into aquatic systems is likely to produce serious ecological damage. Yet on page 229 we discussed uses of hot water for increasing fish yields in aquaculture. Is this a contradiction? Explain.

35. **Hot water.** Discuss some difficulties with use of hot water for agriculture; for aquaculture. Discuss the potential benefits.

36. **Solar energy.** Explain how a solar-operated steam generator can thermally pollute a river without having an effect on the total heat flux of the biosphere.

37. **Waste heat and climate.** What types of energy sources do not add any additional heat to the environment?

38. **MHD.** Explain the operation of an MHD generator. Why is it a desirable form of electrical generation?

39. **Economics.** Explain how "cowboy economics" in the United States is responsible for the fact that: (a) Recy-

cling operations are, in general, not economical; (b) the automobile is the most popular form of transportation; (c) most power plants use once-through cooling; (d) merchant tankers discharge their ballast into the ocean. If the concept of "cowboy economics" were displaced by a more conservationist doctrine, the prices of many goods would rise, but the quality of our environment could improve. How would you go about making decisions in instances where the two systems are in opposition? Do you think it would be helpful to set a dollar value on the environmental improvement and then match that against the price rise? If so, how would you set such a value? If not, why not?

The following problems involve calculations.

40. **Work and energy.** A man drives a 4000 pound truck with a 2000 pound load 40 feet up a hill. The vehicle is 25 per cent efficient. How many foot-pounds of fuel energy are consumed? How many foot-pounds of useful work are done?

41. **Energy.** How many calories are needed to heat 1000 grams of water 30°C? How many kilocalories? Would you need the same amount of heat to warm 10,000 grams of water 3°C?

42. **Cost of thermal pollution control.** A typical home consumes 1200 kwh per month. Use Table 6.9 to determine the monthly cost to a family of (a) a cooling pond, (b) a mechanical wet tower, and (c) a mechanical dry tower.

(*Answers*: 40. 960,000 ft-lbs; 80,000 ft-lbs. 41. 30,000 cal or 30 kcal for both. 42. 9.6¢; 12¢; 96¢.)

BIBLIOGRAPHY

Four recent and comprehensive books on energy and the environment are:

Neil Fabricant and Robert M. Hallman: *Towards a Rational Power Policy: Energy, Politics, and Pollution.* New York, George Braziller, 1971. 292 pp.

John Holdren and Philip Herrera: *Energy.* San Francisco, The Sierra Club, 1971. 252 pp.

Hoyt C. Hottel and Jack B. Howard: *New Energy Technology.* Cambridge, Mass., M.I.T. Press, 1971. 363 pp.

Richard S. Lewis and Bernard I. Spinrad: *The Energy Crisis.* Chicago, Educational Foundation for Nuclear Science, 1972. 148 pp.

A fine book on some physical aspects of power production, thermal pollution and climate is:

Theodore L. Brown: *Energy and the Environment.* Columbus, C. E. Merrill, 1971. 141 pp.

The most comprehensive book devoted to climate modification is:

Inadvertent Climate Modification. Hosted by the Royal Swedish

Academy of Sciences and the Royal Swedish Academy of Engineering Sciences. Cambridge, Mass., M.I.T. Press, 1971.

Two books on special topics are:

Farrington Daniels: *Direct Use of the Sun's Energy.* New Haven, Yale University Press, 1964. 374 pp.

Marvin M. Yarosh, ed.: *Waste Heat Utilization.* Springfield, Va., National Technical Information Service, 1971. 348 pp.

Two books with good articles on strip mining are:

Garry D. McKenzie and Russell O. Utgard, eds.: *Man and His Physical Environment.* Minneapolis, Burgess Publishing Co., 1972. 338 pp.

Fred C. Price, Steven Ross, and Robert L. Davidson, eds.: *McGraw-Hill's 1972 Report on Business and the Environment.* New York, McGraw-Hill, 1972. 505 pp.

Another book with a fine section on thermal pollution is:

H. Foreman: *Nuclear Power and the Public.* Minneapolis, University of Minnesota Press, 1970.

The subject of thermodynamics is covered in many standard texts at various levels, and no specific references need be given here.

The entire issues of *Scientific American* for September, 1971 and of *Science* for April, 1974 were devoted to energy and power.

7

AGRICULTURAL SYSTEMS

7.1 INTRODUCTION

When man first evolved from his ape-like ancestors his diet depended on what he could manage to collect from day to day. He hunted and dragged his food back to his den or was hunted and dragged back to some other predator's lair. He competed with other herbivores for plant foods. Life was very hard during drought, flood, or pestilence. Because his technology was so limited, and his population so small, early man did not appreciably alter the earth's environment. His stone and wooden tools for digging and hunting were competitive with the tusk of the mammoth and the claw of the tiger, but certainly were not overwhelmingly superior.

Even when man first began to cultivate the land, his activities had little impact on global ecosystems. People lived close to their food supplies, and their wastes were returned to the farmlands directly (see Fig. 7.1). Thus, the consumption of nutrients was balanced by a return of nutrients. Of course, even a simple agricultural system is potentially disrupting, for it promotes the growth of a few species, where many species once existed, but the extent of the environmental alteration was small in early cultures. In fact, man was not alone in favoring those plants and animals which benefited him most; bees selectively pollinate the most succulent flowers, ants protect their herds of aphids, and sage grouse scratch the soil around the roots of sagebrush.

When agricultural yields first enabled some individuals to pursue goals other than cultivation of the soil and to live far from sources of food, the cycle was broken, for nutrients were not all returned to the farmland from which they came. The result was an ecological imbalance. As food-growing technology became more efficient, the ecological disruptions became more severe.

Of course, man has understood for many years that his choices were either to refertilize the land or, eventually, to move elsewhere. In some areas, man has been quite successful in keeping farmlands fertile. For example, some regions of China, Japan, and Europe have been farmed successfully for thousands of years. During that period of time the soil has even been enriched by fertilization with human and animal manure and various other materials of biotic origin. However, in some places, as we will see, man still continues to destroy huge areas of previously fertile land.

The story of agriculture includes both successes and failures. While the total harvest of food for man has increased steadily for thousands of years, more than half of the world's people are victims of malnutrition, and famines due to crop failures continue to occur. Man's technological accomplishments, both in agriculture and in industry, have enabled him to rise to his present level of biological ascendancy, but many informed people feel that our present practices,

FIGURE 7.1 A primitive agricultural field in New Guinea. (From Rappaport, R. A.: The Flow of Energy in an Agricultural Society. Sci. Amer., Sept. 1971, p. 116. Photo courtesy of Roy A. Rappaport.)

despite their successes, are altering the earth's ecosystems so severely that the planet's future ability to produce high yields of food for man is being endangered.

7.2 AGRICULTURAL DISRUPTIONS

(Problems 1, 2, 3, 4, 5)[*]

It is generally known that the agriculture of India cannot keep up with the needs of its people and that, consequently, famine is never more than a few dry seasons away. It is less well known that approximately two-thirds of the cropland has been completely or partially destroyed by erosion or soil depletion caused by man. The area of the province of Sind near the mouth of the Indus River is typical of a destroyed cropland. With the exception of areas under irrigation, Sind is now a barren, infertile, semi-desert region. Yet, archeological diggings have uncovered the remnants of highly civilized inhabitants who farmed there over 4000 years ago. Moreover, fossils of the native animals of the area show that elephants, water buffalo, tigers, bears, deer, parrots, wolves, and other similar forest dwellers lived there. The early settlers built temples with fire-baked bricks; undoubtedly, wood fed the fires, and the wood must have come from forests in the nearby areas that are now desert. The details of the transformation from forest to desert are largely conjectural because there are no recorded weather reports over these 4000 years. However, meteorology suggests a plausible mechanism. Remember that a forest maintains a cooler environment than a grassland. Part of this coolness is due to the water-holding ability of the forest ecosystem. Now, rain clouds passing over a hot tropical steppe will generally rise and precipitation will seldom occur, while rain clouds passing over cool jungle will be induced to drop their moisture. Thus, the change in climate which destroyed the land of Sind was probably caused by deforestation. This poses an interesting dilemma. Most farmers grow only those crops that produce an efficient quantity of food. Forests are poor producers of food for man. But the killing of the Sind forest ultimately destroyed the productivity of the land. Had the early inhabitants of Sind understood the importance of natural diversity, and had they been able to reach a workable apportionment of forest and cropland, they might have been able to grow enough food without destroying their ecosystem.

In the light of this hypothetical example it is interesting to observe what is happening in India today. As overgrazing destroys the ground cover, goat herders, using 20-foot-long sickles, cut the limbs of the trees for food for their goats. The trees die and the soil blows away. But try to tell a man who has a hungry baby at home to save the trees for future generations!

Another story of land destruction comes from the area of the fertile crescent, the "cradle of civilization." The Tigris-Euphrates valley gave birth to several great civilizations. We know that highly sophisticated systems of letters, mathematics, law, and astronomy originated in this area. Obviously, then, men had the time and energy to educate themselves and to philosophize. We can deduce that the food supply must have been adequate. Today much of this region is barren, semi-desert, badly eroded, and desolate. Archeologists dig up ancient irrigation canals, old hoes, and grinding stones in the middle of the desert. What must have happened?

Part of the story starts at the source of the great rivers in the Armenian highlands. The forests were cleared to make way for pastures, vineyards, and wheat fields. But croplands, especially if poorly managed, cannot hold the soil and the moisture year after year as well as natural forests or grasslands. As a result, large water run-offs such as those from the spring rains or melting mountain snows tended to flow down the hillsides rather than soak into the ground. These uncontrolled waters became spring floods. As a protective measure, canals were dug in the valley to drain the fields in the spring and to irrigate them in the summer and fall. Later, devastating wars resulted in the abandonment of the canals and croplands. The neglected canals turned to marsh, and the formation of marshes shrunk the rivers, and the diminished rivers could no longer be used to irrigate other areas of land. The water table sank, so that the yields per acre today are less than the yields 4000 years ago.

The story of land destruction can be repeated with monotonous regularity. Ancient Carthage was founded on the shores of the Mediterranean in North Africa amid dry but fertile grasslands. Grain was grown in abundance. Today much of this area has become part of the Sahara Desert. In fact, a large part of the Sahara is a by-product of over-plowing and over-farming fertile land, which led to a depletion of the

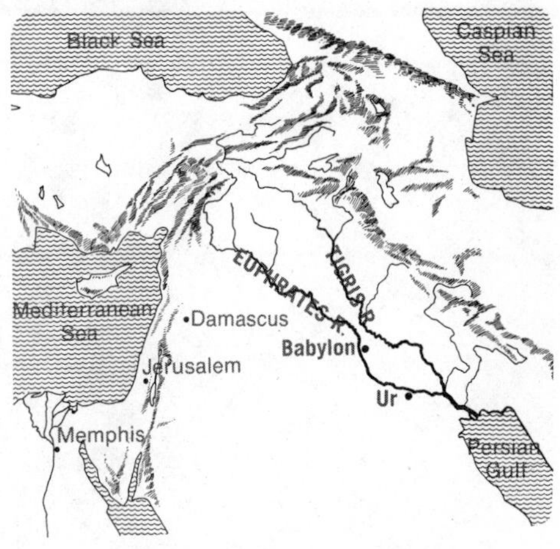

The Fertile Crescent.

amount of available moisture. To study this process in more detail we shall examine a great American fiasco —the Dust Bowl.

The early European settlers found millions of acres of virgin land in America. The eastern coast, where they first arrived, was so heavily forested that even by the mid-eighteenth century, a mariner approaching the shore could detect the fragrance of the pine trees about 180 nautical miles from land. The task of clearing land, pulling stumps, and planting crops was arduous. Especially in New England, long winters and rocky hillsides contributed to the difficulty of farming. It was natural that men should be lured by the West, for here, beyond the Mississippi, lay expanses of prairie as far as the eye could see. Deep, rich topsoil and rockless, treeless expanses promised easy plowing, sowing, and reaping. In 1889 the Oklahoma Territory was opened for homesteading. A few weeks later the population of white people there rose from almost nil to close to 60,000. By 1900 there were 390,000—a people living off the wealth of the soil. In 1924 a thick cloud of dust blew over the East Coast and into the Atlantic Ocean. This dust had been the topsoil of Oklahoma (see Fig. 7.2).

In each of the earlier examples one might contend that the destruction of the land was really caused by

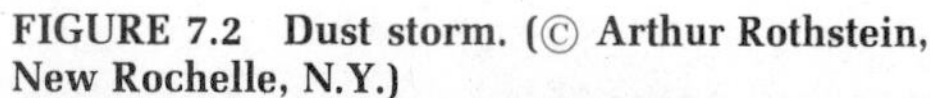

FIGURE 7.2 Dust storm. (© Arthur Rothstein, New Rochelle, N.Y.)

changes in climate rather than by man's mismanagement of the land. Indirect evidence, however, strongly implicates man. For instance, in the areas between the Tigris and Euphrates rivers where the canals were not destroyed, the land remains fertile. Similarly, some areas of North Africa near the Sahara still support trees believed planted by the Romans, and geological evidence indicates relatively constant weather patterns in these areas. However, in the case of the Oklahoma Dust Bowl we *know* that the land was destroyed by man and not by climate.

In Chapter 1 we learned that a natural prairie is a diversified ecosystem with homeostatic mechanisms to protect itself from spring floods and summer droughts. White man's contribution to the prairie was not particularly far-sighted. He planted large fields of single crops, thus destroying the naturally balanced system. He killed the bison to make room for his cattle, then killed the wolves and coyotes to prevent predation of his herds. Moreover, he often permitted his cattle to over-graze. In over-grazed land, the plants, especially the annuals, become so sparse that they cannot reseed themselves. The land itself, therefore, becomes very susceptible to soil erosion during heavy rains. In addition, the water runs off the land instead of seeping in, resulting in a lower water table. Because the perennial plants depend upon the underground water levels, depletion of the water table means death for all prairie grasses. The whole process is further accelerated as the grazing cattle pack the earth down with their hooves and block the natural seepage of air and water through the soil.

Man's introduction of the plow to the prairie had an even more severe effect because the first step in turning a prairie into a farm is to plow the soil in preparation for seeding. At this point, of course, the soil is vulnerable, since the perennial grasses which normally hold the soil during drought have already been killed.

If the spring rains fail to arrive, then the new seeds won't grow and the soil will dry up and blow away. As an emergency measure during droughts, farmers often practice **dust-mulching.** By chopping a few inches of the surface soil into fine granules a thin layer of dust is formed. This dust layer aids the capillary action whereby underground water is brought to the surface. Thus the soil remains moist just below the dust, and the seeds sprout. However, before labeling dust-mulching a success, one must examine several other factors. Fertile soil is more than pulverized rock; it consists of decayed organisms and partially decayed organic matter. Moreover, prairie topsoils have evolved a balancing mechanism whereby the concentration of decayed organic matter remains approximately constant. Dust mulching brings to the surface much of the previously underground organic material. Once on the surface this organic matter reacts with the oxygen of the air (oxidizes) much more rapidly than it would have had it remained underground. The result is a decrease both in the concentration of organic matter in the soil and in the soil's fertility. As these losses continue, a time is finally reached when the soil is so barren that it cannot support even the growth of the hardy prairie grasses. Of course, the farmer can replace lost organic matter by spreading manure or other fertilizer, but the fact is that millions of acres of previously fertile farmland have been ruined because the farmers have not refertilized adequately.

On the other hand, if, after plowing, the spring rains are too heavy, then the soil may easily wash away before the seeds have an opportunity to grow. Even after seeds have sprouted, the practice of pulling

FIGURE 7.3 Farmland in Mississippi ruined by soil-erosion. (Photo courtesy of U.S. Forest Service.)

weeds between the rows leaves some soil susceptible to erosion by heavy rains.

The preceding discussion is relevant to the events in Oklahoma. Over a period of 20 to 35 years the soil fertility slowly decreased. Incomplete refertilization and loss of soil from wind and water erosion took their toll (see Fig. 7.3). Finally, when a prolonged drought struck, the seeds failed to sprout and a summer wind blew the topsoil over a thousand miles eastward into the Atlantic Ocean.

The droughts that killed the Oklahoma farms had no lasting effect on those prairies left untouched by man. In fact, these virgin lands are still fertile. In a few thousand years, perhaps, the wind-scarred Dust Bowl will regain its full fertility. We say "perhaps" because similar destruction of the North African prairie left the land so barren that nothing was left to hold the rain that did fall. Two thousand years after the farms failed one can stand in the center of the ruins of a wealthy country estate and watch the sands of the great Sahara blow by.

One might think that it would be impossible to destroy land by importing water (irrigating). Unfortunately, irrigation, too, can be destructive. When rain water falls on mountain sides, it collects in small streams above and below ground and it filters over, under, and through the rock formations. In the process of flowing into a large river, the water dissolves various mineral salts present in the mountain rock and soil. Usually these salts are concentrated in the oceans. However, if the river water is used for irrigation, man is bringing slightly salty water to his farm. When water evaporates the salt is left behind and, over the years, the salt content of the soil increases. Because most plants cannot grow in salty soil, the fertility of the land decreases. In Pakistan an increase in salinity decreased soil fertility alarmingly after a hundred years of irrigation. In parts of what is now the Syrian desert, archeologists have uncovered ruins of rich farming cultures. However, the land adjacent to the ancient irrigation canals is now too salty to support plant growth.

Irrigation is threatening agricultural production in parts of California, which produces about 40 per cent of the vegetables consumed in the United States. The richest vegetable-growing area in California is the San Joaquin valley where virtually all commercial

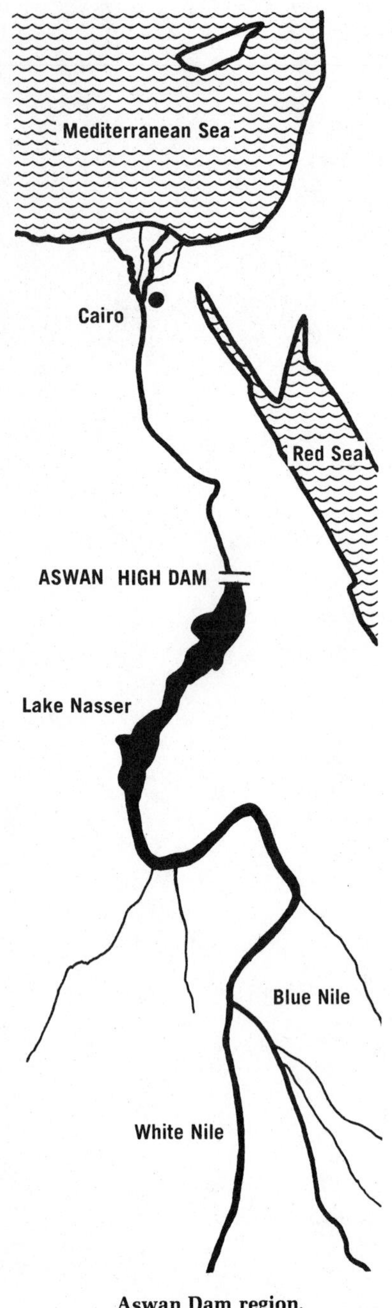

Aswan Dam region.

operations rely heavily on irrigation. As the irrigation intensifies, so does the threat of destructive salination of the soil and the groundwater. Most of the valley is now serviced by a shallow underground drainage system to divert the brackish waters, but the old system now appears inadequate, and new drainage projects are being proposed.

In addition, extensive, poorly managed irrigation can raise the water table of the irrigated land. If the water table is too high, plant roots will become immersed in water and will die from lack of air. In the earlier 1960's waterlogging and salinity problems were causing the loss of 60,000 acres of crop land in Pakistan alone.

Occasionally, even an apparently well-planned upset of the balance of nature has produced disastrous results. For thousands of years the peasants of Egypt farmed the Nile Valley. Every spring the river flooded and brought with it water and fresh soil from the Abyssinian Mountains. When the water level subsided the farmers grew what they could during the long hot summer. To increase the agricultural yield for a growing population, engineers had long considered the feasibility of damming the Nile, thus storing the excess water during flood seasons for more efficient use during the summer. The Aswan High Dam, built over a period of 11 years and inaugurated in 1971, was to make this objective possible as well as to provide abundant hydroelectric power. Water storage started in 1964, when the Nile was diverted into a bypass channel during construction, and power has been produced since 1967. Other benefits obtained have included higher yields of cotton, grain, fruits, and vegetables by irrigation during the summer and by reclamation of previously barren land. But there have been problems. First, the sediment that the flood waters formerly washed out to sea, although useless to farmers, did nourish a rich variety of aquatic life. The absence of this nutrient has resulted in the annual loss of 18,000 tons of sardines. Second, the flood waters previously rinsed away soil salts that otherwise would have accumulated and made the soil less fertile. With the flooding under control, the salts are left in the soil. The consequent rise in soil salinity is a threat to the productivity of the land. Third, the sediment formerly protected the delta land in several ways: it served as an underground sealant to minimize seepage that would otherwise drain off various sweet-water lakes, and it helped strengthen natural sand dikes that protected the coastline against erosion by the powerful currents of the Mediterranean. These buffering actions, too, are gone, as the sediment from the Abyssinian Mountains instead sinks behind the High Dam, and the clear, silt-free waters rush downstream more rapidly than before, eroding the river banks and undermining the foundations of many of the bridges that span the Nile. Fourth, the loss of silt previously deposited on the downstream fields is a loss of nutrient, and must be compensated by chemical fertilizers. Fifth, the dry periods between floods, now eliminated by irrigation, used to limit the population of water snails which carry a worm that spreads easily to man, by depositing its larvae under the skin, whence they invade his intestinal and urinary tracts. The resulting debilitating disease, called **schistosomiasis**, follows along the paths of the irrigation canals, infecting about 80 per cent of the people who work in them. By what reckoning can one balance lost manpower and increased human misery against higher crop yields?

The powerhouse of the Aswan High Dam (Sadd-El-Aali) creating the Lake Nassar that will eventually flood south far into Sudan. Photography from the rim of the dam in either direction was off-limits; however, there were no restrictions in taking powerhouse photographs on the downstream side of the dam (February 1974). (Photo courtesy of Ward W. Wells—DPI.)

Finally, the High Dam seems to be responsible for a serious loss of water, the one essential substance that was never to be in short supply. The lake behind the dam was to have been filled by 1970; it was not yet half full in 1971. Some limnologists (scientists who study the physical phenomena of lakes) predict that it will not be full for a century or more. Losses are suffered by seepage through porous rock and by evaporation in the hot winds of Upper Egypt. Of course, technology may come to the rescue. All the problems cited above are under intensive study; means may be found to control the erosion, the parasites, the seepage, and the salinity, and to make enough power to supply fertilizer plants to manufacture nutrients to replace those formerly carried down as silt from the mountains. There is no law of chemistry or physics that denies the possibility of such technological rescue of technology's disruptions. But thus far this goal is much more elusive than had been expected.

We have cited examples of some of man's past agricultural failures that have led to the loss of large areas of fertile cropland. At the present time, millions of acres are still being lost. In fact, farm soils are presently being destroyed faster than they are being formed by natural processes. In recent times, despite this overall loss of fertile areas, greatly improved farming techniques have increased agricultural yields across the world. However, these increases cannot be expected to continue indefinitely and some future leveling effect can be anticipated.

7.3 ENERGY FLOW IN INDUSTRIAL AGRICULTURE

(Problem 6)

Consider some differences between a wild oat plant growing in an unfarmed prairie and a domestic oat plant in a field. To survive, the wild oat must compete successfully for sunlight and moisture with its neighbors. A tall plant, one that sprouts early, or one with an effective root system, has a competitive advantage. The energy a plant needs to grow a tall stalk or a deep root must come from the sun. On the other hand, a farmer aids the survival of a cultivated oat. He waters it when necessary, removes plant competitors, and loosens the soil to stimulate the growth of root systems. Since all the seeds in the field are planted at the same time and are of the same variety, competition is minimal and the plant does not need a tall stalk or a unique and fast-growing root system to survive. In other words a wild plant must use some of its incident solar radiation for survival and some for the production of seeds or bulbs. A farmer is willing to add auxiliary sources of energy to help the plant survive if he can grow a variety that produces more food for man.

As we mentioned in Chapter 3, there are significant individual differences within a given species. Particularly important among individual plants are the differences in total weight of edible matter. Thus, if a farmer replants seeds only from those plants that produce the most grain, regardless of the viability of that plant in an *untended* field, he should be able to breed high-yielding plants over the course of time. In today's laboratories, different varieties of plants are often cross-bred with the hope that a new combination of existing traits might produce progeny with desirable characteristics (see Fig. 7.4). Alternatively, the existing characteristics may themselves be changed by inducing mutations artificially. These manipulations are often varied and quite sophisticated, but all of them lead to the same kinds of choices: The agronomist must decide whether or not to select a plant for further breeding, and ultimately whether to adopt it for commercial planting. Traditionally, agronomists have chosen the plant varieties that yield the most

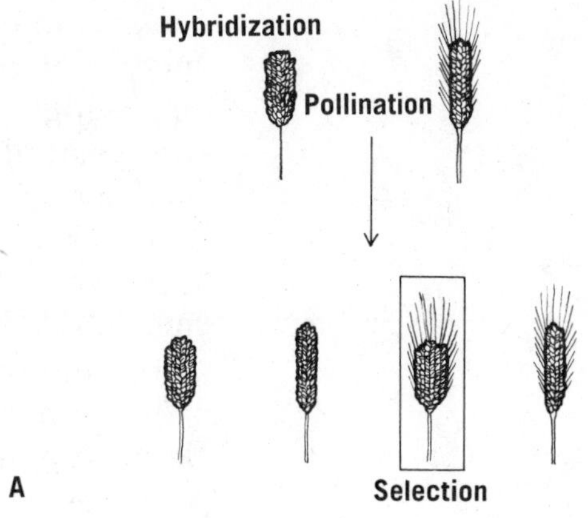

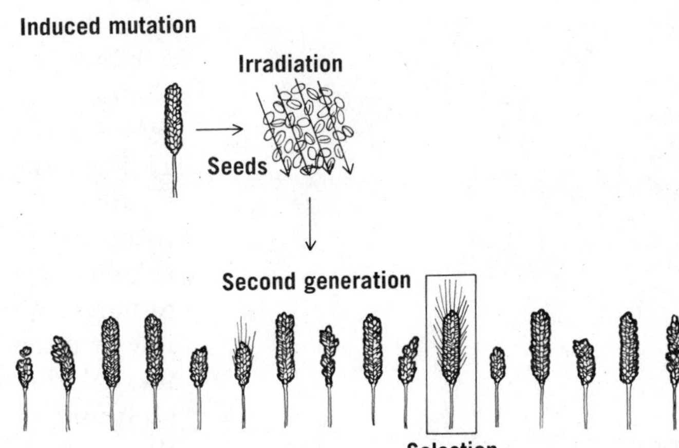

FIGURE 7.4 Scheme illustrating the general procedure for cross-breeding of plants.

desirable crop—usually with emphasis on quantity. As a result of such selections, however, crops have become increasingly dependent on cultivation for survival. Thus, if we compare a variety of Oriental rice developed in the late 1960's with wild Oriental rice, we find that the new variety produces more grain than the old, but is also shorter, less resistant to disease, and more dependent on irrigation than its predecessor. In addition, the new seeds have lost their biological clocks and can sprout any time they are planted, whereas native rices are geared to the seasons, and if left uncultivated, new seedlings will germinate only in the proper growing season. The new seeds require care, and care requires energy—energy to plant, to weed, to control pests and disease, and to irrigate.

Similarly, animal breeders have produced varieties of livestock that are more efficient than prairie grazers in converting grain to meat and to dairy products. But the farmer must invest energy to realize these high yields. A prairie chicken lays 10 to 20 eggs per year, all of them in the spring, then broods them and cares for the chicks. A domestic hen can produce over 200 eggs in a year, but most hens will not sit on a fertilized egg long enough for it to hatch. Therefore, most breeds of domestic chickens require artificial incubations for propagation.

Farmers in industrially developed countries have large supplies of fossil fuels available to them. They use these resources to supplement the energy that the plants get from solar radiation (see Fig. 7.5). Let us examine the overall energetics of a typical modern agricultural system—the mechanized production of beef. Remember that for every 12 calories of energy a plant receives from the sun, only about one-half calorie is available for the production of animal tissue by a herbivore that consumes the plant. Some of the 11.5 calories that are "lost" are required by the plant for its metabolism, some by the animal for its metabolism, and some of the energy is used by the animal in search of more food. In a feedlot, a steer is not required to move, but may stand in front of his feedbin, eating hay or grain that the farmer, with the help of his tractor, baled or threshed and brought in from the field. In addition, food additives and growth hormones, synthesized in factories powered by coal or oil, are used to increase growth. In a feedlot, about six calories of fossil fuel energy are used for every 12 calories that the plant matter receives from the sun. The total weight gain of the animal under these conditions corresponds to about five calories (see Fig. 7.6). The artificial system is more efficient than a natural system in converting both plant and total energy to meat. This enhancement of efficiency is the major benefit of the modern system of producing food. Production of beef in feedlots is based on a new cycle of manmade technology. High food production is needed for the urban workers who provide the technology required to maintain the high food production. The individual components of the cycle are interdependent. Large populations of human beings are dependent on high agricultural yields, while at the same time these high yields are dependent on high industrial outputs. If one link in this chain is cut, there cannot be reversion to the ancient range system without mass starvation.

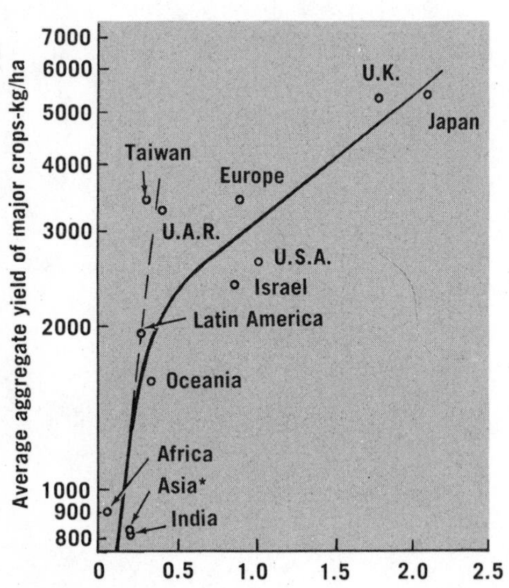

FIGURE 7.5 The relationship between crop yields and energy.

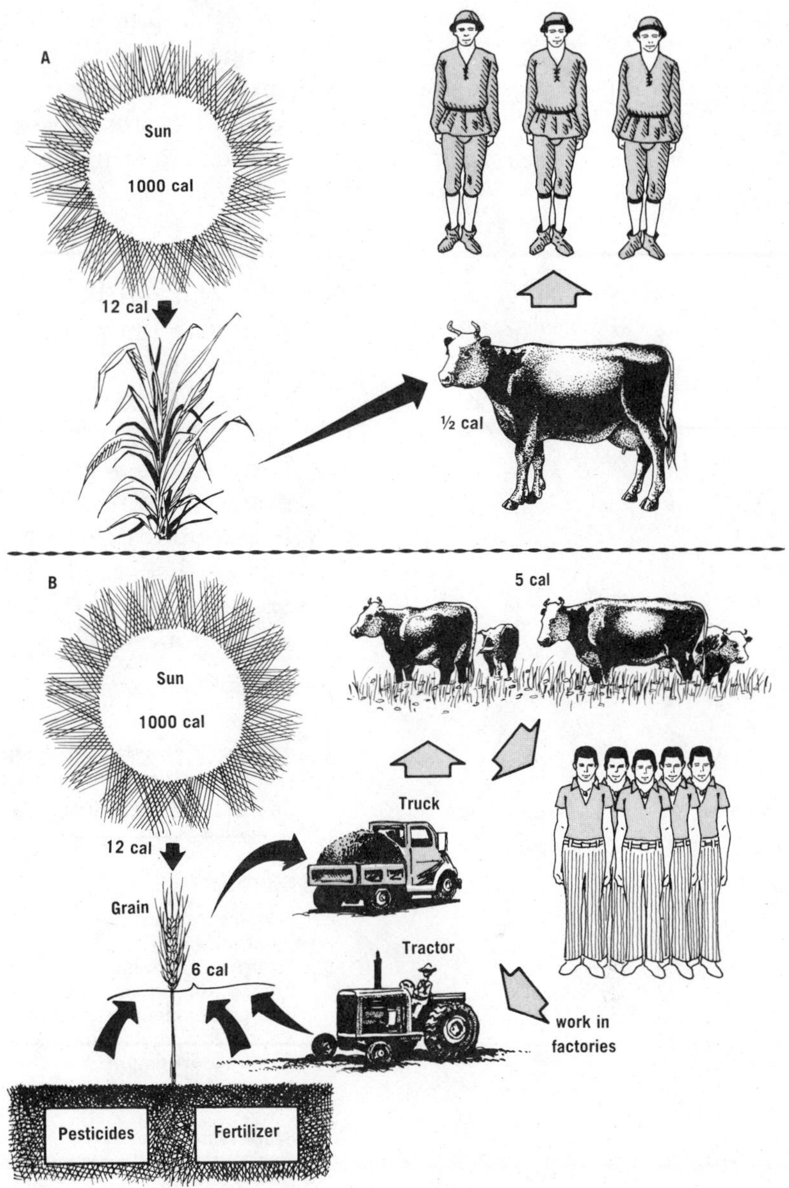

FIGURE 7.6 Comparison between beef production on the range (A) and in a feedlot (B).

Utilization of energy in industrial agriculture has become so intensive in recent years that in the United States in 1970, for example, about nine calories of fossil fuel energy were expended to harvest one calorie of food. Recent gasoline shortages and the prospect of an increasingly severe energy crisis in the years to come have made it important for us to learn how to harvest large quantities of food with a smaller expenditure of energy. To appreciate the complexity of the problem, we must first examine the methods of modern industrial agriculture. They

Note some interesting features of the energy use in the food system of the United States. (See Table 7.1.) More than twice as much energy is used to manufacture cans than to manufacture fertilizer, and the gasoline used in tractors represents less energy than is consumed by the food processing industry.

are: (1) chemical fertilization, (2) mechanization, (3) irrigation, and (4) the chemical control of disease, insects, and weeds. Energy use in agriculture in the United States is summarized in Table 7.1, and the preceding four subjects will be treated in the following sections.

7.4 CHEMICAL FERTILIZATION

(*Problems 7, 8, 9, 10, 12, 13*)

Man fertilized his crops with manure, straw, or dead fish long before he understood the chemistry of fertilization. Today, mined and manufactured fertilizers are used so extensively (see Fig. 7.7) that the total energy requirement of the fertilizer industry constitutes a major portion of agriculture's large demand for fossil fuel. The chemistry of fertilization is essentially the same as the chemistry of the nutrient cycles discussed in Chapter 1. Oxygen and carbon are readily available to all plants that are exposed to air, and are of no concern to the fertilizer industry.

TABLE 7.1 USE OF ENERGY FROM FARM TO TABLE IN THE UNITED STATES, 1940 AND 1970 (MULTIPLY ALL VALUES BY 10^{12} KCAL)*

	1940	*1970*
On Farm		
Fuel for tractors	70.0	232.0
Electricity	0.7	63.8
Energy to manufacture fertilizers	12.4	94.0
Energy to produce agricultural steel	1.6	2.0
Energy to manufacture farm machinery	9.0	80.0
Energy to manufacture tractors	12.8	19.3
Energy to irrigation	18.0	35.0
Subtotal	124.5	526.1
Processing Industry		
Food processing industry	147.0	308.0
Energy to manufacture food processing machinery	0.7	6.0
Energy for paper packaging	8.5	38.0
Energy to manufacture glass containers	14.0	47.0
Energy to manufacture aluminium and steel cans	38.0	122.0
Gasoline for food transport	49.6	246.9
Truck and trailer manufacture	28.0	74.0
Subtotal	285.8	841.9
Commercial and Home		
Commercial refrigeration and cooking	121.0	263.0
Energy to manufacture refrigeration machinery	10.0	61.0
Home refrigeration and cooking	144.2	480.0
Subtotal	275.2	804.0
Grand total	685.5	2172.0

*Adapted from Energy: *Sources, Use, and Role in Human Affairs* by Carol and John Steinhart. © 1974 by Wadsworth Publishing Company, Inc., Belmont, California 94002. Reprinted by permission of the publisher, Duxbury Press.

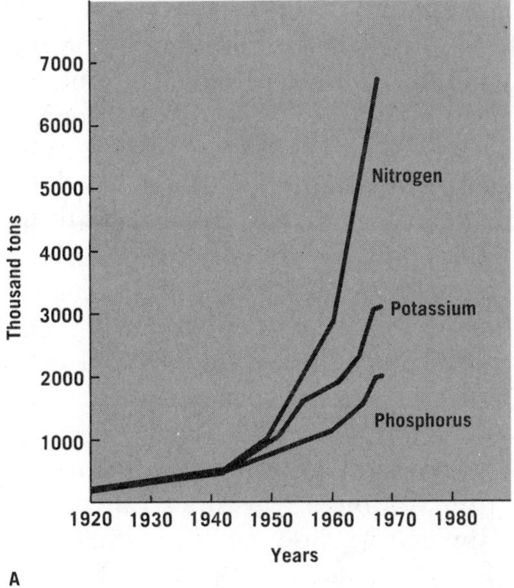

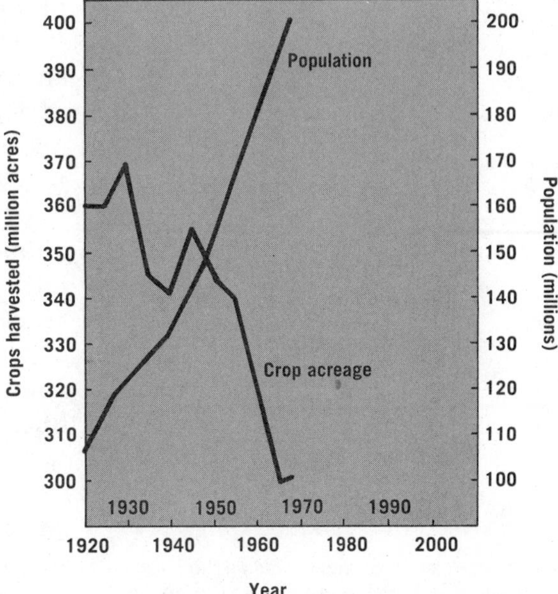

FIGURE 7.7 Fertilizers in U.S. agriculture. *A.* Use of plant nutrients, 1920 to 1968. *B.* Acreage of 59 principal crops harvested plus acreages in fruits, tree nuts, and farm gardens. Total U.S. population, including persons in the armed forces. (Reprinted by permission from Agricultural Practices and Water Quality, edited by Ted Willrich and George Smith, © 1970 by Iowa State University Press, Ames, Iowa.)

As discussed in Chapter 1, man has recently learned to convert atmospheric nitrogen to synthetic plant fertilizers by producing ammonia:

$$N_2 + 3\ H_2 \rightarrow 2\ NH_3$$

Nitrogen makes up about 78 per cent of the atmosphere by volume, and the supply is therefore practically unlimited. We cannot run out of hydrogen either, because it is obtained from water; however, the synthesis of ammonia requires large expenditures of energy. As we will see in Section 7.8, poor farmers in less developed nations are seriously affected by the rising price of fuel, since the price of fertilizers reflects the cost of crude oil.

Mineral fertilizing is qualitatively different from nitrogen fertilizing. If the minerals cannot be recycled by the reuse of dead plant matter, or by such material as bone meal, man must mine the needed minerals from some geological deposit. Thus, phosphorus, potassium, calcium, magnesium, and sulfur, all important constituents of a well-balanced fertilizer, are extracted from the earth. Sometimes these minerals are simply pulverized and dusted onto the soil, while sometimes they are chemically treated in order to enhance the speed or efficiency of uptake by plant roots. Of these five minerals, all but phosphorus are plentiful in the earth's crust. Known geological deposits of phosphate rock (an oxidized form of phos-

Hydrogen is produced by the reaction of steam with hot carbon:

$$C + H_2O \rightarrow CO + H_2$$

This chemical change consumes energy (42 kcal per mole of hydrogen). The reaction of H_2 with N_2 to produce ammonia releases energy (11 kcal per mole of ammonia). Since 3 moles of hydrogen (consuming 126 kcal) are needed to produce 2 moles of ammonia (releasing 22 kcal), the net change is energy-consuming. However, this consumption is modest compared with the energy requirements for establishing and maintaining the high temperatures required for all the reactions and the high pressure needed for the ammonia synthesis, as well as the energy for processing and shipping the raw materials and final product.

Incidentally, hydrogen can also be made by the electrolytic decomposition of water, but this reaction consumes much more energy and is rarely used except when extremely pure hydrogen is needed.

phorus) will be depleted early in the twenty-first century if current mining rates are continued. Without an adequate supply of phosphorus, highly productive agriculture will be impossible. New discoveries are not expected to increase known reserves appreciably.

When we speak of the depletion of a mineral deposit such as phosphate, we do not mean that all the elemental phosphorus on the earth will disappear, but rather that it will be dispersed. Our phosphate reserves will be considered to be depleted when the concentrated deposits have been mined, sold, and spread over agricultural systems. At this time, some fraction of the total phosphorus will have been absorbed by plants, consumed by humans, collected in sewage systems, and ultimately washed into rivers and from there to the oceans. Some will remain in the soil in chemical forms unusable by plants, and will slowly pass into groundwater and eventually return to the oceans. Mining phosphate dissolved in the ocean would be prohibitively expensive, for one would have to process several thousand pounds of sea water to extract one pound of phosphorus. In certain regions, various combinations of ocean currents and of gradients of chemical concentrations can bring about the deposition of phosphorus in mineral form. The long-term recycling of phosphorus has always been dependent on these geological processes, but the accumulation of phosphate deposits on the ocean floor and the uplifting of continental shelves to form dry land takes millions of years.

"Organic" originally meant "derived from a living organism." According to this definition, composted straw, manure, offal from a slaughterhouse, and bone meal would all be considered to be organic materials. After it was shown in a series of discoveries from 1828 to 1845 that some components of these substances could also be synthesized from nonliving sources, "organic" came to refer to compounds containing C–C or C–H bonds, with or without other atoms. (Marble, $CaCO_3$, is not considered to be organic.)

By this new definition, composted straw, manure, and slaughterhouse offal are all complex mixtures of many different organic compounds (with a few inorganic ones in the mixture, too), but bone meal, which consists mostly of calcium salts and other minerals, and contains little carbon and hydrogen, is largely inorganic.

Very recently, in usages such as "organic foods" and "organic gardening," the word has recaptured some of its old meaning and has added new overtones. In these contexts "organic" implies a relationship to, or a derivation from, biotic systems that are unadulterated by human technology. Special emphasis is given to the absence of synthetic or other additives that are foreign to natural food webs, and some importance is also placed on the avoidance of excessive refining that greatly alters natural compositions. Thus, DDT-sprayed beans are not organic, and pure-white sugar (sucrose) is questionable, but sea salt qualifies. In this context, therefore, "organic" may be briefly defined as "naturally constituted." Thus, an organic gardener would include pulverized phosphate rock as well as bone meal in his list of organic materials.

In spite of the fact that manufactured fertilizers have raised agricultural yields across the world, and in spite of the general health and vitality of peoples raised on fertilized crops, the use of inorganic fertilizers is being questioned, and there are many outspoken advocates of "organic" gardening and farming. Exactly what is meant by "organic" gardening?

The major argument against "nonorganic" farming is based on the chemistry and physical nature of the soil. Plants need soil with properly regulated nutrients, density, moisture, salinity, and acidity. Soil conditions have traditionally been regulated by the **humus**, a very complex mixture of compounds resulting from the decomposition of living tissue. A given piece of tissue, such as a leaf or a stalk of grass, is considered to be humus rather than debris when it has decomposed sufficiently in the soil system so that its origin becomes obscure. Compared to inorganic soil, humic soil is physically lighter, it holds moisture better, and it is more effectively buffered against rapid fluctuations of acidity. Additionally, certain chemicals present in humus aid the transfer and retention of nutrients. For example, calcium ions, Ca^{++}, can exist in water solutions from which they are usable by plants. However, atmospheric carbon dioxide reacts with water to form carbonate ion, $CO_3^{=}$, which reacts with calcium in water to form the sparingly soluble compound, calcium carbonate, $CaCO_3$. A striking consequence of its meager solubility has been the deposition of large masses of $CaCO_3$ on earth in the form of limestone and marble. But it is not *completely* insoluble; moreover, its solubility depends largely on the soil acidity, and under certain conditions, the calcium may be liberated. Under other conditions, it is very insoluble and the calcium may not be readily available to plants, even though it is present in the soil. Moreover, if dissolved calcium is not used immediately by plants, it may travel with water droplets down below the root zones where it becomes unavailable. This movement of free ions into the subsoil and underground reservoirs is called **leaching.**

$$CO_2 + H_2O \rightleftharpoons HCO_3^- + H^+$$
$$\downarrow\uparrow$$
$$CO_3^{=} + H^+$$
$$\downarrow Ca^{++}$$
$$CaCO_3$$

$$CaCO_3 + 2H^+ \rightarrow Ca^{++} + CO_2 + H_2O$$

To avoid confusion among the various definitions of the words *organic* **and** *chemical*, **the following conventions will be used:** *organic* **means containing carbon-hydrogen or carbon-carbon bonds, and** *"organic"* **will be taken to mean naturally constituted; also,** *inorganic* **means not containing carbon-hydrogen or carbon-carbon bonds, and** *"nonorganic"* **or** *"chemical"* **will mean not** *"organic."*

Richly humic soils provide chemical reaction pathways for metal ions that are not available in simple inorganic solutions. Certain chemicals in the humus, which are known as chelating agents, react with inorganic ions such as Ca^{++} to form a special class of compounds known as chelation complexes. Ions bonded in chelation complexes are held tightly under some conditions, but are easily released under others. A calcium ion chelated by humus will not react readily with carbonate to form $CaCO_3$, nor will it leach easily. Rather it will tend to remain bonded to the chelating agent and thus be retained in the humic matter.

The word comes from the Greek *chele*, meaning "claw." Chelates bind atoms from two or more different directions, much as a lobster grabs on to its prey.

Plants and soil microorganisms have evolved mechanisms whereby they can release chelated ions readily and incorporate them into living tissue, as shown schematically in Figure 7.8. Naturally, the specific chelation chemistry is different for each soil nutrient, but in general, humus maintains a nutrient reservoir and enhances the ease with which nutrients are used by the soil organisms.

When soil is plowed, the humus which normally remains underground becomes directly exposed to air

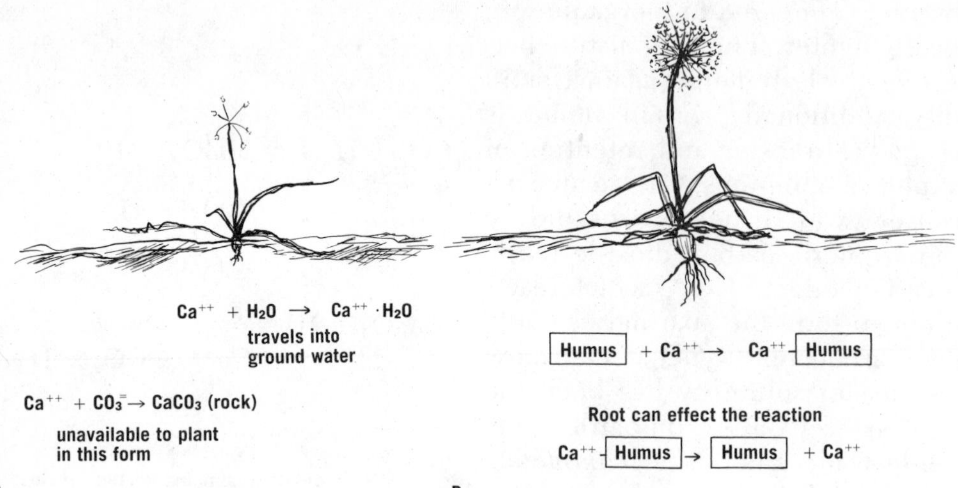

FIGURE 7.8 Schematic illustration of pathways for transfer of inorganic nutrients. *A.* Without humus. *B.* With humus.

where some of it oxidizes. Under these conditions the complex chemicals which contribute to the unique chemistry of humus decompose more rapidly than they do in undisturbed soil. Thus, when a virgin area is plowed and converted into cropland, the humus content of the soil decreases. In many areas of the world, organic fertilization has allowed land plowed for thousands of years to remain fertile.

In recent years, some potentially serious problems have arisen from the depletion of humus.

One hundred years ago the area that is now North and South Dakota, eastern Montana, and southern Saskatchewan was buffalo prairie; now it is cattle and wheat country. Currently, it is common to plow fields, leave them idle during the spring and early summer, then plant winter wheat in the late summer or early fall. The practice of leaving soil exposed to spring rains and summer heat has rapidly accelerated the decomposition of humus. Soil with little humus and no roots is vulnerable to leaching. The last ice age had deposited large quantities of salts in the subsoil of this region. Between the time of the last glacier and the twentieth century, the prairie roots and the humus had served as a barrier against the seepage of large quantities of rainwater down into the subsoil. At the present time, however, such filtering and leaching occurs more rapidly and the water table is rising each year. This rising water is carrying the glacial salts toward the surface, and if the salt water reaches the root zone, present agriculture will fail in this area.

When soil humus is not maintained, the efficiency

of converting fertilizer to plant tissue is low and the ability of soils to store reserves of nutrients is poor. As a result, large quantities of nitrogen or mineral fertilizers are leached into surface and ground waters or tied into the soils in chemically unusable forms. This loss is a relatively minor problem to the farmer, for despite the inefficiency, inorganic fertilizers represent one of his best investments, returning about $1.20 worth of yield increase for every $1.00 spent. However, lost fertilizers collect in the waterways and significantly add to water pollution in agricultural areas. Compounds that are nutrients on land are also nutrients to aquatic life, and serve to feed algae and plankton, thereby upsetting the food webs in rivers and lakes. Nitrogen fertilizers are also potentially harmful in that the nitrate ion, NO_3^-, may pose a health hazard. (Note the sequence of equations in the margin, showing conversion of nitrates to carcinogens.)

NO_3^- (nitrate ion)
↓ biological reduction
NO_2^- (nitrite ion)
↓ R_2NH (amines in foods, etc.)
$R_2N—NO$
cancer-producing nitrosamines

Another argument against the practice of chemical farming is related to the nutritional value of the crops. Several studies have shown that the food value of many grains and vegetables decrease when high concentrations of fertilizers are used. In one study, alfalfa yields were increased dramatically with high applications of chemical fertilizers; however, cattle fed on that alfalfa became ill from cobalt and copper deficiencies. These trace minerals, necessary to the animal's health, were not present in the fertilizer, and were therefore lacking in the alfalfa. In a study examining the relationship between oats and phosphate fertilizer, both the total yield of grain and the phosphorus content of each oat plant increased with increasing quantities of phosphate fertilizer, but the ratio of weight of phosphorus per unit weight of oat grain decreased.

Nothing in the preceding discussion implies that the use of synthetic "inorganic" fertilizer, *per se*, results in nutritionally poor crops. Plants cannot tell whether a given molecule of nutrient came from a bag of "inorganic" fertilizer or from a bag of compost. The major deficiency of commercial "inorganic" fertilizers is that they usually contain no more than three nutrients: nitrogen, potassium, and phosphorus. Trace elements could be incorporated into manufactured fertilizers, but the expense is high and is not compensated for by a large increase in yield. As a result, the continued use of commercial "inorganic" fertilizers may engender nutritional imbalances. On the other hand, one doesn't render a soil nutritionally perfect simply by tossing in a little manure. Only by using the proper mixture of fortifying materials such as manure, wood ashes, bone, and blood meal does the "organic" farmer grow prime foods.

The lowering of nutritional values caused by improper fertilization does not present a serious problem to most inhabitants of the developed nations. Indeed, many foods do not appear to be harmed by large fertilizer applications, and the decreased food value of many others is insignificant to people who can afford a rich and varied diet. Secondly, transportation and refrigeration enable shoppers to purchase foods from many different geographical areas. Because some soils in different regions are naturally endowed with an overabundance of certain trace minerals, a person eating foods raised in many different soils is likely to ingest an adequate supply of minerals. The problem of the nutritional value of individual foods is serious in areas where families eat a subsistence diet grown in one field.

Although commercial fertilization has been quite successful in the past, the problems associated with these practices are now demanding increased attention. Perhaps the most serious problem is that the humus content of soils is continuing to drop. One solution would be to formulate more sophisticated artificial fertilizers including synthetic chelating agents. A second solution would be to initiate a massive program of naturally constituted fertilization through recycling. There is no doubt that the soils in the United States, for example, could be improved if the two billion tons of animal feces, slaughterhouse offal, and other organic residues produced each year were composted and returned to the land. Currently, some of this material is used as fertilizer. It is often cheaper, however, to buy and spread chemical fertilizers than to handle the natural materials. While the yield per dollar in a single growing season may often be maximized by discarding manure, the wastes wash into streams, thereby polluting them, and the cost of this pollution must be considered. Additionally, the agricultural yields for future generations must be assured, and many economists consider that we should pay our share of the costs for such provisions. In many countries dried and pasteurized sewage sludge from municipal sources is marketed as fertilizer. However, short-sighted economic criteria have often favored disposing of wastes into nearby rivers.

A second problem, which we have mentioned previously, stems from the need for external sources of energy to manufacture synthetic "chemical" fertilizers. Since naturally constituted materials, by contrast, are produced by natural biological processes, one might think that a return to "organic" farming would conserve energy. However, the economic differences are not so sharp. In many feedlots in the United States, for example, hundreds of thousands of cattle are concentrated within small areas. Much gasoline would be needed to transport the tons of manure generated in these feedlots to distant farms. Therefore, more widespread use of manure would conserve energy only if feedlots were broken up into smaller units and dispersed widely. Agricultural systems in developed countries, however, are designed to produce high yields, and any basic change would be economically disruptive and costly. Perhaps there are other ways to reduce energy consumption and to realize some of the objectives of "organic" farming without disrupting the entire agricultural system. For example, the farmer could plant more legumes in rotation with corn or grain. Even such changes, however, pose complex economic problems in planting and marketing. Some farmers report that "organic" techniques are profitable on farms of 300 acres or more.

7.5 MECHANIZATION IN AGRICULTURE

(Problem 11)

High agricultural yields depend on specific varieties of seeds, an active fertilizer program, and pest control. The use of tractors, on the other hand, does not improve yield per acre. For example, most Japanese farms vary in size from about one to five acres. Many farmers own small, two-wheeled rotary tillers, but very few own four-wheeled tractors, and the large combine is virtually unknown. Yet the yields per acre in Japanese agriculture are the highest in the world. However, these yields depend on great quantities of fertilizers and much labor. Therefore, the total auxiliary energy requirement of Japanese farm-

Combine. (Photo courtesy of Rorer-Amchem, Inc., Ambler, Pennsylvania.)

ing is high even though the use of heavy equipment is minimal.

The fact that a tractor can out-perform any animal in pulling a plow is not the only advantage of mechanization. Another advantage is versatility. Farm equipment that can perform many intricate operations simultaneously is now available. A single device can plow, fertilize, and plant in one pass through a field, while another can pick the fruit, sort it according to size, bag it, weigh the bags, and release the labeled sacks for delivery to market. As a result of such extensive mechanization of agriculture, farms in the United States now use more petroleum than does any other single industry.

As mechanization becomes more versatile and more intricate, and as the individual devices get larger, agricultural yields per acre actually *decrease.* A machine that processes 10 rows of crops will necessarily be less precise than one that handles only two or three rows, because of unevenness of ground levels and other natural variations. The result will be that the larger machine causes more damage and hence more waste. A mechanical tomato picker requires a variety of tough-skinned tomatoes that all ripen at the same time, because the machine has no eyes to differentiate between red and green. Such varieties have in fact been developed, but their simultaneity of ripening is not perfect, and hence, again, the machines make waste. There are machines that shake apples from trees into collecting frames, but they bruise some apples. Furthermore, all of these problems are aggravated by the crowding of plants that is made possible by the greater productivity of fertilized soils.

According to the data in Table 7.1, in 1970 gasoline for tractors accounted for almost 45 per cent of the energy used on farms in the United States. What would happen if American farmers were to revert to more manual labor and fewer machines? In other words, suppose that the factory workers who build mechanized tomato pickers were to move to the country and pick tomatoes. This type of system would be similar to Japanese farming in which high inputs of fertilizer, pesticides, and labor give rise to high yields. A labor-intensive agricultural system could operate in the United States and other industrial nations only if major changes were made in social structure and living standards.

Tomato picker. (Photo courtesy of Rorer-Amchem, Inc., Ambler, Pennsylvania.)

7.6 IRRIGATION
(Problem 16)

Although irrigation has often destroyed land, it continues to be important in highly productive agriculture. Former success in irrigation has led many planners to look toward the deserts as future sources of food. The conversion of deserts into agricultural systems depends on huge quantities of water and, in many cases, on large amounts of energy to desalinate ocean water. Chapter 6 discussed the availability of energy sources in the future and the proliferation of nuclear reactors which could be used to desalinate seawater. Advocates of reactor-based irrigation systems point to the fact that if all the deserts that lie within 300 miles of the ocean were to be developed and planted to grain, the total grain-producing acreage of the world would rise from the present 1.6 billion acres to a phenomenal 6.8 billion acres. However, the required quantities of energy are not presently available, for now desalinated water is often too expensive to be practical, but, as energy technology improves, and as agricultural practices use water more conservatively, the economics of the system is likely to change.

In traditional irrigation systems water is pumped through ditches in open fields. About 98 to 99 per cent of the water is lost to evaporation either directly from the canals or indirectly through transpiration from the plant leaves. Most of this loss can be eliminated if crops are grown in plastic greenhouses under careful-

FIGURE 7.9 Flood-water irrigation in the valley of Teotihuacán, Mexico, May 1955. (From Eric R. Wolf: *Sons of the Shaking Earth.* **Phoenix Books, University of Chicago Press, 1959, p. 77. William T. Sanders, photographer.)**

FIGURE 7.10 A desert greenhouse project. (Photo courtesy of Environmental Research Laboratory, University of Arizona.)

ly controlled conditions. In the most advanced greenhouse technology, the atmosphere is maintained at close to 100 per cent humidity to minimize transpiration, and irrigation water is carried directly to the plant root system through plastic pipes to eliminate evaporative losses that occur in open systems. Additionally, carbon dioxide gas is pumped into the artificial environment to accelerate photosynthesis, and the soil is heavily fertilized. When the high incident solar radiation of desert areas is coupled with the enhanced water, CO_2, and nutrient inputs, some rather striking yields can be achieved. In a pilot project on the Arabian Peninsula, in one year experimenters have grown about a million pounds of vegetables on five acres (see Fig. 7.10). This harvest represents about a twenty-fold increase over the average production of the same crops in open fields. These yields must be balanced against the high capital investment for the desalination plant and the costs of the greenhouse operation, but the project appears to be economical for high-priced crops such as tomatoes and cucumbers.

The failures of some past irrigation projects (see Section 7.2) serve as a warning that future dependence on mammoth uses of imported water may not be reliable. As described in Section 7.2, heavily irrigated lands tend to become salty, and eventually unfit for agriculture.

7.7 CONTROL OF PESTS AND WEEDS

Perhaps the most persistent and difficult problem in the agricultural production of large quantities of food has been competition from small herbivores. If a cow wanders into a field, the farmer's son can chase her away. An insect, a field mouse, the spore of a fungus, or a tiny root-eating worm (a nematode) is more difficult to deal with. Since these small organisms reproduce rapidly, their total eating capacity is very great. In addition to their voracity, these pests may be carriers of disease. The bubonic plague, carried by a flea which lives on rats, swept through medieval Europe, killing as much as one-third of the total population in a single epidemic. Malaria and yellow fever, spread by mosquitos, have killed more people than have all of man's wars.

However, not all insects, rodents, fungi, and

nematodes are pests. Most do not interfere with man and many are directly helpful. Millions of nematodes live in a single square yard of healthy soil. Most are necessary to the process of decay and hence to the recycling of nutrients. Fungi, too, are essential to the process of decay in all the world's ecosystems. The role of insects in the maintenance of the biosphere is extensive and versatile. For instance, bees are essential to the life cycle of most flowering plants. In their search for food, bees inadvertently transfer pollen from flower to flower, and thereby ensure fertilization. Many insects, such as species of springtails, are part of the process of decay.

Insects are the prime food source of many animals that are vital, in turn, to the maintenance of natural balance. For example, the diet of many species of birds includes both insects and fruit. Fruit seeds transferred intact through the bird's digestive system and deposited at distant locations have been an important contributing factor in the continuing existence of certain plants. Thus, insects are necessary to the survival of many species of birds that are necessary in the life cycle of many wild fruit trees, which, in turn, help to support wildlife. In addition, many carnivorous or parasitic insects feed on insects that eat man's crops. The periodic invasion of some African villages by driver ants is a fascinating illustration of insect ecology. Many disease-carrying rodents and insects live in the village houses, posing a constant threat to the human population. At periodic intervals, however, millions of large driver ants invade the villages, chase away the inhabitants and eat everything that remains. When the people return, they find that their stored food supply is gone, but so are all the cockroaches, rats, and other pests—everything has been eaten.

An insect is an animal; therefore, an insect-eater is a carnivore.

Therefore, the problem in pest control has been to destroy the harmful species while sparing beneficial ones. During most of man's history, insect control was not particularly effective, and species of pests have lived side by side with man for millions of years. Several factors have made this accommodation possible, if not pleasant. First, during most of man's existence, the total human population, and hence the total human food requirement, has been much smaller than it is now. Also, many species of food plants produce insecticides for their own protection. For instance, the roots of some East

Crop duster dusting sulphur to retard mildew on grapevines 20 miles south of Fresno, California. (Courtesy of EPA-DOCUMERICA, photographer, Gene Daniels.)

Indian legumes contain the insecticide **rotenone**; and many wild cereal plants such as wheat, corn, rye, and oats are naturally resistant to fungal attack. Such natural accommodations are now insufficient for man. One reason is that the natural controls do not always work too well, especially in the short run; thus throughout history natural periodic cycles of pest populations have occurred, resulting in blooms of these species and famine for man. Perhaps even more critical is the rapid growth of the human population and hence its increasing food requirements.

Systematic spraying of crops was initiated in the early twentieth century when natural insecticides such as rotenone, pyrethrum, nicotine, kerosene, fish oil, and compounds of sulfur, lead, arsenic, and mercury were in common use. These formulations did not change appreciably until the 1940's. The most significant breakthrough occurred around the start of World War II, when the insecticidal properties of a chemical called DDT were discovered. DDT was far cheaper and more effective against almost all insects than the previously known control methods. The use of DDT led to dramatic early successes: it squelched a threatened typhus epidemic among the Allied army in Italy; anti-mosquito programs saved millions from death from malaria and yellow fever; and pest control, leading to increased crop yields all over the world, saved millions more from death by starvation.

Dichlorodiphenyltrichloroethane (DDT)

$$\mathrm{Cl{-}C_6H_4{-}C(H)(CCl_3){-}C_6H_4{-}Cl}$$

Enthusiastic supporters of DDT predicted the complete destruction of all pest insects within the foreseeable future; the chemist who first discovered its insecticidal properties received a Nobel Prize. But within 30 years the promise of insect-free abundance had been broken, and the "miracle" chemical that was to have achieved it had fallen from grace. On January 1, 1973, all interstate sale and transport of DDT in the United States was banned except for use in emergency situations in which life is immediately threatened.

In the past 30 years, a whole new array of powerful chemical insecticides, in addition to DDT, has been developed. However, problems arise because they are also active against a much wider range of organisms. Thus, DDT not only poisons insect pests, but it can also kill insect-eating birds, carnivorous insects, domestic cattle, wild animals, house pets, and even man. Therefore, spray applications destroy the natural food webs within agricultural ecosystems. The resulting simplified ecosystems are potentially unstable, and if one species somehow becomes immune to the poison, its population can grow at a nearly exponential rate. In many instances, sprayed fields have, in fact, suffered serious pest outbreaks. Insects are extremely adaptable, and an increasing number of pest species have evolved immunity to certain insecticides. As an example, when DDT and two other insecticides were used extensively in pest control in a valley in Peru, the initial success gave way to a delayed disaster. In only four years, cotton production rose from 440 to 650 pounds per acre. However, one year later, the yield dropped precipitously to 350 pounds per acre, almost 100 pounds per acre less than before the insecticides were introduced. Studies indicated that the pests had become immune, but predator insects and birds had not. Thus, with natural controls eliminated, the pest population thrived better than ever before.

"There I was, coming in low at a hundred and seventy-five m.p.h. with forty-two acreas of broccoli to the left of me, eighteen acres of asparagus to the right of me, eighty acres of carrots straight ahead! Power lines all around! My target: seven acres of badly infested garlic smack in the center. . . ." (**Drawing by Dedini; © 1972,** *The New Yorker Magazine, Inc.)*

We have shown how destruction of predator populations by indiscriminate spraying may cause upsurges in pest populations. It is reasonable to assume that the opposite treatment—importation of predators—may be an effective control measure.

Excellent success with such techniques has been achieved in several cases. For example, the Japanese beetle, a native of the Orient, was inadvertently imported to the United States with a shipment of some Asiatic plants. In the absence of any effective natural control, the beetles thrived on the eastern seaboard of the United States and gradually became a major pest. Scientists then searched for natural predators and imported several likely species. One of these, an Oriental wasp, provides food for its young by paralyzing the Japanese beetle grub and attaching an egg to it. When the young wasp hatches, it eats the grub as its first food. The life cycle of the wasp is dependent upon the grub of the Japanese beetle; it does not naturally breed on the grubs of other insects. Therefore, this type of control is species-specific and does not seriously affect the rest of the ecosystem.

Other natural control methods which have proven effective in certain instances include the following:

1. **Sterilization techniques.** Male screwworm flies were sterilized in the laboratory and released in the field. They mated with but did not fertilize wild females. Unfertile eggs were laid, and a pest epidemic was controlled.

2. **Hormone control.** Hormones are chemicals which regulate bodily functions. Specific insect hormones trigger the metamorphosis from juvenile stages, such as the caterpillar, through various phases until the adult is reached. Insect hormones can be synthesized in the laboratory and sprayed on agricultural fields. These sprays disrupt the pest insects' life cycle, usually killing them.

3. **Use of resistant strains of crops.** Plant breeders have been able to develop strains of crops which are naturally resistant to certain pests. Occasionally, these resistant crops produce smaller yields than the nonresistant crops do, but if loss to pests and disease can be eliminated, the eventual yield is likely to be greater. As an added bonus, the materials and energy needed to manufacture and spread pesticides can be conserved.

7.8 AGRICULTURE IN THE LESS-DEVELOPED NATIONS AND THE GREEN REVOLUTION

(Problems 14, 15)

There has never been widespread famine in the United States. Barring a major war, a famine in the United States within this century is highly unlikely despite the potential problems discussed in the previous section which may make us worry about the durability of our agricultural systems. Agricultural difficulties could easily cause increases in the price of food, but there should be enough produce to supply the people of this nation in the near (and maybe far) future with adequate diets. Conversely, most school-age children in India have already experienced famine and might again in their lifetimes. In the less-developed countries of Asia, Africa, the Middle East, and Central and South America, the race between population growth and food production has been close; sometimes most of the people have enough food to survive, sometimes many don't. Even in "good" years, malnutrition and related illnesses are common among the living. It is a fallacy to believe that hunger exists in the less-developed nations solely because the population density is high. Looking at Figure 7.5, we see that India, Asia, Africa, and Latin America are four of the five least productive agricultural areas in the world, and it is this lack of productivity which is primarily responsible for famine. If rice farmers in India could achieve the success of their Japanese counterparts, there would be no hunger in India. Thus far, attempts to increase the efficiency of agriculture in these areas have been unsuccessful. Why so?

Oceania (Central and South Pacific island groups including Australia and New Zealand) is unproductive in terms of yields per acre, but the population density is so low that people are generally well fed.

The poorest people in the world are primarily concerned with survival, and often the decision to eat today jeopardizes the food supply for tomorrow. For example, Indian women follow cattle, collect the dung, dry it, and use it as fuel for cooking. The poor often have no alternative sources of heat, for there are virtually no trees in parts of India, and coal or oil is too expensive. But at the same time, few farmers can afford to purchase chemical fertilizers, and most of Indian agriculture is dependent on manure and other organic wastes. When the fertilizers are used as fuels to cook this year's crops, the yield of next year's harvest is being jeopardized, so a continuous downward spiral develops. Agriculture in these depressed areas must somehow reverse the current trends.

Many attempts by wealthy nations to aid the less-developed countries have met only limited success.

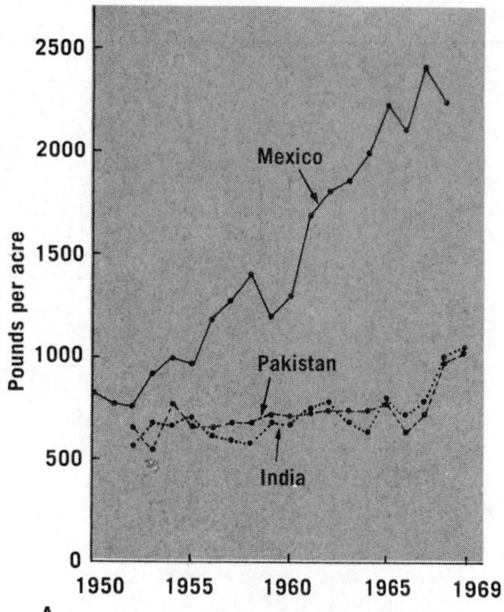

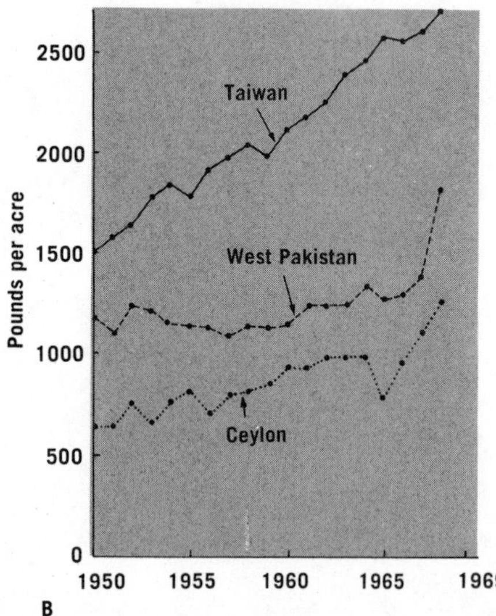

FIGURE 7.11 The yield takeoff of the Green Revolution. *A.* **Wheat yields in Mexico, Pakistan, and India.** *B.* **Rice yields in West Pakistan, Ceylon, and Taiwan. (From L. R. Brown:** *Seeds of Change.* **New York, Praeger Publishers, Inc., 1970, pp. 37, 39. Copyright 1970 by Praeger Publishers, Inc.)**

Gifts or loans of machinery are seldom used efficiently for several reasons: (a) While almost all American or European farmers can service their own equipment, few farmers in the less-developed nations can fix a tractor. (b) The distribution and supply system for spare parts and for petroleum in many countries is practically non-existent. (c) Poor farmers often cannot afford petroleum or spare parts, even if they are available, and interest rates in some villages approach 50 per cent per year. Consequently, tractors often lie idle soon after they arrive.

Gifts or loans of fertilizers have been more useful, but problems have developed with these imports, too. The varieties of grains grown in most of the world 10 years ago were closely related to wild grasses, and are characterized by long, thin stalks and small grain clusters on top. If these varieties are fertilized heavily, the young plants will produce such abundant clusters of seed that the stalk cannot support the top, and the plant bends over (lodges) and falls to the ground before it matures.

In the mid-1960's American scientists working in Mexico and the Philippines developed new varieties of wheat and rice that were adaptable to tropical climates and were capable of producing higher yields than any native grains. These seeds were heralded as the means to make most countries self-sufficient, to free millions of farmers from their downward spiral and to buy time to solve the ultimate problem of population expansion. The era of the new varieties, called the **Green Revolution**, has opened with some truly spectacular successes. One such instance was the reversal of the crop failures and famine that occurred in Pakistan in 1965 and 1966, when the monsoon rains failed to fall on schedule. Massive shipments of new varieties of wheat seeds during the next two growing seasons, coupled with good weather, raised the total wheat harvest almost 60 per cent in two years, and Pakistan made strides toward self-sufficiency. Similarly, India's wheat crop increased by 50 per cent between 1965 and 1969. The incidence of famine in that country, too, has decreased. The Philippines had been importing rice for half a century and suddenly, with the cultivation of large acreages of new rice varieties, the trends have been reversed and Filipinos became rice exporters. The development of these high-yielding seeds has had more impact on

Old way of harvesting wheat in India by cutting off stalks at the base by hand. (© Food and Agricultural Organization of the United Nations. Via delle Terme di Caraculla 00100, Rome, Italy.)

the daily lives of millions living in poor countries than any other technological development of the past 15 years (see Fig. 7.11).

On the other hand, figures of national grain production do not always reflect the fate of groups of poor farmers within that nation. To understand the social impact of the Green Revolution, we must first understand that the new grain varieties, planted in an impoverished soil and dependent on variable rainfall for growth, produce equal or smaller yields than the native grains which have been cultivated in poor areas for centuries. The major advantage of the imported seeds is that they are more responsive to fertilizers than are the old seeds. The new varieties of wheat and rice have short, thick stalks and do not lodge when fertilized heavily. For example, the "miracle" rice variety IR-8 can absorb 120 pounds of fertilizer per acre economically, while native rices can only absorb 40 pounds of fertilizer per acre before the stalks bend or break. Moreover, IR-8 plants require fewer nutrients to build stalks, and hence the conversion of fertilizer to grain is twice as efficient as the conversion in tall-stemmed plants. The dependence of new seeds on high fertilizer applications means that a relatively poor farmer who can afford somehow to buy fertilizers or to obtain them on credit can gradually extract himself from poverty when his high yields are harvested. The truly poor, marginally subsistent farmer, however, cannot afford the investment that is required to rebuild his worn-out land and free himself from the threat of hunger.

Unfortunately, fertilizers alone do not ensure the success of the miracle seeds. These grain varieties are not so resistant to insect pests and fungal diseases as the traditional plants. As a result, farmers who invest in the seed and the fertilizer must also invest in pesticides. But even fertilizers and pesticides are not sufficient. Successful production of high yields with new grains is dependent on the availability of water and on a carefully controlled and regularly maintained irrigation system. The necessary equipment requires time, knowledge, and capital for investment in pipes, pumps, and fuel.

In reviewing the situation, one observer remarked that the tenant farmer in India has gone from secure poverty to insecure poverty, for he is now faced with a host of new problems. Many landlords, realizing that for the first time farming can be profitable, have simply evicted their tenants, invested in fertilizers, pesticides, and irrigation systems, and gone into business themselves. The evicted tenants suddenly have found themselves without homes or jobs. Current tenants live in fear of eviction, so that many are reluctant to invest in an irrigation system, even when low government loans are available.

Even the poor landowner is facing problems. The man who decides to continue his old secure system finds that because the total grain production of the country has increased, the market is flooded, the price of grain has dropped, and the value of the small quantity of wheat or rice that he has traditionally sold in good years has been deflated. Many of those farmers who are capable and willing to invest seed and the auxiliary inputs nevertheless also face insecurity. Grain prices in countries like India and Pakistan are notoriously unstable. Therefore, the price of grain at harvest time may not be sufficient to repay investments for fertilizers, sprays, and irrigation equipment. Moreover, few storage facilities are available in these countries, so a farmer cannot store his grain until its market value increases.

The use of the new seeds has led to many other problems, such as:

1. The taste and texture of hybrid rice IR-8 differ from those of traditional rice, and it commands a lower price. In Pakistan, many wholesalers refuse to handle this variety.

2. In many rice cultures, paddy farmers have raised fish in irrigation canals. Now fertilizer and pesticide pollution of the canals have killed fish. Thus, increased total caloric content of many people's diets has been associated with decreased protein content.

3. The problem of protein deficiencies has been augmented by the curtailment of production of soy beans and other protein-rich legumes, which are now less profitable for some farmers than grains.

4. New grain varieties mature earlier than the old varieties. While this enables the prosperous farmer to grow two crops a year instead of one, many other farmers are finding that since harvest time now coincides with rainy weather, the grain can no longer be dried in the sun. Instead, mechanical driers are often needed.

5. National economic problems have also arisen because of frequently insufficient foreign exchange to purchase imported pesticides and chemical fertilizers. Even if the less-developed nations were to build their own chemical factories, many would have to buy fuels, phosphate rock, and other raw materials.

The consequences of the Green Revolution have been mixed. Food production in the less-developed nations has increased. Starvation has decreased, and millions are living with a new-found freedom from hunger and want. On the other hand, many rural people have sunk into deeper poverty, and homeless and untrained rural poor are migrating in large numbers to the city slums.

7.9 FOOD SUPPLIES FOR A GROWING POPULATION

(Problems 17, 18, 19, 20)

Worldwide food production and population growth have been in a nip-and-tuck race in recent decades. Widespread use of pesticides and fertilizers caused food production in the 1950's to increase faster than population. During most of the next decade, however, the situation was reversed; high birth rates and some poor crop years in many areas caused population growth to outstrip gains in agricultural yields. The latter part of the 1960's saw some tangible benefits of the Green Revolution and, once again, food production forged ahead of population growth. Gains were slight, however, and when a serious drought struck India, Pakistan, and the grazing lands just south of the Sahara Desert between 1972 and 1974, millions again became famished. Moreover, the costs of fertilizers, pesticides, and machinery have risen with the rising price of fuel, so that poor farmers will find it extremely difficult to grow large crops even if the drought eases. The situation is serious, and world planners fear that unprecedented famines might occur by the late 1970's.

Demographers predict that population levels in the less-developed nations will grow at about 2.6 per cent per year through the decade 1970–1980. During the same time span grain production will be expected to grow at a rate of between 3.1 and 3.9 per cent. If both predictions hold, and the weather improves, the standard of living for many people will rise slightly in the near future. However, all such predictions are subject to large errors. Agricultural planners are now examining several different sources of future food supplies.

Continued Increases in Crop Yields

Japanese farms are six times as productive per acre as Indian farms. Yet, agronomists believe that with proper techniques many impoverished farms along the Ganges flood plain in India could be made at least as fertile as those of Japan. Even if yields in India rise at 3.5 to 4 per cent per year (an optimistic view), food production at the end of 25 years would still be well below the potential maximum. To reach this potential, the Indian farmer would have to multiply his energy and chemical inputs many times (see Fig. 7.5). At the moment, India lacks sufficient foreign exchange for this investment. Furthermore, the energy and raw material reserves of the planet may not be sufficient to support maximum production throughout India and the rest of the less-developed nations. If virtually unlimited supplies of energy become available from fusion reactors within a short time, and if agricultural techniques improve further, perhaps agricultural yields will rise significantly. The ultimate productivity of croplands is difficult to predict accurately.

Increases in Crop Acreage

Most of the prime, fertile land in the world is currently being farmed, so expansion of crop acreage will have to be in areas which were uneconomical to cultivate in the past, such as deserts, forests, and jungles. The farming of deserts will be feasible if massive quantities of inexpensive energy become available from nuclear sources and if salinity problems can be solved (see Section 7.4). Unfortunately, those nations which are technically and financially capable of entering the field of desert farming are not the countries that are facing immediate food shortages.

Many poor countries, especially those in South America and Africa, are looking hungrily at their jungles. However, jungles are not easily converted into fertile fields. They are ecologically peculiar, for decay and growth of plant tissue is so rapid that nutrients do not remain in the soil very long. Thus, the large biomass that exists is dependent on rapid recycling of tissue and not on large nutrient reservoirs in the soil. Moreover, simple fertilization of cleared land does not solve all the problems, for jungle soils harden when exposed to air, and they cannot be farmed with the techniques of temperate agriculture. When a farmer in these areas cuts down the trees and fights back the brush, he finds that the soil underneath is infertile and incapable of producing high-yield grain fields.

Most of the forests that lie between the tropics and the arctic have been cut back so that today only the mountain sides and the high plateaus are used for timber. If these areas are ever cleared for agriculture, alternative sources of fiber for construction and for writing material will be needed.

In summary, it is possible to increase crop acreage by exploiting marginal land, but the required energy and chemicals are too costly for the people who need new sources of food most urgently.

Food From the Sea

Man has often regarded the seas as a boundless source of food, especially protein. In truth, most of the central oceans are biological deserts (see p. 26), while the more productive continental shelf areas are already heavily fished. In 1967, approximately 60 million metric tons of fish were harvested from the seas. Marine ecologists believe that the oceans could support a sustained catch of about 100 to 150 million metric tons. Larger catches would dangerously deplete fish populations. Thus, if the seas were fished wisely and not exploited, the total food protein from the sea could be doubled.

Unfortunately, even the hope of a sustained yield of 100 to 150 million metric tons might not be realized, for the oceans are even now being threatened with overexploitation of many edible species. For example, since 1960, marine biologists have annually recommended to the International Whaling Commission to limit the catch of various species of whales. The biologists have warned that if whaling were intensified, the numbers of whales killed would exceed the reproductive capacity of the existing population, and the number of whales would decline. Since 1960, the International Whaling Commission has ignored the biologists' suggestions and set limits which greatly exceed the sustained catch levels. The Japanese and Russian whaling industries have practiced particularly intense whaling. Predictably, the total food yield from the whaling industry has decreased from 1960 to 1970. Recently, Russian whalers, finding that it is now unprofitable to hunt the big mammals, have started netting the Antarctic shrimp, called **krill**, which form the staple of the whale's diet. However, whales prove to be more efficient krill harvesters than man, and the protein harvest from the krill operation is less than that which would have been possible from sustained whaling a decade ago.

If the precedent set by the whaling industry is adopted by fishermen, the short-term catch may exceed the 100 to 150 million ton limit, but the future protein yield from the sea would then necessarily decrease. Unfortunately, efforts at international marine cooperation have not been particularly successful, and stocks of sardines, salmon, tuna, flatfish, and other sources of marine protein are presently being overexploited.

Some nations, notably Japan and the United States, have practiced **aquaculture**, or fish farming. Proponents of this method point to already successful ventures with oysters and yellowtail fish and claim that aquaculture will revolutionize ocean-based food supplies just as animal husbandry revolutionized food gathering on land. In aquaculture, the animals are housed in underwater cages, fed, and slaughtered at maturity. Food production is maximized because every mature fish can be "caught," and slaughtering can be done at a time which maximizes the ratio of weight gain to food consumed by the fish. If certain problems, such as the control of disease epidemics in artificially concentrated and caged schools can be solved, aquaculture will probably expand rapidly for a short time. Ultimately, however, these ventures will be limited by the food requirements of the animals. If the fish are fed seaweed and plankton, then man must grow or collect these plants, and these operations are difficult (see p. 279). If the feed is manufactured from terrestrial vegetation, then once again man is dependent on his croplands for food.

The Japanese have demonstrated the feasibility of seaweed culture in shallow waters, and undoubtedly an extension of coastal farming could increase man's worldwide food supply if care were taken not to disrupt estuary nurseries. One serious drawback to seaweed farming is that fertilizing the fields efficiently is difficult, for ocean currents often carry the nutrients away. The coastal areas suitable for efficient farming are small, and seaweed culture, though helpful, is not expected to yield prodigious quantities of food. The prospect of collecting or cultivating plankton is discouraging. First, it is difficult, expensive, and energy-demanding to collect billions of small organisms, concentrate them, and somehow render them palatable. Second, plankton form the base of salt-water food webs, and plankton harvests would necessarily reduce wild fish populations. If man first over-exploits fish populations, and then turns to the small organisms for food, it is likely that plankton harvests would yield less total food protein than present fishing harvests do, just as krill collection is now less efficient than whaling once was. Finally, any attempt to fertilize the central oceans to increase plankton yields is an undertaking well beyond our present technology.

Judicious regulation of the fishing industry, coupled with some expansion of seaweed and fish farming, could increase to some extent the productivity of the oceans. However, it is also possible that man will pollute the ocean beyond its ability to support even the present life systems.

Another serious threat to the oceans is the urbanization of coastlines and the destruction of estuaries. In the United States, especially on the eastern seaboard, it is often cheaper to purchase a salt-water marsh and convert it to solid land than to purchase prime industrial sites. Ecologically, the destruction of an estuarine marsh is disastrous, for, as was mentioned in Chapter 1, estuaries are nursery grounds for many deeper water fish. It has been estimated that one acre of estuary produces enough fry per day to grow into 240 pounds of marketable fish. Legally, the problem of conversion of marsh to factory site is quite complex. One such dispute involves the Ryker Industrial Corporation of Bridgeport, Connecticut, which purchased 700 acres of marsh in 1948, paid taxes on them for 25 years, and now wants to develop the coastline. Opposition lawyers argue that because coastal waters are under the jurisdiction of the Federal Government, and a salt marsh is an extension of coastal waters, Ryker never had a legal claim to the estuary, does not own it now, and cannot develop it.

The Cultivation of Algae

In Figure 1.7, we saw that terrestrial plants, which operate at about one per cent efficiency, are poor converters of sunlight to leaf tissue. Underwater algae are much more efficient converters of solar energy to organic matter. If the underwater efficiencies could be maintained in concentrated cultures at full sunlight, a farm half a square meter in size could feed a man on a sustained basis. While this calculation

FIGURE 7.12 The relation between food yields and auxiliary energy.

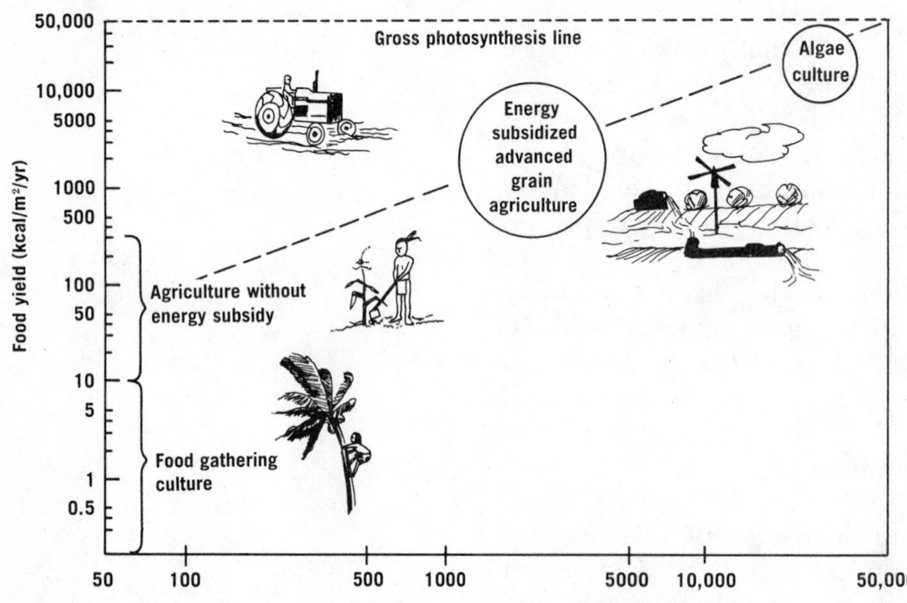

is accurate, it does not follow that algae farming will solve man's food problems in the future, for algae do not grow as well in concentrated surface cultures as they do under water. As light intensity is increased, the conversion rate decreases, so that at full sunlight, algae are only about four times as efficient as wheat or rice. Furthermore, the advantage of this efficiency is countered by the fact that algae are not well adapted to concentrated growing conditions, and tremendous work inputs are required to maintain high yields. Figure 7.12 compares the overall efficiencies of algae cultures and other means of producing food.

Processed and Manufactured Foods

Food chemists have pursued three different concepts in developing new food products: (1) alteration of good natural foods so as to make them more acceptable or more marketable; (2) conversion of agricultural wastes to edible food; and (3) the chemical synthesis of food. Let us discuss each in turn.

Alteration of Natural Foods

Perhaps the most familiar example to us is the conversion of vegetable oils into the butter substitute, oleomargarine. Food conversion is becoming increasingly important in both the developed and the underdeveloped nations. In some areas such high quality protein sources as soybeans, lentils, and other legumes, though readily available, are unpopular. One solution is to use concentrated vegetable protein as a base for manufactured foods that taste like something else. Thus, soy protein "hamburger," soy "milk," soy "bacon," and other simulated products are readily available in retail stores in many parts of the world. In some areas, undernourished people spend their money on soft drinks rather than on protein-rich foods. Recently, soy protein has been used as an ingredient in soft drinks, and various sweetened, carbonated, soy beverages are now sold in Asia, South America, and Africa.

Conversion of Agricultural Wastes

When cooking oils are extracted from peanuts, soybeans, coconuts, cottonseeds, corn, or any other vegetable seed, the residue is a waste mash high in vegetable protein. Presently, most of these residues are fed to cattle or discarded, but it is relatively easy

```
      CH2OH
 H    C------O     O
  \  /H       \   /
   C           C
  /  \OH   H/   \
 O    C-----C     H
      H     OH       n
```

cellulose

Simulated ground meat made from soybean flour. (Courtesy of Archer Daniels-Midland Company.)

to incorporate the proteins into synthetic foods. Cottonseed mash contains a natural insect repellent which is toxic to humans. Research is under way to find an economical process to remove this poison.

Sawdust, straw, and other inedible plant parts also represent a major source of waste materials. The major chemical component of sawdust and straw is cellulose. Cellulose consists of very large molecules which, when heated with dilute acid and water, will decompose into a mixture of various sugars. Yeasts can thrive on this sugar mixture if other nutrients such as urea (a source of nitrogen) and mineral salts are added to the culture medium. In turn, yeasts are sources of high-quality protein and can be eaten directly. Alternatively, yeast protein extract can be used as an additive in manufactured or processed foods. For example, whole wheat is deficient in the amino acid lysine which is necessary for protein synthesis in the human body. Lysine can be extracted from yeast cultures and sold for about $1.00 per pound, and in India it is added to dough in government-owned bakeries for the production of enriched bread. Lysine fortification requires a much larger quantity of yeast than is needed to make bread rise.

Yeast culture, like algae culture, requires large quantities of auxiliary energy to synthesize the fertilizers and to maintain the growth chambers. Yeasts are heterotrophs, while algae are autotrophs and can fix energy from the sun. Thus, yeasts do not represent a primary source of food, but rather are organisms which can convert low-quality organic compounds into high-quality foods if sufficient additional energy is supplied. The acids and microorganisms present in the digestive system of cattle are also capable of converting cellulose and urea into protein, and future livestock feed may contain only sawdust, urea, and inorganic chemicals.

Chemical Synthesis

The high costs of maintaining yeast organisms or feedlot cattle are prompting chemists to completely bypass the heterotrophic organisms and synthesize amino acids directly from coal and petroleum. Presently, lysine synthesized chemically costs about the same as lysine extracted from yeast cultures, and requires about as much energy for production. Other synthetic foods include vitamins used for food additives and artificial cellulose derived from petrochemicals used as a pie filler.

It is encouraging to know that bread in New Delhi is more nutritious than ever before, but we must recognize that such dietary advances are concomitant with the trend from the range cow to the feedlot cow to synthetic amino acids, and that man's existence is thus becoming increasingly dependent on his technology.

PROBLEMS

1. **Agricultural ecosystems.** Explain why agriculture is more disruptive to natural ecosystems than are hunting and gathering of food.
2. **Destruction of ecosystems.** Chemical defoliation was introduced as a tactic of war in Southeast Asia in the 1960's, and a study group of the American Association for the Advancement of Science concluded that "it is to be expected that in any future wars of this nature, more extensive use will be made of it (defoliation)." Mangrove trees that grow in the river delta areas can be killed by a single application of defoliant, and many sections of mangrove forest have been rendered barren by this tactic. Outline the ecological factors that will determine whether or not such forests

The destruction of Indochina. (Photograph courtesy of AAAS Herbicide Assessment Commission, Harvard University, Cambridge, Massachusetts.)

will eventually reestablish themselves. Do you think such reestablishment is certain, doubtful, or hopeless? Defend your answer.

3. **Destruction of ecosystems.** A common logging practice in the United States today is **clearcutting,** the removal of all trees and bushes to make room for logging roads across hillsides (see below). Conservation groups attack clearcutting because it destroys wilderness recreation areas. Will clearcutting affect people other than those who enjoy wilderness? Explain.
4. **Longevity of agriculture.** If a given land area is farmed for many years, does its productivity necessarily decrease? Justify your answer.
5. **Natural vs. agricultural ecosystems.** Briefly discuss the relative importance of each of the following characteristics to a plant species existing (a) in a natural prairie; (b) in a primitive agricultural system; (c) in an industrial agricultural system: (i) resistance to insects, (ii) a tall stalk, (iii) frost resistance, (iv) winged seeds, (v) thorns, (vi) biological clocks to regulate seed sprouting, (vii) succulent flowers, (viii) large, heavy clusters of fruit at maturity, (ix) ability to withstand droughts.
6. **Energy.** List the energy inputs that are added to a wheat field in industrial agriculture.

"Happy?"

(Drawing by Koren; © 1973 The New Yorker Magazine, Inc.)

7. **Phosphorus reserves.** The depletion of both natural gas and of phosphorus reserves is likely to be a serious problem in the early part of the twenty-first century. Explain how the meaning of the word depletion varies when discussing these two different resources.

8. **Definition of "organic."** (a) Write all the definitions of the word "organic" that you can find in a good dictionary and that were not given in this text. (b) Obtain definitions of the word "organic" from organic gardening publications or from the proprietors of health food stores, and compare them critically with the definitions given in this text.

9. **"Organic" foods.** "Organic" food is grown and processed without using any unnatural product. Reliable commercial sources contend that 10 times as much "organic" food is sold as is produced. (a) "Organic" food can be sold to the consumer at a relatively high price. How is that relevant to the above contention? (b) Suppose a farmer grows a crop of tomatoes "organically." He can show his field to several potential buyers and promise each he will deliver tomatoes from that field. What if the farmer promises more than the field delivers? How can the buyer be certain he has received the organically grown tomatoes and not produce from some other field? (How would you test whether a tomato has been grown organically?) (c) How can a consumer be certain that his purchased "organic" tomatoes are genuine? (d) If you wanted to buy a jar of organic honey, what would convince you that the product was unadulterated? Its appearance? taste? the reputability of the store you bought from? the label? the price? (e) "Organic" granola is a cereal made from many ingredients; the more elaborate formulations include oatmeal, wheat germ, other grains, honey and other natural sweeteners, almonds, sesame seeds, and dried fruit. At what points in the growing of the raw materials and in the processing of the granola itself might unnatural foods or contaminants enter the product? Do you feel contamination would be likely to be accidental or deliberate? Do you think the FDA or some other governmental body should be charged with the duty of insuring that the label "organic" refers to an uncontaminated product? Consider the cost of policing and the relevance of natural foods to nutritional health in the United States.

10. **Humus.** Briefly discuss some of the chemical and physical properties of humus.

11. **Industrial agriculture.** What are the major techniques of industrial agriculture? Under what circumstances of use or misuse may each of these methods engender a loss of fertility of the land?

12. **Fertilizers.** Assess the advantages and disadvantages of using manure as a fertilizer. Do you feel manure

should be used to fertilize crops? (Remember that a field fertilized with manure stinks; however, such odors are not harmful to health.)

13. **Manure.** (a) A father and two sons walk along a plain in India, driving several cows before them. The mother follows, carrying a basket on her head. When a cow defecates, the mother scoops up the droppings into her basket. Later, she plasters these droppings onto a wall to dry. Most of the dried dung is then traded for rice; some is kept by the family for use as a fuel to cook the rice. Outline a possible sequence that might have led to the use of cow dung as fuel. Does the situation described above represent the lowest condition of poverty of the land, or does it lead to still further disruptions? Explain. (b) Many of the buffalo-hunting Plains Indians of North America used buffalo chips (dried buffalo dung) as fuel. Would you describe the situation of those Indian tribes as being comparable to the situation of the Indian family described in part (a)? Defend your answer.

14. **Agriculture in India.** Which of the following inventions would you believe has had significant effects on agriculture in India? Which has had a moderate effect? No effect? (a) pesticides, (b) the tractor, (c) inorganic fertilizers, (d) the electric motor, (e) the high-pressure irrigation pipe.

15. **The Green Revolution.** Examine the curve for Pakistani wheat yields in Figure 7.11. If no significant new advances occur in agricultural sciences, how would you expect the curve to grow in the near future? (*Hint:* Recall some of the discussion of graphical extrapolation in Chapter 5.)

16. **Agriculture in marginal lands.** Discuss some problems inherent in farming deserts; jungles; temperate forest hillsides; the Arctic.

17. **Food from the sea.** Jacques Cousteau has said, "The shores of the rivers are the roots of the oceans." What do you think he meant?

18. **Aquaculture.** Aquaculture, or seaweed farming, is likely to be practiced along shorelines and in coastal bays or marshes. Outline a series of events which might lead to a situation where successful aquaculture would *decrease* the total food harvest from the sea. Do you think that this set of events is likely to occur?

19. **Synthetic foods.** It is reasonable to believe that food chemists will be able to synthesize economically a complete and balanced diet from coal in the near future. Outline some advantages and disadvantages of the chemical synthesis of foods.

20. **Synthetic foods.** How does the probable energy requirement necessary to make a soy drink compare to the direct use of soybean soups?

21. **Sources of protein.** Mushrooms are high in protein and some varieties can be grown to maturity in 10 days to two weeks. Discuss the advantages and disadvantages of mushrooms as a source of protein. (*Hint*: Mushrooms and yeasts are both varieties of fungi.)

22. **Synthetic food.** Discuss the advantages and disadvantages of the food production sequence outlined below.

$$CO_2 + H_2O \xrightarrow[\text{using nuclear energy}]{\text{factory process}} C_6H_{12}O_6 \text{ (sugar)}$$

$$\text{sugar} + \text{urea} + \text{yeasts} \longrightarrow \text{protein}$$

$$\text{protein} + \text{sugar} \xrightarrow{\text{industrial processes}} \text{imitation bread, meat, string beans, etc.}$$

BIBLIOGRAPHY

An excellent book which discusses the power base of industrial agriculture is:

Howard T. Odum: *Environment, Power, and Society.* New York, Wiley-Interscience, 1971. 331 pp.

The reader who wishes to investigate the Green Revolution and food in the future should refer to:

Lester R. Brown: *Seeds of Change: The Green Revolution and Development in the 1970's.* New York, Encyclopaedia Brittanica, Praeger Publishers, 1970. 205 pp.

Willard W. Cochrane: *The World Food Problem: A Guardedly Optimistic View.* New York, Thomas Y. Crowell Co., 1969. 331 pp.

Rene Dumont and Bernard Rosier: *The Hungry Future.* New York, Praeger Publishers, 1969. 271 pp.

Francine R. Frankel: *India's Green Revolution.* Princeton, N.J., Princeton University Press, 1971. 232 pp.

Two recent excellent books on agriculture and food resources are:

Kusum Nair: *The Lonely Furrow: Farming in the United States, Japan, and India.* Ann Arbor, University of Michigan Press, 1970. 336 pp.

N. W. Pirie: *Food Resources, Conventional and Novel.* New York, Penguin Books, 1970. 208 pp.

A delightful book about human nutrition from prehistoric to modern times is:

Lloyd B. Jensen: *Man's Foods.* Champaign, Ill., Garrard Publishing Co., 1953. 278 pp.

Some of the pollution problems associated with fertilizer applications are discussed in:

Barry Commoner: *The Closing Circle.* New York, Alfred A. Knopf, 1971. 326 pp.

Ted Willrich and George E. Smith, eds.: *Agricultural Practices and Water Quality.* Ames, The Iowa State University Press, 1970. 415 pp.

GLOSSARY

Abortion, induced – The artificial termination of pregnancy.

Age-specific vital rate – A vital rate defined as the number of vital events occuring in the lives of individuals of a specific age group divided by the total number of individuals in that age group.

Amensalism – A two-species interaction in which the growth of one species is inhibited, while the growth of the second is unaffected.

Arithmetic growth – In population studies, growth characterized by the addition of a constant number of individuals during a unit interval of time. For instance, if there are x individuals in year 0 and $x + a$ in year 1, arithmetic growth implies $x + \mathrm{n}a$ in year n.

Autotroph – An organism that obtains its energy from the sun, as opposed to a heterotroph, which is an organism that obtains energy from the tissue of other organisms. Most plants are autotrophs.

Benthic organism – A plant or animal that lives at or near the bottom of a lake, river, stream, or ocean.

Biogeography – The study of the geographical distribution of living things.

Biomass – The total weight of all the living organisms in a given system.

Biome – A group of ecosystems characterized by similar vegetation and climate, and which are collectively recognizable as a single large community unit. Examples include the arctic tundra, the North American Prairie, and the tropical rain forest.

Biosphere – That part of the Earth and its atmosphere which can support life.

Biotic potential – The maximum rate of population growth of a species that would result if all females bred as often as possible, and all individuals survived past their reproductive age.

Birth cohort – All individuals born during a specified interval of time.

Birth rate – The number of individuals born during some time period, usually a year, divided by an appropriate population. For example, the crude birth rate in human populations is the number of live children born during a given year divided by the midyear population of that year.

Bloom — A rapid and often unpredictable growth of a single species in an ecosystem.

Calorie — A unit of energy used to express quantities of heat involved in chemical changes, especially in biochemistry and nutrition. When calorie is spelled with a small c, it refers to the quantity of heat required to warm 1 gram of water 1 Celsius degree. (This definition is not precise, because the quantity depends slightly on the particular temperature range chosen. The gram calorie, normal calorie, mean calorie, and thermochemical calorie are all precisely defined, but they are nearly the same, and their differences need not concern us.) When Calorie is spelled with a capital C, it means 1000 small calories, or one kilocalorie, the quantity of heat required to warm 1000 grams (1 kilogram) of water 1 Celsius degree. Food energies for nutrition are always expressed in Calories.

Carnivore — An animal that eats the flesh of other animals.

Carrying capacity — The maximum number of individuals of a given species that can be supported by a particular environment.

Chelating agent — A molecule that can offer two or more different chemical bonding sites to hold a metal ion in a claw-like linkage. The bonds between chelating agent and metal ion can be broken and reestablished reversibly.

Climax system — A natural system that represents the end, or apex, of an ecological succession.

Cohort measure — In demography, rates and measures pertaining to specific birth cohorts.

Commensalism — A relationship in which one species benefits from an unaffected host.

Competition — An interaction in which two or more organisms try to gain control of a limited resource.

Contraception — Prevention of conception.

Critical level — In general, a critical condition relates to a point at which some property changes very abruptly in response to a small change in some other property of the system. In ecology, a population is said to be reduced to its critical level when its numbers are so few that it is in acute danger of extinction.

Crude rate — A vital rate with the entire population of some area as the denominator.

Crude reproductive rate — See Rate of natural increase.

Death rate — The number of individuals dying during some time period, usually a year, divided by an appropriate population. For example, the crude death rate in human populations is the number of deaths during a given year divided by the midyear population of that year.

Demographic transition — The pattern of change in vital rates typical of a developing society. The process can be outlined briefly as follows. Birth and death rates in preindustrial society are typically very high; consequently, population growth is very slow. Introduction, or development, of modern medicine causes a decline in death rates and hence a rapid increase in population growth. Finally, birth rates fall, and the population grows slowly once more.

Demography—That branch of sociology or anthropology which deals with the statistical characteristics of human populations, with reference to total size, density, number of deaths, births, migrations, marriages, prevalence of disease, and so forth.

Density-dependent components of environmental resistance—The components of the environmental resistance that are influenced by the population density of a species in a specific environment. Examples include disease, competition for space, and control by species-specific predators.

Density-independent components of environmental resistance—The components of the environmental resistance which act upon an entire population irrespective of its size. Examples include flood, drought, and windstorm.

Desiccate—To deprive thoroughly of moisture.

Detritus—Traditionally, it is the accumulation of small particles of rock worn away by weathering. In ecology, the word has recently been used to describe all the nonliving organic matter in an ecosystem.

Detritus feeders—Organisms which consume organic detritus.

Doubling time—The time a population takes to double in size, or the time it would take to double if its annual growth rate were to remain constant.

Dust mulching—An agricultural practice of pulverizing the surface of the soil to enhance its capillary action and thereby draw underground water up to the root zone.

Ecological niche—The description of the unique functions and habitats of an organism in an ecosystem.

Ecology—The study of the interrelationships among plants and animals and the interactions between living organisms and their physical environment.

Ecosystem—A group of plants and animals occurring together plus that part of the physical environment with which they interact. An ecosystem is defined to be nearly self-contained, so that the matter which flows into and out of it is small compared to the quantities which are internally recycled in a continuous exchange of the essentials of life.

Ecosystem homeostasis—The control mechanisms within and ecosystem that act to maintain constancy by opposing external stresses.

Ecotone—A transitional area between two different types of ecosystems, containing characteristics of both, yet having a unique character of its own.

Ecotypes—Separate populations of a single species which have become adapted to local conditions and therefore have different limits of tolerance to various factors such as temperature, moisture content, or wind speed.

Energy—The capacity to do work or to transfer heat.

Environmental resistance—The sum of various pressures,

such as predation, competition, adverse weather, etc., which collectively inhibit the potential growth of every species.

Estuary – A partially enclosed shallow body of water with access to the open sea and usually a supply of fresh water from the land. Estuaries are less salty than the open ocean but are affected by tides and, to a lesser extent, by wave action of the sea.

Euphotic zone – The surface volume of water in the ocean or a deep lake which receives sufficient light to support photosynthesis.

Expectation of life – The number of years an infant can be expected to live under a specified schedule of age-specific death rates.

Extrapolation – The prediction of points on a graph outside the range of observation.

Fertility rate – See birth rate.

First Law of Thermodynamics – See Thermodynamics.

Food chain – An idealized pattern of flow of energy in a natural ecosystem. In the classical food chain, plants are eaten only by primary consumers, primary consumers are eaten only by secondary consumers, secondary consumers only by tertiary consumers, and so forth.

Food web – The actual pattern of food consumption in a natural ecosystem. A given organism may obtain nourishment from many different trophic levels and thus give rise to a complex, interwoven series of energy transfers.

Gaia – The ancient Greek goddess of the Earth. This word has recently been used to describe the biosphere and to emphasize the interdependence of the Earth's ecosystems by likening the entire biosphere to a single living organism.

Generator – A device that converts mechanical power to electrical power. Generators must be powered by some external source of energy. In most commercial applications, steam or flowing water drives a turbine, and the generator is connected to the spinning shaft of the turbine.

Geometric growth – In population studies, growth such that in each unit of time, the population increases by a constant factor. For instance, if there are x individuals in year 0 and ax in year 1 (where a is greater than 1), geometric growth implies anx individuals in year n.

Geothermal energy – Energy derived from the heat of the Earth's interior.

Green Revolution – The realization of increased crop yields in many areas due to the development of new high-yielding strains of wheat, rice, and other grains in the 1960's.

Habitat island – A particular area of a continental land mass that is ecologically similar to an aquatic island because it is isolated from other similarly constituted ecosystems. Immigration and emigration are slow among habitat islands.

Heat engine – A mechanical device that converts heat to work.

Heterotroph – An organism that obtains its energy by consuming the tissue of other organisms.

Homeostasis – See Ecosystem homeostasis.

Home range – The area in which an animal generally travels and gathers its food.

Humus – The complex mixture of decayed organic matter that is an integral part of healthy soil.

Hydroelectric power – Power derived from the energy of falling water.

Hypolimnion – The lower levels of water in a lake or pond which remain at a constant temperature during the summer months.

Inbreeding – The mating of closely related individuals. Inbreeding generally weakens a population.

Infant mortality rate – One thousand times the number of deaths to infants less than one year old in a given time period divided by the number of live births during the same period.

Law of limiting factors – A law which states that, under steady state conditions, the growth of a population is limited by that essential material which is least available.

Leaching – The extraction by water of the soluble components of a mass of material. In soil chemistry, leaching refers to the loss of surface nutrients by their percolation downward below the root zone.

Legumes – Any plant of the family Leguminosae, such as peas, beans, or alfalfa. Bacteria living on the roots of legumes change atmospheric nitrogen, N_2, to nitrogen-containing salts which can be readily assimilated by most plants.

Magnetohydrodynamic generator – A type of electrical generator which operates by passing ions through a magnetic field. MHD systems are more efficient than conventional mechanical generators because there are fewer moving parts, and hence fewer frictional losses.

Malthusian theory – A social theory predicting that human population growth will continue until it is checked by human misery.

MHD generator – See Magnetohydrodynamic generator.

Microclimate – The local climate of a very small area, such as the underside of a rock, the inside of a rotten log, or the center of a pile of organic debris.

Mortality rate – See Death rate.

Mutualism – An interaction beneficial and necessary to both interacting species.

Natural selection – A series of events occurring in natural ecosystems which eliminates some members of a population and spares those individuals endowed with certain characteristics that are favorable for reproduction.

Natural succession – The sequence of changes through which an ecosystem passes during the course of time.

Neutralism – The inconsequential case of very little interaction between two species.

Niche – See Ecological niche.

Omnivore – An organism that eats both plant and animal tissue. Common omnivores include bears, pigs, rats, chickens, and man.

Parasitism – A special case of predation in which the predator is much smaller than the victim and obtains its nourishment by consuming the tissue or food supply of a larger living organism.

Period measures—In demography, rates and measures which are the result of viewing all persons alive in the year of study. A period measure can be viewed as a cross-section of many cohort measures.

Photosynthesis—The process by which chlorophyll-bearing plants use energy from the sun to convert carbon dioxide and water to sugars.

Photovoltaic cell—A semiconductor device that converts sunlight directly into electrical energy.

Phytoplankton—Any microscopic, or nearly microscopic, free-floating autotrophic plant in a body of water. There are a great many different species which exist in a community of phytoplankton; these plants occur in large numbers and account for most of the primary production in deep bodies of water.

Plankton—Any small, free-floating organism living in a body of water. See Phytoplankton and Zooplankton.

Pleistocene Age—The geologic age encompassing the time span from one million to ten thousand years ago. The ice ages, the emergence of man, and the extinction of many species of mammals all occurred during this time.

Population bloom—See Bloom.

Population density—The size of a population divided by the area in which the members live.

Population distribution—The composition of a population categorized by several variables, often age and sex.

Population ecology—The branch of ecology dealing with the size, growth, and distribution of populations of organisms.

Power—The amount of energy delivered in a given time interval.

Prairie—An extensive area of fairly level, predominantly treeless land. Prairies are characterized by an abundance of various types of grasses.

Predation—An interaction in which some individuals eat others.

Predator—An animal that attacks, kills, and eats other animals. More broadly, an organism that eats other organisms.

Primary consumer—An animal that eats plants.

Protocooperation—A relationship between two individuals that is favorable to both of them but is not essential for the survival of either.

Rate of natural increase—The difference between the crude birth and crude death rates. Also called the crude reproductive rate.

Replacement level—The level of the total fertility rate which, if continued unchanged for at least a generation, would result in an eventual population growth of zero.

Saprophyte—An organism, usually a plant such as a mold or fungus, that consumes the tissue of dead plants or animals.

Savanna—A type of tropical or subtropical prairie that is subject to seasonal patterns of rainfall. Savannas are common in central Africa.

Second Law of Thermodynamics—See Thermodynamics.

Sigmoid curve—A mathematical function which is roughly S-shaped, characterized by an initially slow rate of increase, followed by rapid increase, and followed again by a slow, near zero rate of increase.

Solar energy—Energy derived from the sun.

Species—A group of organisms that interbreed with other members of the group, but not with individuals outside the group.

Stable population—A population with constant age-specific vital rates and an unchanging age-sex distribution. The stable population is useful as a demographic concept; real populations are not stable.

Sterilization—A procedure which renders a person incapable of fathering or conceiving a child.

Strip mining—Any mining operation which operates by removing the surface layers of soil and rock, thereby exposing the deposits of ore to be removed.

Survivorship function—In demography and population ecology, a graph depicting the proportion of survivors of a given hypothetical birth cohort from birth until the last member of the cohort dies.

Taiga—The northern forest of coniferous trees which lies just south of the arctic tundra.

Thermal pollution—A change in the quality of an environment (usually an aquatic environment) caused by raising its temperature.

Thermodynamics—The science concerned with heat and work and the relationships between them.

First Law of Thermodynamics—Energy cannot be created or destroyed.

Second Law of Thermodynamics—It is impossible to derive mechanical work from any portion of matter by cooling it below the temperature of the coldest surrounding object.

Total fertility rate—The average number of infants a woman would bear, if she lived through age 49, under a specified schedule of age-specific birth rates.

Transformer—A device that changes the voltage of a circuit but does not by itself produce electrical power.

Transpiration—The controlled evaporation of water vapor from the surface of leaf tissues.

Trophic levels—Levels of nourishment. A plant that obtains its energy directly from the sun occupies the first trophic level and is called an autotroph. An organism that consumes the tissue of an autotroph occupies the second trophic level, and an organism which eats the organism that had eaten autotrophs occupies the third trophic level.

Tundra—Arctic or mountainous areas that are too cold to support trees and are characterized by low mosses and grasses.

Turbine—A mechanical device consisting of fanlike blades mounted on a shaft. When water, steam, or air rushes past the blades, the shaft turns, and this mechanical energy can be used to generate electricity.

Vital event—In demography, a birth, a death, a marriage, a termination of marriage, or a migration.

Vital rate – The number of vital events occurring in a population during a specified period of time divided by the size of the population.

Work – The energy expended when something is forced to move.

Zooplankton – Microscopic or nearly microscopic free-floating aquatic animals that feed on other forms of plankton. Some zooplankton are larvae of larger animals, while others remain as zooplankton during their entire life cycle.

INDEX

Key to abbreviations:

ff = following pages
fig = figure
mrg = margin
prb = problem
tbl = table